Information, Entropy, Life and the Universe

What We Know and What We Do Not Know

Information, Entropy, Life and the Universe

What We Know and What We Do Not Know

Arieh Ben-Naim

The Hebrew University of Jerusalem, Israel

NEW JERSEY • LONDON • SINGAPORE • BEIJING • SHANGHAI • HONG KONG • TAIPEI • CHENNAI

Published by

World Scientific Publishing Co. Pte. Ltd.
5 Toh Tuck Link, Singapore 596224
USA office: 27 Warren Street, Suite 401-402, Hackensack, NJ 07601
UK office: 57 Shelton Street, Covent Garden, London WC2H 9HE

Library of Congress Cataloging-in-Publication Data
Ben-Naim, Arieh, 1934–
Information, entropy, life, and the universe : what we know and what we do not know / Arieh Ben-Naim, The Hebrew University of Jerusalem, Israel.
pages cm
Includes bibliographical references and index.
ISBN 978-9814651660 (hardcover : alk. paper) -- ISBN 9814651664 (hardcover : alk. paper) -- ISBN 978-9814651677 (pbk. : alk. paper) -- ISBN 9814651672 (pbk. : alk. paper)
1. Information theory--Mathematics. 2. Entropy. 3. Physical biochemistry. I. Title.
Q360.B46 2015
003'.54--dc23
2015000737

British Library Cataloguing-in-Publication Data
A catalogue record for this book is available from the British Library.

Printed in Singapore

This book is dedicated
to readers of popular-science books
who are
baffled, perplexed, astonished
and discombobulated
by reading about
Information, Entropy, Life and the Universe.

Contents

Prologue

The Lost Information

Once upon a time Crytos Cipherus III, mighty ruler of Cipher Island, wrote a small book in a language he had invented and only he knew. After his death, his sons found the book in one of his rooms in the keep of the castle. The sons tried to decipher what their father had written but to no avail. Not one of them understood what he had written. Therefore, they commissioned six world-renowned cryptologists to unravel the mystery of their father's messages encapsulated in his incomprehensible language, tentatively called Cryptese. The first paragraph of the book is quoted below:

The scrolls of Crytos Cipherus III

Upon seeing the text, one of the cryptologists immediately noticed that it conveyed *information*. One other professor declared that each letter in the unknown language conveyed *distinct information*. Another commented that it looked as if the *pairs* of letters conveyed information. "No," yet another interjected, "I think each *triplet* of letters constitutes a word which contains meaningful information."

The fourth professor interrupted the lively, almost competitive exchange of opinions and claimed that the *sequence* of symbols as a whole tells a story. "If that is the case, then how do we know in which order the sequence was written? Was it written from left to right, or right to left, or perhaps from top to bottom?" The fifth professor, a linguistics expert, suggested, "Why don't we first apply Shannon's measure of information to the text? By doing so, we will know *how much information* it contains. Perhaps it is redundant, and many symbols are synonyms, and convey the *same information*. The sixth professor, an expert in thermodynamics who had listened in silence, said, "Perhaps we could first measure its entropy and this would tell us something about its content."

They examined, re-examined, argued, debated, and put forward conjectures which did nothing but get in the crosshairs of the others in the group.

Can you help these professors in answering the following questions?

Does the text convey information?
What is the information conveyed by the text?
How much information is conveyed by the text?
How can one calculate Shannon's measure of information for this text?

How many *bits* are there in the message?
What is the entropy of the message?

Whatever the information in the text is, do you think that it will ever be destroyed by the ravages of entropy?

I will let you think about these questions for some time. If you cannot provide any answer, I suggest you read the first two chapters, and perhaps by the time you reach the end you will be better equipped to answer these questions. My answers will be given at the end of Chapter 2.

List of Abbreviations

BH	black hole
C-information	colloquial information
ln	natural logarithm
log	logarithm with base 2
SMI	Shannon's measure of information
20Q	20 questions
i	initial
f	final

Preface

The title of this book encompasses everything. I had initially thought of using the subtitle "What we can and what we cannot say." However, upon realizing that it could be misconstrued as "what we are allowed and what we are forbidden to say," I decided to change the subtitle to a more appropriate one.

Information is a very general concept. Everyone knows what it is, but defining information is as elusive as defining many other abstract concepts. I could have chosen to write anything I wanted about *information*, and this book would hold (and hopefully also transmit) a great deal of *information*, some of which might be true, some false, some neither true nor false, some meaningful, some meaningless, and some neither meaningful nor meaningless. Perhaps I should refer to them as *meaningful-less*. Some might be important, some trivial, and some objective, some subjective. Some could be exciting, while some might be boring.

I could have gone on and on in describing what *kind* of information is contained in this book, and I am sure that the copious aggregation of words all sums up into *information*. However, I will not write here about what I could have written but instead describe what I have written in this book.

The main objective of this book is twofold: first, to inform the curious reader about what is known and what is not known about *information* and *entropy*, and their relationships to *life* and to the *universe*; second, to help the reader in sorting out the meaningful from the meaningless statements, made in many popular-science books dealing with these subjects.

In short, I will try to sort out for you what we know and what we do not know; how one can use and how one cannot use these concepts (perhaps also how some people misuse and sometimes abuse these concepts).

In doing so I am well aware of the risk of ruffling feathers, especially those of some authors who have come up with all sorts of claims that are either unfounded or outright meaningless.

The book is written in reader-friendly English, without highfalutin words. I have made an effort to limit the use of technical terms which are employed in thermodynamics, biology and cosmology, lest the uninitiated reader should get intimidated. For the most part the style will be simple and accessible to anyone who is curious about some of the statements made by famous authors. On some occasions, when mathematics is necessary, I will describe in plain and simple terms the main mathematical argument and relegate the details to the "Notes" section at the end of the book.

This book is organized into four chapters:

Chapter 1 contains two parts. The first part discusses the general concept of information. We all know what information is although its precise definition is elusive. There are no prerequisites for you to read and understand this chapter. In this part, I will take the liberty to discuss at random several topics which are related to the general concept of information.

We will taste some of the fruits from the "tree of knowledge." The second part of this chapter is devoted to what is currently known as *information theory*. In order to understand this part you need to know what probability distribution is, as well as the meaning of the logarithm of a number. No other mathematical knowledge is required to understand the contents of this part. The central concept of information theory is *Shannon's measure of information*. We will see in what sense this concept is a *measure of information*. Although the two parts are related to each other, they are very different in their styles. We will see that although information theory contains a *measure* of *information*, it is not a *theory of information*!

In this book, we will make a distinction between *colloquial information* and Shannon's measure of information. As was stated explicitly by Shannon, the meaning or the significance of the information transmitted is irrelevant to what is called "information theory." There were several attempts to construct a "semantic" theory of information. Probably, the first publication on this was the article by Bar-Hillel and Carnap (1954). To the best of my knowledge, these theories do not have much in common with Shannon's measure of information, and are certainly irrelevant to thermodynamic entropy.

Chapter 2 is devoted to the topic of entropy. It contains three parts. The first part is a brief discussion of entropy and the Second Law as introduced by Clausius. We will use the familiar concepts of energy, heat, temperature, etc., but we will not discuss the mathematics involved in the development of thermodynamics. It should be noted that what is referred to as Clausius' definition is not a definition of entropy, but only changes in entropy in a very specific process. To understand this part, no mathematical prerequisite is called for. The second part of this chapter discusses the *microscopic* interpretation of entropy and the Second Law of Thermodynamics. This part consists of two subsections: one presenting Boltzmann's entropy, and the other a brief, nonmathematical outline of the derivation of entropy from Shannon's measure of information. We will show how the latter is superior to both the Clausius and the Boltzmann definition of entropy. The third part dwells on some examples of changes in entropy in some simple processes, how we interpret those changes and why they might or might not occur. We will also discuss some processes for which we do not know how to calculate the changes in entropy, and perhaps we will never know, or perhaps there is no sense in assigning changes in entropy to such processes.

In Chapter 3, we will discuss entropy and information in relation to life. This chapter will be shorter than the previous two. It will contain "information" and "entropy," just enough to examine its relevance to life. We will show that neither entropy nor the Second Law applies to living systems. This fact has a profound effect on all life questions, from the existence of the Maxwell demon to the survival of Schrödinger's cat, and to the fate of life in the universe.

Chapter 4 will touch on the entire universe but nothing regarding the many theories about the universe, such as the big bang theory. These theories are based on a highly sophisticated mathematical language. We will focus only on those aspects of the universe for which the concepts of information and entropy have been involved. In fact, most of what we will say about the applicability of information and entropy to life also applies to the universe.

Of course, "life" and the "universe" can each easily fill many books. The last two chapters will be the shortest — inversely proportional to their scope and magnitude. We will discuss only those topics which are relevant to information theory and entropy.

Many books have already been written on this topic. Yet, as we will see, not much can be said at the moment (and perhaps never) on the relationship between information theory and entropy on one hand, and life and the universe on the other.

The reader should be aware of the fact that much of what we know about the world that surrounds us is acquired through our senses: hearing, seeing, touching, tasting and smelling. However, a larger part of what we know about the world which is not directly accessible to us through our senses is acquired by the tools of mathematics. If you like, you can view these mathematical tools as an extension of our senses. Perhaps we can refer to these tools as our *common sense.*

In recent years, there has been a flood of popular-science books on information, the theory of information and entropy. The confusion in this field is unprecedented. The main confusion is between the three major concepts: *information*, *Shannon's measure of information* (SMI) and *entropy*. It is difficult to overexaggerate the confusion between these concepts (and, in addition, between "information" and "bits"). I am not aware of any concept in physics that is so misinterpreted, misused and abused.

Following von Neumann's suggestion to call SMI "entropy" (see Chapter 1), people have confused entropy with SMI. Furthermore, SMI contains the word "*information*." Therefore, it is easy to confuse SMI with *information*, and hence also identify entropy with information. The latter has led to a very popular and grave misconception. Since information can be subjective, some scientists refuse to accept the informational interpretation of entropy. The reason is that people believe that SMI is the same as information, and therefore accepting the SMI interpretation of entropy would make entropy a subjective quantity. On the other hand, those who *accept* the informational interpretation of entropy, but confuse SMI with information, reach the (erroneous) conclusion that entropy must be a subjective quantity.

My aim in writing this book is to put some *order* into the existing *disorder*, and hopefully lessen the confusion. Even the words "order" and "disorder" appear in the literature in a variety of forms and combinations, as varied as shades and hues: "order from order," "order from disorder," "disorder from order" and many others. Before we put some order into this disorder, let me describe briefly the three main terms. *Information* is the most general term. No one has a definition of "information," and yet everyone knows what it means. Next comes information theory, which is *not* a theory of information. There exists no theory of information! The central quantity in information theory is *Shannon's measure of information*. This is a well-defined concept. It will be discussed in Section 1.2. For now, suffice it to say that SMI is a *measure* of information. It is not information, and it has nothing to do with the meaning or value of information. The third quantity is *entropy*. Again, this is a well-defined quantity. Chapter 2 is entirely devoted to this well-defined quantity.

People talk about the *laws of entropy*, which is another way of referring to the Second Law of Thermodynamics. But then, confusing entropy with SMI, and SMI with information, leads one to talk about the "laws of information." What they mean is the Second Law, applied to information where it does not apply. To the best of my knowledge these "laws of information" do not, and cannot, explain anything. Yet, such incredible claims appear in the literature and abundantly so.

I would like to quote a paragraph from Bricmont's article (1996) on "Science of Chaos or Chaos in Science":

> Popularization of science seems to be doing very well: the Big Bang, the theory of elementary particles or of black holes are explained in countless books for the general public. The same is true for *chaos, theory, irreversibility* or *self-organization*. However, it also seems that a lot of *confusion exists concerning* these latter notions, and that at least some of the popular books are *spreading misconceptions*. The goal of this article is to examine some of them, and try to clarify the equation.

The goal of the present book is similar, though the focus is on *confusion* and misconceptions regarding theories about information, entropy, life and the universe.

This book may be classified as a *popular-science* book. It is addressed to the *general public* and yet its contents are far from being popular. It is not popular in the sense that it does not contain fancy statements that one finds in many textbooks and popular-science books. You can find examples of such statements on the dedication page, as well as throughout the entire book.

Arieh Ben-Naim

Department of Physical Chemistry
The Hebrew University of Jerusalem
Jerusalem, Israel
Email: ariehbennaim@gmail.com

Acknowledgments

I have benefitted from the many discussions with Jacob Bekenstein, Patrick Fahey, Zvi Kirson, Alex Levitski and Naftali Tishby. I am also grateful to John Anderson, Jordan Bell, Frank Bierbrauer, Diego Casadei, David Deutsch, Claude Dufour, Ivan Erill, Robert Hanlon, Douglas Hemmick, Bernard Lavenda, Azriel Levy, Robert Mazo, Mike Rainbolt, Andres Santos, Jens Smiatek, David Thomas and Brian Wood for reading parts or the entire manuscript and offering useful comments.

Special thanks to my friend Alex Vaisman for the beautiful illustrations in the book.

As always, I am very grateful for the gracious help I got from my wife, Ruby, and for her unwavering involvement in every stage of writing, typing, editing, re-editing and polishing the book.

Chapter 1

Information

This chapter consists of two main parts. In the first part we discuss some aspects of the general concept of *information*. The second part is devoted to *specific* information to which a *specific measure* is assigned. More precisely, we will introduce *Shannon's measure of information*, its definition, its various interpretations, and a few of its applications.

Information is a very general concept. Everything written in this book is information. Information can be true or false, or even either true or false. It can be meaningful or meaningless, or neither. It is usually about something, but sometimes it is about itself. It can be redundant or ambiguous. It can be objective or highly subjective. It may be short or long. Some information can be assigned a measure, or even a few measures. We may refer to this measure as the "size" or the "length" of the information. In Section 1.2, we will introduce a *specific measure* of information which is measured in units of "bits." Some information can have the same *meaning* but different *lengths*. Some can have the same *length* but different *meanings*. Information can flow from one place to another. It can be created and destroyed.[1]

At this point, I will allow the reader to continue with the list of adjectives that can be attached to *information*. In this section we will focus only on some of those attributes of information which are needed in order to grasp the ideas constituting what is referred to as information theory.

1.1 The General Concept of Information

In any dictionary or encyclopedia, you will find a few "definitions" of information:

The communication or reception of knowledge
Knowledge obtained from investigation, study, experience or instruction
A collection of facts or data; statistical information
A numerical measure of the uncertainty of an experimental outcome

The first three "definitions" associate the word "information" with knowledge of facts and data. None of them are *bona fide* definitions of information. In fact, there is no definition of information, and yet we use this word in connection with whatever we hear, see, read, etc. The fourth "definition" is not a definition of information but a measure of certain types of information. In this section, we will discuss the general concept of information, which we will refer to as colloquial information or, for short, C-information. We will devote Section 1.2 to a measure of information and refer to it as Shannon's measure of information (SMI).

We sometimes use the words "information" and "knowledge" interchangeably. We might say, "I have *information* about the weather in Jerusalem tomorrow," or, equivalently, "I have knowledge about the weather in Jerusalem tomorrow."

However, there are some subtle differences between "information" and "knowledge." Although we do not have a precise definition of each of these two concepts (independently of each other), we do distinguish between them. In an article titled "Musing on Information and Knowledge," Aumann attempted to clarify the distinctions between these two concepts.[2]

Some suggest that "information" is the raw material from which "knowledge" is produced. One distinction (in common usage) is that *information* can be impersonal whereas *knowledge* is personal. Knowledge does not exist on its own, which is independent of people's minds. On the other hand, *information* exists outside people's minds, and the distinction (in common usage) is that knowledge has a more precise connotation than information. For example, a detective might say, "I have *information* on the person who committed the crime, but I do not *know* who did it." Or a weather forecaster might say, "We have *information* on the probability of rain tomorrow, but we do not have that *knowledge*."

Aumann suggested, "Information can be measured, knowledge cannot." As we will see in Section 1.2, not every piece of information can be measured, in the sense of SMI. This kind of information can be measured in *bits*. On the other hand, *knowledge* is something that one either knows or does not know.

We will not attempt to define information in this book. We will assume that we have an intuitive understanding of what information is, much as we have an intuitive understanding of temperature and energy. In the rest of this section, we will discuss some aspects of C-information. In particular, we will describe those aspects of C-information which will have their counterparts in *information theory*, discussed in Section 1.2.

In this book, we will make a clear-cut distinction between the abstract concept of *information*, and *physical entities*. This is in contrast to the many statements in the literature claiming, "Information is physical." I will discuss my views about such statements throughout the book.

In fact, there is another extreme view of reality, which we will not discuss here. Some philosophers think that even *reality* itself is all in our mind. This philosophy is called *solipsism*.

Solipsism maintains that everything we perceive by our senses is only an illusion or a dream. In other words everything exists "only in the mind." It does not specify which mind — mine or yours or ours! There is a nice story about a professor of philosophy who lectured on solipsism [quoted in Deutsch (1997)]. During the lecture the professor said, "Everything I see, hear, or perceive with my senses is only an illusion. Nothing really exists outside my own mind." The lecture was so persuasive that, soon after, a student came up to the professor, applauded him for a very convincing lecture, and said, "I agree with every word you have said in this lecture." The professor was ostensibly satisfied with the compliments of the student.

But what exactly did the student agree with? Did he agree that all that the professor saw, including the student, were *only* in the professor's mind? This is tantamount to agreeing that he, the student himself, did not really exist.

1.1.1 *Private and common knowledge*

The term "common knowledge" is used in game theory, when all the players know some information, and also each of them knows that all the others know it, and that all know that everyone knows it, and so on. The famous example is the problem of a group of people wearing hats with different colors:

Ten people were gathered in a party. The host organized a social game. He stuck a colored label on the back of each person. Specifically, he placed eight blue stickers on the backs of eight people, and two red stickers — one on Bob's back and one on Linda's back. The guests were allowed to look at the stickers on the backs of the others, but not those on their own backs. In addition, they could not reveal to each other the color of the stickers they saw on the others. Then the host announced that each time he rang the bell, and someone found out, by pure reasoning, the color of his or her *own* sticker, he or she should raise his or her hand, and would get a million dollars. The guests were warned that if they guessed their own color incorrectly they would have to pay a million dollars. Of course, no one would risk making the wrong guess and coughing up a million dollars. The host rang the bell, but no one raised his or her hand. After the bell had rung sixty times, everyone gave up. Clearly, Bob and Linda had the following *information*: they knew that eight of their friends had blue stickers and one had a red sticker. The other eight people knew that seven of their friends had blue stickers and two had red stickers. But no one could deduce the color of his or her own sticker. Equivalently, we could say that no one had *information* on the color of his or her own sticker.

Before the party ended, the host offered a variation of the game. He would ring the bell every minute. If a person found out correctly the color of his or her own sticker, he or she would get a bottle of wine. This time he offered his guests a "hint." He declared that there was "at least one sticker which is red" and then he rang the bell. A minute passed and no one had raised his or her hand. After the second ring both Bob and Linda raised their hands and got a bottle of wine each.

How did they know the color of their stickers? Obviously, the host gave them *information*: "There is at least one red sticker." But this information was already *known* to all the guests, even before the host gave them the "hint." Clearly, Bob saw that there was one red sticker on the back of Linda, and Linda saw one red sticker on the back of Bob. All the other eight people saw the red stickers on Linda's and Bob's backs. Therefore, each of the guests saw *at least* one red sticker (some saw one, some saw two, but each saw *at least* one). It seems therefore that the "hint" given by the host was useless since everyone already *knew* that there was at least one red sticker.

Yet, after the second ring of the bell, Bob and Linda *knew* the color of their stickers, and got their bottle of wine. How did they get that information? (Remember the game's first run, and why the host dared to risk one million dollars?)

Indeed, the *information* conveyed by the host, "at least one sticker is red," was known to all the guests. Before the host announced that there was at least one red sticker, everyone had already known that *information*. However, after the host's announcement, this information was made *public*. Now each one not only knew this *information*, but also knew that all the other guests knew that *information*. Let us see how Bob and Linda could "activate" their common sense to take advantage of this additional information.

After the first ring of the bell, Bob thought, "I see that Linda has a red sticker; I do not know what she sees. There are two possibilities — either Linda sees one red sticker, or she sees no red sticker. If she sees only one red sticker, this is consistent with what the host said but she cannot reach any conclusion regarding her own sticker. On the other hand, if she sees no red sticker, and she just heard that there is at least one red, then she must have concluded that her own sticker is red. Therefore, she must have raised her hand after the first ring. But she did not raise her hand after the first ring." Exactly the same argument went on in Linda's mind, and therefore both could not have concluded that their own sticker was red. Therefore, both did not raise their hands after the first ring.

Then they heard the second ring. Bob argued as follows: "I know that Linda knows that there is at least one red sticker. If she does not see any red sticker she must have concluded that it is her own sticker which is red, and therefore she would raise her hand after the first ring. The fact that she did not raise her hand after the first ring means that she sees one red sticker, and that the red sticker must be mine." Therefore, Bob concluded that his sticker was red, and after the second ring he raised his hand and got the prize. The same argument went on in Linda's mind, and she also concluded that her sticker must be red, and therefore after the second ring she raised her hand and got the prize.

If we were to *measure* the information given by the host, we could say that this was the *same information* they had already possessed before the announcement. However, the act of announcing made that information public. Now Bob knew that Linda also knew that there was at least one red sticker. Bob also knew that Linda knew that he knew the same thing. This *additional* knowledge could be exploited to get the prize.[3] One cannot avoid the conclusion that the host did provide *new* information which was not available to the players.

Having concluded that *additional* information was provided by the host, let us turn to discussing the "quantity" of that additional information.

Initially, each person knew the information that "there is at least one red sticker." Call it for short "one RS." After the host's announcement, it seemed that each of the players in the party already knew the "one RS" information. Therefore, we could have concluded that no one *gained* any new information from the announcement. Yet, from the announcement of "at least one RS," each person could *deduce* (or perhaps create) new information. Bob knew, in addition to the information on "one RS," that Linda *knew* about "one RS." Furthermore, Bob knew that Linda knew that he knew that she knew about "one RS." The same was true of all the other players. The question is now: Can we quantify the *added information*, followed by the announcement? We also have to decide what we mean by "added information" — is it just the *new* "added information?" Or do we count the added information for each person? — Or the added information for all the players? Note also that Bob and Linda *made use of* this additional information to deduce not only that "Bob knows that Linda knows that Bob knows etc.," but also the information about what Linda had concluded after the first ring, and also what she had concluded after the second ring, and also what he himself had concluded after the second ring, specifically that his own sticker must be red. This was certainly *new* information that neither was known to Bob before nor was transmitted to him by the host. It was information Bob had obtained by *logical deduction*. The same applies to Linda, who obtained the information on her own sticker by logical deduction. This additional information was specific to Bob and Linda, and was not shared by the other players. Also, note that the information deduced by Bob on the color of *his* sticker was *different* from the information deduced by Linda on *her* sticker. It is intriguing to speculate on the following situation: Suppose that Bob or Linda did not care to make the *logical deduction* described above. Did the host provide new information to Bob or to Linda?

This kind of *additional* information which one can (or cannot) deduce is not measured by information theory. The situation is similar to many other cases where one piece of information provides other information which one may or may not deduce. The example of the Pythagoras theorem is discussed in Subsection 1.1.2.

1.1.2 *Knowledge obtained by mathematical reasoning*

Much of the knowledge we acquire depends on our senses; hearing, sight, touch, taste and smell. In general this information is quite crude and vague, and sometimes cannot be expressed in words. As the French philosopher Descartes put it, "Sense perception is sense deception."

However, most of what we know about the physical world, both in the realm of the microscopic world of atoms and molecules, and of the macroscopic world of remote planets and galaxies, is not obtained directly by our senses. Instead, we rely on mathematical deductions, some of which we can refer to as our "common sense." We *see* that the sun rises in the east and sets in the west. We can deduce, based on our sense of sight, that the sun rotates around the earth. This was certainly the conclusion reached by the ancient people (referred to as the *geocentric theory*). In modern times mathematical deductions tell us that the earth revolves around the sun, and not the other way around (referred to as the *heliocentric theory*).

Likewise, we *see* different colors. We do not *see* the electromagnetic waves. It is only with the help of mathematical deductions (in this case Maxwell's theory of electromagnetism) that we know about the behavior and properties of electromagnetic waves — knowledge which has been used to transmit information by radio, television and other devices.

Among the most dramatic knowledge we obtain by mathematical deduction is the meaning of temperature. Our senses tell us the difference between *hot* and *cold*. Nothing in our senses reveals to us the deep connection between *temperature* and the molecular motion of the atoms and molecules. We can feel when a moving rock is cold. We can also feel when the same rock is very hot, yet it is not moving. These statements apply to macroscopic objects such as a rock. The situation is very different in the microscopic world of atoms and molecules.

One of the greatest achievements of scientists in the late 19th century was to find out that *temperature* (which we feel with our sense of touch) is a measure of the average kinetic energy of the atoms and molecules. This knowledge could not have been obtained without the tools of mathematics. Note that the temperature cannot be assigned to individual molecules. The speed and the kinetic energy of the atoms can be deduced by mathematical reasoning.

A large part of Kline's book (1985) *Mathematics and the Search for Knowledge* is devoted to the following question: How do we acquire knowledge about the physical world? He summarizes the role of mathematics in acquiring knowledge as follows:

> Contrary to the impression students acquire in school, mathematics is not just a series of techniques. Mathematics tells us what we have never known or even suspected about notable phenomena and in some instances even contradicts perception. It is the essence of our knowledge of the physical world. It not only transcends perception but outclasses it.

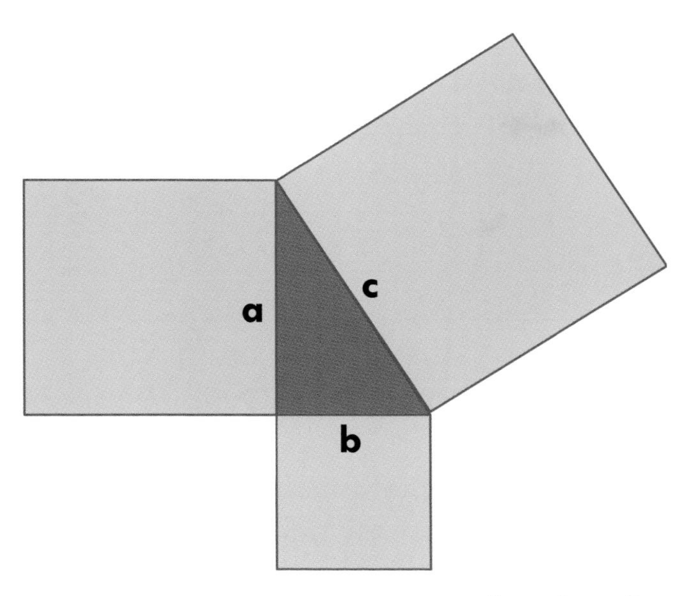

Fig. 1.1(a) Pythagoras' theorem, $c^2 = a^2 + b^2$.

I should add to this, that knowledge of mathematics is information in itself, and yet it can convey additional information. As an example, Pythagoras' theorem states that in a right-angled triangle, the area of the square on the hypotenuse is equal to the sum of the areas of the squares of the two sides of the triangle, a and b, in Figure 1.1, i.e. $a^2 + b^2 = c^2$. The proof may be found in any book on elementary geometry. In Figure 1.1(b), we provide a simple "visual" proof. The two large squares of edge $a + b$ in Figure 1.1(b) have equal areas. Now, remove the four triangles in both of these squares. What remains? One square of edge c on the right hand side, and two squares of edges a and b on the left hand side.

This theorem is *information*. Knowing this theorem allows us to deduce *new information*. If you give me the information about the lengths of the two sides a and b in Figure 1.1, I can calculate,

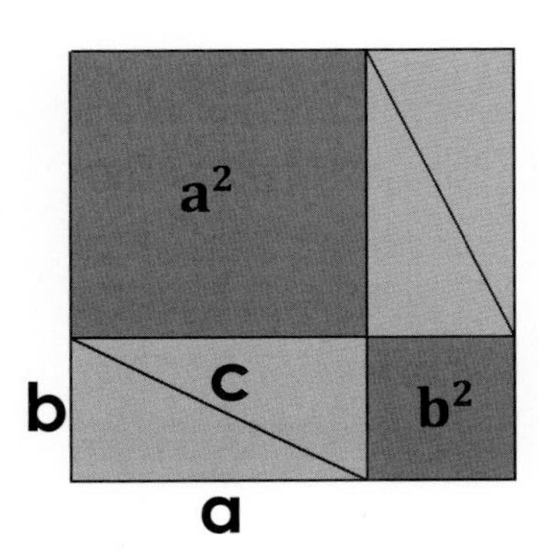

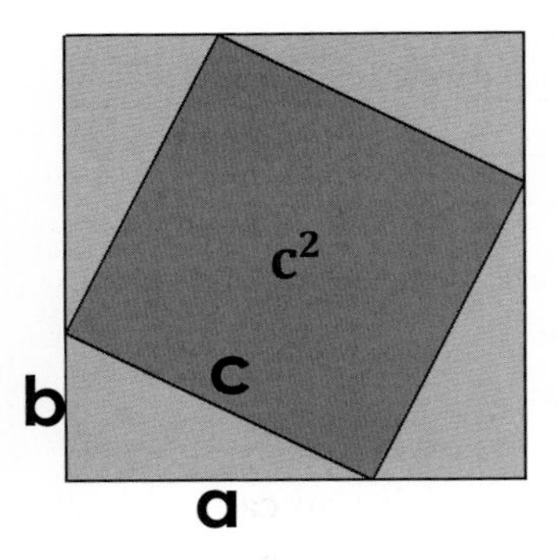

Fig. 1.1(b) The visual proof of Pythagoras' theorem.

and give you in return the information about the length of the hypotenuse side, *c*. This is new information we have deduced from the knowledge of Pythagoras' theorem.

1.1.3 *Information is always about something — sometimes it is about itself*

My first encounter with self-referential sentences was in 1981, when I read Douglas Hofstadter's article in *Scientific American*.[4]

While reading this fascinating article, I had an awesome feeling that these self-referential sentences were "talking" to me directly, informing me not about some other topics as most sentences do, but rather about themselves. I immediately tried to construct some of my own self-referential sentences. In Table 1.1 I list some of my favorite ones taken from Hofstadter's article and book,[5] and some of my own. There is not much to say about self-referential sentences, as they speak for themselves. They certainly convey some *information*, and perhaps also information about the one who wrote them.

There is one self-referential statement which is particularly outstanding. This will be discussed in the next subsection.

Table 1.1 Self-referential sentences (my reactions are in parentheses). The first five sentences were taken from an article by Hofstadter (1981).

This is a sentence with "onions," "lettuce" and "tomato."
(Indeed, it is.)
i should begin with a capital letter.
(Thanks.)
.siht ekil ti gnidaer eb d'uoy, werbeH ni erew ecnetnes siht fI.
(Of course, if you can read Hebrew.)
What would this sentence be like if it were not self-referential?
(It could be anything.)
Are you reading me at this moment?
(Yes, indeed.)
Please quote me next time you write a book.
(I promise to do so.)
I am a unique sentence, no one has ever seen me before.
(Are you sure?)
Thanks for reading me.
(You are welcome.)
This sentenc contains three erors.
(Yes, I found them.)
This sentence contains no errors.
(Indeed, you are perfect.)
Cn y pls dd th mssng vowels in ths sntnc?
(Yes, I can do that.)
Does reading me distract your attention from reading the book?
(To some extent, yes.)
I realize that I am the last, but I hope not the least of your list of favorite self-referential sentences.
(No dear, you are my most favorite one!)

1.1.4 *True and false information*

In many cases we can tell whether some information is true or false. Clearly, the statement "Five plus one equals six" will be immediately qualified as a *true* statement. On the other hand, the statement "Five minus two equals ten" will be qualified as a *false* statement.

However, there can be statements which we cannot qualify as either true or false:

"Tomorrow, it will snow in Jerusalem."
"If I draw a marble from an urn containing ten white and ten red marbles, I will get a red marble."
"There are other living creatures in a distant galaxy."

Clearly, we cannot tell whether these statements are true or false. In some cases we can assign *probabilities* to express the extent of our belief in the truth of a statement. We can say that there is a 30% probability that it will snow tomorrow in Jerusalem. But we are not sure about this. We say that there is a probability of one half to draw a red marble from an urn containing ten white and ten red marbles, but we cannot be sure that the first draw will yield a red marble. Regarding the existence of other living creatures in a distant galaxy, we can only guess its probability, and our guesses will be highly subjective and unreliable.

1.1.4.1 *The liar paradox*

There are cases where we cannot say whether a statement is true or false. Perhaps it is neither true nor false. The most famous example is the so-called liar paradox.

Consider the following sentence: "This sentence is false." If the sentence is true, then it is false. On the other hand, if the sentence is false, then it follows that it is true.

We see that this sentence is neither true nor false. This is known as the liar paradox, and is associated with the name of Epimenides, who lived in Cretan about 600 BC. He declared that "all Cretans are liars." This statement is known as the Epimenides paradox. But it is not completely equivalent to the above statement.[6] For a thorough discussion of this paradox, see Hofstadter (1985). In the context of this book we can say that the statement "This sentence is false" conveys *information*. Although this information is neither true nor false, we shall see that this statement can be assigned a *measure of information*, a measure which is independent of the meaning of the information.

The liar paradox is related to Russell's paradox, which is somewhat more abstract and involves the set of all sets that are not members of themselves.[7]

Here, we present a simpler version of the Russell paradox which may be referred to as the self-referential catalogs.

1.1.4.2 *The catalog that does not exist*

Each of the public libraries was asked to compile a catalog of all its books and send them to the National Library.

After receiving a few thousand catalogs the librarian at the National Library decided to separate them into two groups and put them on two separate shelves (Figure 1.2).

On the left shelf, he put all the catalogs that *included* reference to themselves. That shelf was called the *self-referential* catalog shelf, or the *self-ref* for short.

On the right shelf, the librarian put all the catalogs that did not mention themselves. That shelf was called the *non-self-referential* catalog shelf, or the *non-self-ref* for short.

Realizing that there were too many catalogs on each of those shelves, the librarian decided to compile two Super Catalogs. One contained all the catalogs which were *self-ref*, and the other contained all the catalogs which were *non-self-ref*.

Catalogs

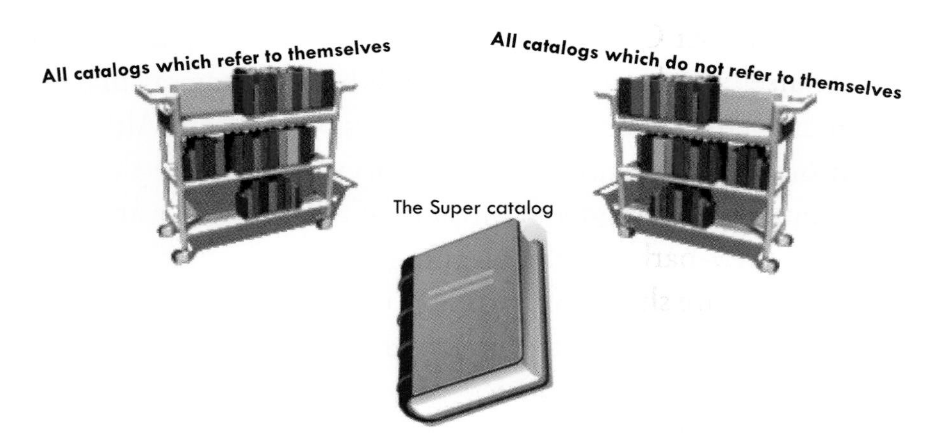

Fig. 1.2 The catalog of all catalogs.

Having been done with the writing and editing of the *first* Super Catalog, the librarian had pondered whether or not to mention the Super Catalog itself. If he included a reference to the Super Catalog of the *self-ref* in the Super Catalog, then this catalog would itself be a self-ref catalog. If he did not include reference to itself, it was still going to be a proper *catalog* of all catalogs that mentioned themselves. So he could decide either way.

Next, he considered the case of the Super Catalog of all the catalogs which were *non-self-ref*, i.e. all the catalogs that did not list themselves. Should that catalog list itself?

He went on to ponder endlessly but never reached a decision. If he decided to have reference to itself in this Super Catalog, then this Super Catalog would list itself, and therefore could not be a Super Catalog of those catalogs that did not list themselves. If he decided not to have reference to itself in this catalog, then it would not be a Super Catalog of *all* the catalogs that did not list themselves.

Thus, a Super Catalog of all the *non-self-ref* catalogs cannot refer to itself, and cannot avoid reference to itself. Therefore, such a catalog does not exist. Likewise, the set of all sets that do not include themselves does not exist.

A related paradox is the barber paradox. In a town there is only one male barber. All the males in this town either shave themselves or are shaved by the barber. Who shaves the barber?[8] We shall leave it to the reader to answer this question.

Sometimes, knowing the liar paradox can be useful to your survival. The following story demonstrates this.

1.1.4.3 *Admitting "I am a liar" can save your life*

A man lost his way in the jungle. A tribe of cannibals spotted him and immediately made a wild dash to catch their meal-to-be. The tribe's leader gave the captive two options. He said, "You should say something meaningful. If it is true, we will eat you alive. If it is not true, we will cook you and then feast on you."

[Can you help the poor captive to craft the best sentence which could at the very least buy him time?]

Quivering, the man said, "I am a liar."[9]

Upon hearing this, the chieftain immediately declared, "He is a liar," and ordered his tribesmen to light the fire.

"Wait!" said the deputy chief, his booming voice breaking the deafening silence. "If he is a liar, then what he said is true, and therefore we must eat him alive."

"No," the chieftain said, defiantly. If what he said is true, then he is a liar and therefore we should cook him first, and then eat him."

Before the chieftain could even finish his sentence, the deputy chief said, "No, he admitted that he is a liar and therefore his statement is false, and therefore we must. . . ."

They continued to argue. . . .

Clearly, the hapless captive provided *information*. But what kind of information was it? Was it true or false, was it registered somewhere? Was it created and could it be destroyed? Whatever your answers are, you must admit that the "information" provided by the captive bought him some time.

1.1.5 *Can information flow, or be created or destroyed?*

Of course, we say almost every day that "one person transmitted information to a second person." *Information* is transmitted through communication channels. The efficiency of such transmission was the main concern of Shannon when he developed the "mathematical theory of communication," which later became known as information theory. We say that information *flows* from the transmitter to the receiver or from one person to another. The verb "flow" is oftentimes used in C-information; also, the verbs "create" and "destroy" are used in connection with C-information.

A writer who writes a novel certainly *creates* information (which might be true, false, or imaginary). On the other hand, we say that a criminal *destroys* all the evidence, or all the information which links him to the crime. If we burn all the books and all the records on the history of the Mayan civilization, we destroy this information, and perhaps this information will never be recovered. The question whether or not this information is still "out there" in the ashes and smoke of the burnt book is in my opinion more of a philosophical question.

I raise the matter of these verbs that we attach to information in our daily usage to contrast it with their usage in connection with information theory, which we shall discuss in the next section.

Some people claim that since information must be conveyed by some physical means — a book, a magnetic tape or in our neural network — it cannot be destroyed or created. If you burn a book, they claim that the information is still there in the ashes and smoke. Unfortunately, this kind of information can never be retrieved from the ashes and smoke of the burnt book.[10]

Another question sometimes asked is: Where does the information reside? Colloquially we say that information resides in a book, in a tape, on a CD, or in our brain. However, *information* as an abstract concept does not need to reside in any physical object. Suppose that a crime was committed, and an unidentified body was found with clear signs of murder. The person who committed the crime has not been found. Obviously, he has the information about the criminal. We can say that the information resides in him. After hundreds of years, people would say that they do not know who the criminal was, and that the criminal "took the information to his grave." Does that information reside in the grave, or still in the criminal?

1.1.6 *Can information be measured?*

Most popular books on information theory would say yes, information is measurable, and that is the greatest achievement of Shannon's measure of information theory.

In my view the *measure* (in bits) of information as used in information theory is very different from the measure (or measures) that we associate with the colloquial information.

Two messages can have different "sizes," but convey the same information. On the other hand, two messages can have the same "size," but convey very different information.

Consider the following examples:

(1) "The population of Xland is one million, nine hundred and fifty-five thousand, and seven hundred and twenty-two people.
(2) The population of Xland is 1,955,722 people.

Clearly, these two messages convey the same information. However, the second is *shorter* (in the sense that it contains a smaller number of symbols), and thus will cost less to transmit than the first.

Now, consider the following messages:

(3) Bob loves Linda.
(4) Linda loves Bob.
(5) Lov bebo adinl.

These three messages have exactly the same *size*, but the first two convey different information. The third does not convey any information.

Another interesting example is:

(6) Only Bob loves Linda.
(7) Bob only loves Linda.
(8) Bob loves only Linda.

Clearly, these three messages have the same *size*, but they convey different information. Here, we refer to "size" in the sense of the number of letters. In Section 1.2, we shall introduce Shannon's measure of information, which depends not only on the number of letters but also on the frequencies of the various letters in a particular language.

1.1.7 *Precision of information*

I tell you that the length of the circumference of a circle having a radius of 1 cm is 2π cm. If you know elementary geometry you will know that this information is *true*. However, if I tell you that the length of the circumference of a circle having a radius of 1 cm is 6.2831853071795864769 cm, there is no way you can tell whether this answer is *exact* or not. To verify this information you would have to measure the length with a precision that you cannot achieve with any given measuring device (presuming that you "know" that the radius is *exactly* 1 cm).

In Section 1.2 we will play various 20-question games. I tell you that a dart hit a point of a line of length 1 cm, and you have to guess where exactly the dart hit [Figure 1.3(a)]. To find out you will need to ask an infinite number of questions. The reason is that there are an infinite number of points between zero and one.

In practice we are always interested in a finite precision. So, we divide the length of 1 cm, say, into n small intervals, each of

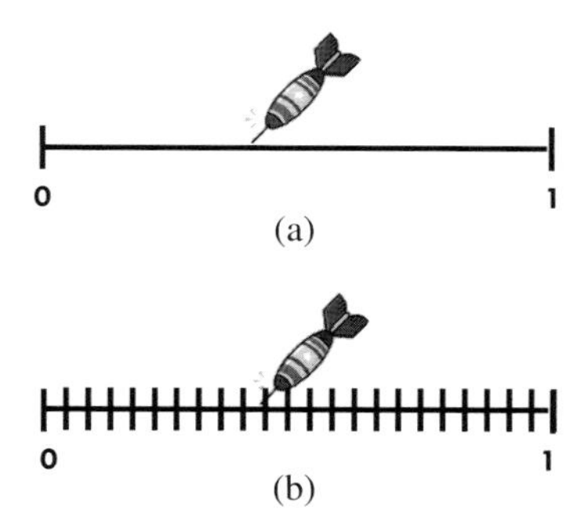

Fig. 1.3 (a) A dart hitting a point on the segment (0, 1), (b) the segment is divided into n equal small segments.

length $1/n$ cm, and ask where the dart hit [Figure 1.3(b)]. In this case you can find where the dart hit by asking a finite number of questions. No matter into how many intervals we divided the length between zero and 1 cm, for as long as it is finite, you can find out the exact interval by asking yes/no questions. However, if I tell you that the dart hit a point which is a rational number (a ratio of two integers), then you will have to ask an infinite number of questions. Thus, since there are an infinite number of rational points between zero and one, you would need an *infinite* number of questions to get the *information* on where the dart hit the segment (0, 1).

If I tell you that the dart hit the exact point $\sqrt{2}$, then you might need "more" than an infinite number of questions. The reason is that the number of points in the segment (0, 2) is *larger* than all the rational numbers between zero and two. $\sqrt{2}$ is an *irrational* number, and as such cannot be expressed as a ratio of two integers. The proof is in the Notes.[11]

1.1.8 *Independent information and mutual information*

The concept of dependence and independence between information is important in information theory. We shall present here only a qualitative discussion of dependence between various pieces of information. A quantitative measure of dependence between pieces of information is discussed in Section 1.2.

A quick example is the following:

Consider the following two items of information:

Today it is raining in Jerusalem.
Today it is raining in New York.

Clearly, we feel that knowing the information on the weather in Jerusalem (or in New York) does not provide any information on the weather in New York (or in Jerusalem). On the other hand, consider the following:

Today it is very cloudy in Jerusalem.
Today it is rainy in Jerusalem.

In this case, we *feel* that information on the amount of clouds in Jerusalem provides some information on whether or not it is raining in Jerusalem.

The examples given above are very qualitative. Here is an example where the dependence between pieces of information can be quantified:

You are shown an urn with four marbles: two blue and two red. Blindfolded, you draw one marble, peek at its color and return it to the urn, and draw another marble while still blindfolded.

Clearly, knowing the marble's color in the first draw does not provide any new information about the probability of drawing a red or a blue marble in the second draw.[12]

Now, we modify the experiment. As before, you draw one marble and record its color. You *do not* return the marble to the urn, and draw another marble. In this experiment it is clear that the information on the color of the marble in the first draw affects the information (in probability terms) on the color in the second draw.

For example, the probability of drawing a blue marble in the first draw is $\frac{1}{2}$. Knowing that a blue marble was picked in the first draw affects the probabilities of the outcomes blue and red in the second draw (here $\frac{1}{3}$ and $\frac{2}{3}$, respectively).

It is interesting to know that the same dependence holds if we *know* the outcome of the *second* draw, and we ask about the probability of the outcomes for the *first* draw.[13]

Here is another example where we can estimate the extent of dependence between two events:

Suppose that you throw two fair dice at a very large distance from each other. We assume that the probability distribution of the outcomes {1, 2, 3, 4, 5, 6} is uniform, i.e. each outcome has the same probability of occurrence. Clearly, knowing the outcome for one dice does not tell us anything about the probability of the outcome for the second dice. We say that the two experiments are independent.

Now, we add little magnets to each of the dice. If we throw the two dice far apart from each other, the outcomes will be independent. However, if the dice are thrown very close to each other, then we expect that the two magnets will affect the relative orientations of the two dice. The smaller the distance between the dice, the larger the dependence between the outcomes for the two dice.

In probability theory we measure the extent of the dependence by a correlation function. We talk about a positive correlation and a negative correlation. In the first case, given an outcome in one experiment *enhances* the probability of an outcome in the second experiment. In the second case, given an outcome in one experiment *diminishes* the probability of an outcome in the second experiment.[14]

In Section 1.8, we will discuss the extension of the idea of independence between pieces of information which measures, on average, how much information is obtained in one experiment knowing the outcome of another experiment.

Another example involving dependence between different pieces of information is the following:

Suppose that we have three diseases: X, Y and Z.
The symptoms of the disease X are $\{S_1, S_2, S_3, S_4\}$.
The symptoms of the disease Y are $\{S_4, S_5, S_6\}$.
The symptoms of the disease Z are $\{S_7, S_8, S_9, S_{10}\}$.

Now, in a given population it is known that the percentages of people with the diseases X, Y and Z are 20%, 30% and 10%, respectively. Also, assume that there is some distribution of the symptoms $S_1, S_2, \ldots, S_{10}$ in the entire population. Without specifying the conditional probabilities connecting the various symptoms and the disease, it is clear, in a qualitative manner, that given that a person has a set of symptoms might give us some *information* on the probability of having a specific disease — say, X. Again, we may say that given the information on the symptoms provides some information on the disease. It is also true that given information on the disease provides information on the symptoms. In Section 1.8, we shall introduce the terms of conditional information and mutual information to quantify the extent of dependence between the two experiments.

1.1.9 *Redundancy in information*

Another concept that is used in information theory is the redundancy of a message. Here, we introduce the concept of redundancy in a qualitative manner. Suppose that a store sells three different fruits; apples, oranges and bananas. A person enters the store, looks at the fruits and asks the vendor:

Please give me 1 kg of each fruit you have.

Please give me 1 kg of apples, 1 kg of oranges and 1 kg of bananas.

Clearly, the two messages are equivalent. The vendor will weigh 1 kg of each fruit. It is also clear that the second message is longer; the additional specification of each of the fruits is redundant.

Another example is the following:

At one TV station the weather forecast for next week is: It will rain every day in the coming week. At another TV station, the same forecast is stated as: On Sunday it will rain, on Monday it will rain, on Tuesday it will rain, on Wednesday it will rain, on Thursday it will rain, on Friday it will rain, and on Saturday it will rain.

Obviously, the two forecasts provide the same *information.* It is clear that the specification of each of the days in the second is redundant.

Redundancy is sometimes used in daily life to avoid any misunderstanding. At times it involves the repetition of the same message again and again. In other cases one repeats the same message using slightly different words but the overall message is the same. The famous biblical story of Jacob and Laban is well known:

Jacob met Rachel near a well and fell in love with her. He then went to her father, Laban, who happened to have another daughter who was older than Rachel, named Lea.

Jacob said to Laban, "I will serve you seven years for Rachel, your younger daughter." (Genesis 29:18[15])

It is believed that Jacob wanted to confirm the accuracy of the message, and therefore he added the redundant information "your younger daughter."

Today, the phrase "Rachel, your younger daughter" is used as an idiomatic expression in Hebrew, especially among lawyers who want to remove any doubt or ambiguity in their statements. Of course, everyone notices that there is redundancy in this phrase. Laban knew that Jacob was in love with Rachel, and there was no need to add that Jacob was referring to Laban's *daughter*, and that he was referring to Laban's *younger* daughter. Perhaps Jacob knew about the dubious character of Laban and in order to avoid any misunderstanding he added "your younger daughter" to ensure that Laban would get the right message and could not claim any misunderstanding.

The irony of the story is that in spite of Jacob's unambiguous message, Laban did cheat Jacob and under the cover of darkness he gave him his older daughter, Lea, instead of Rachel.

Sometimes lack of redundancy might lead to ambiguity. Sometimes lack of explicit information is filled up implicitly by the receiver. I took the following story from Hofstadter[16]:

A father and his son were on their way to a ball game when their car stopped right in the middle of the railroad tracks. What began as a distant rumbling became increasingly loud, as the approaching train made its way to the crossing. The train's deafening whistle blew ominously just meters away from the stalled car. The father tried to start the engine but panic overtook him, numbed his fingers and he could not turn the key. The other motorists could only watch in shock as the speeding train smashed into the car, dragging it and its hapless occupants at death's doorstep. Shortly after the crash, an ambulance came and picked them up, but before they could even get to the hospital the father expired. The son, although still alive, was in a very serious condition and needed immediate surgery. As soon as they arrived at the hospital, the father was declared dead and was taken to the morgue while the son was immediately brought to the operating room. Responding to a code blue, the surgeon walked into the emergency operating room calmly. However, upon seeing the boy lying on the gurney, the surgeon's face seemed to have been drained of blood and said, "I cannot operate on this boy — he is my son."

What exactly just happened? Had a kept secret been unlocked? Was the dead father the son's real father?

What, then, is the explanation? Think it through until you have figured it out on your own — I insist! You'll know when you've got it, don't worry.

You didn't explicitly ponder the point and ask yourself, "What is the more plausible sex to assign to the surgeon?" Rather, you merely let your past experience assign the sex for you. Default assumptions are by their nature implicit assumptions. You never were aware of having made any assumption about the surgeon's sex, for if you had been, the riddle would have been easy to solve!

This ability to ignore what is very unlikely — *without even considering whether or not to ignore it!* — is part of our evolutionary heritage, coming out of the need to be able to size up a situation quickly but accurately. It is a marvelous and subtle quality of our thought processes; however, once in a while, this marvelous ability leads us astray. And sexist default assumptions are a case in point.

This story demonstrates how people "fill in" information that is not given explicitly. This would not occur in a language such as Hebrew, since the statement "I can't operate. . . . " would instantly reveal the gender of the surgeon.

Sometimes the whole information given is redundant. This is when everyone already knows that information.

Once, I visited a friend in Bergen, Norway. While I was there, it had been raining for a whole week. My friend told me that it rains in Bergen almost every day of the year, and everyone is aware of it and thus they never fail to take the umbrella when they leave home. He added that the weather forecaster on the radio is proud that his "formula" for predicting the weather is the best in the whole world. "What is the formula?" I asked.

"Very simple: If today it rains, he predicts that tomorrow it will rain. If today it does not rain, he predicts that tomorrow it will rain. Admirably, he was correct 99% of the time. . . . "

That is really an impressive rate of success in predicting the weather. I was wondering why the forecaster bothers to "predict" and inform the people of Bergen about what they already know.

Exercise: Name a US president who is not buried in North America.[17]

Hofstadter (1985) quotes from some poems which were intentionally written to be nonsensical poems. It is clear that the ultimate redundant message is a nonsensical message — or a message that conveys no information at all. As we will see in Section 1.2, even nonsensical messages can be assigned a measure. In the same section, we will also discuss redundancy as used in information theory. This redundancy is only qualitatively related to the redundancy of information as used in daily life. On the one hand, a message can be highly redundant in a colloquial sense, but not having high redundancy in information theory. On the other hand, a message can have zero redundancy in a colloquial sense, but high redundancy in the information-theoretical sense. It is unfortunate that recent books on information theory muddle up the two kinds of redundancy.

1.2 Shannon's Measure of Information

The best way to start is with a quotation from Shannon's article "A Mathematical Theory of Communication."[18] In 1948 he published that landmark paper. In Section 6 of the paper he writes:

> Suppose we have a set of possible events whose probabilities of occurrence are $p_1, p_2, \ldots, p_n$. These probabilities are known but that is all we know concerning which event will occur. Can we find a measure of how much "choice" is involved in the selection of the event or how uncertain we are of the outcome?
>
> If there is such a measure, say $H(p_1, p_2, \ldots, p_n)$, it is reasonable to require of it the following properties:
>
> 1. H should be continuous in the p_i.
> 2. If all the p_i are equal, $p_i = \frac{1}{n}$, then H should be a monotonic increasing function of n. With equally likely events there is more choice, or uncertainty, when there are more possible events.
> 3. If a choice be broken down into two successive choices, the original H should be the weighted sum of the individual values of H.

Then Shannon proved that:

> The only H satisfying the three assumption above is the form:
>
> $$H = -K \sum p_i \log p_i.$$

In this book, we will not be interested in the *derivation* of the quantity H, which we will refer to as Shannon's measure of information (SMI). In this chapter, we will only be interested in the *properties* of SMI, and its *meaning* as a measure of information. The relevance of H to thermodynamics will be discussed in Chapter 2. At the moment, we will study the quantity H defined above without any reference to thermodynamics. Let us quote another paragraph from Shannon's paper:

> This theorem, and the assumptions required for its proof, are in no way necessary for the present theory. It is given chiefly to lend a certain plausibility to some of our later definitions. The real justification of these definitions, however, will reside in their implications.
>
> Quantities of the form $H = -K \sum p_i \log p_i$ (the constant K merely amounts to a choice of a unit of measure) play a central role in information theory as measures of information, choice and uncertainty. The form of H will be recognized as that of entropy as defined in certain formulations of statistical mechanics where p_i is the probability of a system being in cell i of its phase space. H is then, for example, the H in Boltzmann's famous H theorem. We shall call $H = -K \sum p_i \log p_i$ the entropy of the set of probabilities $p_1, \ldots, p_n$.

Note carefully that Shannon describes H as a "measure of information, choice and uncertainty." All these are valid interpretations of the quantity H, as defined above. Shannon goes on to say that "the *form* of H will be recognized as that of *entropy* as defined in certain formulations of statistical mechanics where p_i is the probability. . . . "

We quoted above a small section of Shannon's article. Most of his article is concerned with the theory of *communication*: problems of coding and decoding, efficiency of transmission of information, etc.

The reader is urged to read carefully the above quotation. You do not need to understand the details of the various statements. Note, however, a few points that will be crucial for understanding the relevance of SMI to thermodynamics.

First, note that Shannon formulated his problem in terms of a *probability distribution* $p_1, \ldots, p_n$. He sought a measure of how much "choice" or "uncertainty" there is in the outcome, and he later referred to the quantity H as a measure of "information, choice and uncertainty."

Shannon did not seek a measure of the general concept of information — only a *measure* of *information contained in* or *associated with* a probability distribution. This is an important point that one should remember whenever using the term "information" either as a measurable quantity or in connection with the Second Law of Thermodynamics.

Second, Shannon proposed three plausible properties of such a measure, *presuming* that such a measure exists. We shall discuss these properties and their plausibility in the following sections of this chapter. Here, the attention of the reader is drawn to the "methodology" of seeking and finding a quantity whose existence is not even clear. Again, we note that the properties specified by Shannon do not apply to the *general concept* of *information* (referred to as C-information in Section 1.1) — only to a specific class of information. Try to apply the properties (1), (2) and (3) to "information" of your choice and see if it makes sense.[19]

Finally, note carefully that Shannon was not interested in thermodynamics in general, or in *entropy* in particular. However, he noted that "the form of H will be recognized as that of entropy as defined in certain formulations of statistical mechanics. . . ." Therefore, he suggested calling H "the entropy of the set of probabilities $p_1, \ldots, p_n$."

Indeed, the *form* of the function H is the same as the *form* of the entropy as used in statistical mechanics. However, the fact that the *form* of H is the same as the form of the entropy in statistical mechanics *does not* imply that H is entropy. We will further discuss this point in Chapter 2, after we learn some of the properties of SMI and entropy. For the moment we will study SMI without any reference to entropy. However, the reader should be aware of the fact that in many applications of the concept of SMI, the concept of entropy has also been involved. This fact has caused great confusion in both information theory and thermodynamics.

SMI is a very general concept. It is defined on *any discrete distribution function*. Examples are the outcomes of throwing a dice and the frequencies of the appearance of letters of the alphabet in certain languages. There is a vast range of fields in which the quantity H is definable, and as such SMI became a very useful tool in so many fields of research.

As we will see in Chapter 2, the entropy is defined only on tiny small sets of probability distributions. Thus, when H is applied to those distributions used in statistical mechanics, it is identical with the statistical mechanical entropy. Thus, the statistical mechanical entropy is a particular case of SMI, but SMI is in general not the entropy. Unfortunately, confusion of the two concepts abounds. The source of this confusion is probably von Neumann's suggestion to Shannon to name the quantity H "entropy." The story is told by Tribus and McIrvive (1971):

> What's in a name? In the case of Shannon's measure the naming was not accidental. In 1961 one of us (Tribus) asked Shannon what he had thought about when he had finally confirmed his famous measure. Shannon replied: "My greatest concern was what to call it. I thought of calling it 'information,' but the word was overly used, so I decided to call it 'uncertainty.' When I discussed it with John von Neumann, he had a better idea. Von Neumann, told me, 'You should call it entropy, for two reasons. In the first place your

> uncertainty function has been used in statistical mechanics under that name, so it already has a name. In the second place, and more important, no one knows what entropy really is, so in a debate you will always have the advantage."

In my opinion, naming Shannon's measure of information "entropy" was a mistake. This has caused a great amount of confusion in thermodynamics, as well as in many other fields of science. I do not agree either with the statement that "no one knows what entropy is."

It should be noted here that many authors confuse entropy (thermodynamic entropy) with SMI (referred to as Shannon entropy), and SMI with information. A typical example may be found in Gleick (2011):

> Indeed, H is ubiquitous, conventionally called entropy of a message, or Shannon entropy, or simply, the information.

In this book, we will refer to the quantity H defined above as *Shannon's measure* of *information* (SMI). We will not be interested in the formal proof of the uniqueness of this function. Instead, we will survey the properties and the meanings of the quantity as defined above.

Several comments are in order before we discuss the meaning, the properties and the applications of SMI. The first thing to note is that both in the title of the article and in the introduction, Shannon was interested in *communication* theory, not information theory. More specifically, he was interested in transmitting a message from a source to a receiver. The message might or might not have any meaning. The meaning of the *information* carried by the message is irrelevant to the theory he sought to develop.

The second point is that after finding the quantity H which fulfills certain requirements, Shannon says that this quantity "plays a central role in information theory as a measure of information, choice and uncertainty." It should be emphasized that the meaning of H as a "measure of information" might be misleading. We shall discuss shortly in what sense H is a "measure of information."

In this book, we will reserve the term "entropy" for the quantity defined in thermodynamics. The quantity derived by Shannon will be referred to as Shannon's measure of information (SMI). In Section 2.2 we will see how the thermodynamic entropy may be obtained as a special case of SMI. It should be stressed, however, that neither *entropy* nor SMI is *information* in its colloquial sense.

In the following subsections we will discuss the meanings of SMI, some of its properties, and some examples of its applications.

1.2.1 *Interpretations of SMI*

In this subsection we discuss three interpretations of SMI. The first is an average of the uncertainty about the outcome of an experiment; the second, a measure of the unlikelihood; and the third, a measure of information. We will use the letter H for the quantity defined above, and refer to it simply as SMI. It should be noted again that SMI is a measure of information in a very restricted sense. The meaning, the value, the importance or the content of the information is irrelevant to SMI.

1.2.1.1 *The uncertainty meaning of SMI*

The interpretation of H as an average *uncertainty* is very popular. This interpretation is derived directly from the meaning of the probability distribution.

Suppose that we have an experiment yielding n possible outcomes with probability distribution $p_1, \ldots, p_n$. If, say, $p_i = 1$ we are certain that the outcome i occurred or will occur. For any other value of p_i we are *less certain* about the occurrence of the event i. *Less certainty* can be translated to *more* uncertainty. Therefore, the larger the value of $-\log p_i$, the larger the extent of uncertainty about the occurrence of the event i. Multiplying $-\log p_i$ by p_i, and summing over all i, we get an *average* uncertainty about all the possible outcomes of the experiment.

1.2.1.2 *The unlikelihood interpretation*

A slightly different but still useful interpretation of H is in terms of *likelihood* or *expectedness*. These two are also derived from the meaning of probability. When p_i is small, the event i is unlikely to occur, or its occurrence is less expected. When p_i approaches 1, the occurrence of i becomes more likely or more expected. Since $-\log p_i$ is a monotonous increasing function of p_i, we can say that the larger the value of $-\log p_i$, the larger the likelihood or the larger the expectedness of the event. Since $0 \leq p_i \leq 1$, we have $-\infty \leq \log p_i \leq 0$. The quantity $-\log p_i$ is thus a measure of the *unlikelihood* or *unexpectedness* of the event i. Therefore, the quantity $H = -\sum p_i \log p_i$ is a measure of the *average unlikelihood* or *unexpectedness* of the entire set of events.

1.2.2 *The meaning of SMI as a measure of information*

As we have seen, both the uncertainty and the unlikelihood interpretation of H are derived from the meaning of the probabilities p_i. The interpretation of H as a measure of information is a little trickier and less straightforward. It is also more interesting, since it conveys different *information* on the Shannon measure of *information* (I hope you do not mind my repetitions of the word "information" in this sentence). It should be emphasized from the outset that SMI is not *information*. Also, it is not a measure of any type of information, but of a very particular kind of information. The confusion of SMI with information is almost the rule, not the exception, for both scientists and nonscientists.

Some authors assign to the quantity $-\log p_i$ the meaning of information (or self-information) associated with the event i.

The idea is that if an event is rare, i.e. p_i is small and hence $-\log p_i$ is large, then one gets more information when one knows that the event has occurred. I do not agree with this interpretation. When I know that an event i has occurred, I have got the *information* on the occurrence of i. I might be surprised to learn that a rare event has occurred, but the size of the *information* one gets when the event occurs is not dependent on the probability of that event.

Both p_i and $\log p_i$ are measures of the uncertainty about the occurrence of an event. They do not measure *information* about the events. Personally, I prefer not to refer to $-\log p_i$ as information (or self-information) associated with the event i. Hence, H is not interpreted as average information associated with the entire experiment. Instead, I prefer to assign "informational" meaning directly to the quantity H, rather than to the individual events.[20]

In the following subsections, we will develop the interpretation of H as a measure of information. We will start with the familiar 20-question (20Q) game. In this game the player is seeking *information* by asking questions. This leads to a new interpretation of the quantity H in terms of the *number* of questions one needs to ask in order to obtain that information. The reader is urged to read carefully the following subsections. In addition to their entertaining aspects as "games," we will see in the next chapter that these games led to an interpretation of entropy.[21]

1.2.3 *The 20-question game*

To discover the informational meaning of SMI, join me in playing a few 20-question games (20Q). Suppose that I am thinking of a person, and you have to ask *binary* questions (i.e. questions answerable by "yes" or "no" only). Your task is to find the person with as few questions as possible. You might start by asking questions like:

Is it Einstein?
Is it Gandhi?
Is it Mozart?
And so on.

Of course, there is a chance that you might guess the right person in the first or second question. That is a possibility. However, the chance is quite low. A better strategy is to ask questions like:

Is the person a male?
Is the person alive?
Is the person a scientist?
And so on.

With this *strategy* of asking questions, you *cannot* possibly win after one, two or three questions. However, you might feel that by choosing this strategy the chances of winning the game are much better than for the previous strategy. The last sentence may sound paradoxical at first: in the first strategy you *can* win with one, two or three questions, yet I claim that the second strategy is the better choice.[22] Note that "winning" means finding the chosen person. We shall quantify this game soon and make it more precise.

All that has been said so far is very qualitative. We have all played this kind of games as children do, without giving much thought to the rules of the game.

It is clear that many assumptions are made implicitly. For instance, we must limit the size of the group of people from which I am allowed to choose. Also, we have to agree that I will choose a person who exists or existed, and that we both know him or her. There are other rules that we have to specify if we are going to analyze this game in a scientific manner. These rules will be made soon.

Besides the rules of the game, the players may use all kinds of information, which may reduce the number of questions. Therefore, in order to *formalize* the game we must define it in a precise and objective manner.

Before we *formalize* our 20Q game, pause to ponder: Why 20 questions — why not a hundred or thousand questions?

When we play the 20Q game, we usually do not specify the rules of the game. However, it is implicitly assumed that we have about a hundred, a thousand or ten thousand objects. You might be surprised to discover that with 20 questions you can find an object (or a person) from a set of about a million objects. More precisely, if you choose one out of 1,048,576 objects, you can find the chosen object by asking 20 binary questions (presuming that you are smart enough to ask the right questions). Simply divide the entire set of all possibilities into two equal halves at each step and at the 20th question you will have the answer. (See Table 1.2.)

1.2.4 *A well-defined 20Q game*

Let us now define the 20Q game more precisely, and in such a way that it will be devoid of any subjective elements.

Suppose that I show you a rectangular board divided into eight equal squares (Figure 1.4). I throw a dart onto this board and I tell you that it hit one of the squares. Your task is to find out where the dart hit the board by asking binary questions, i.e. I can answer "yes" or "no." I also tell you that I threw the dart at *random*, which means that each of the eight squares is equally likely to get the dart (we ignore the possibility that the dart hit

Table 1.2 Reduction in the number of possibilities left after each question.

Question 1:	1,048,576
Question 2:	524,288
Question 3:	262,144
Question 4:	131,072
Question 5:	65,536
Question 6:	32,768
Question 7:	16,384
Question 8:	8,192
Question 9:	4,096
Question 10:	2,048
Question 11:	1,024
Question 12:	512
Question 13:	256
Question 14:	128
Question 15:	64
Question 16:	32
Question 17:	16
Question 18:	8
Question 19:	4
Question 20:	2

any borderline between the squares, and we assume that one of the squares did get the dart). To make this game a little more dramatic, suppose that you have to pay a dollar for each answer you get, and when you find out where the dart hit, you get five dollars. Assuming that you want to maximize your earnings, how would you plan to ask questions? Of course, as in the 20Q game we discussed above, there are many ways or *strategies* of asking questions. Two extreme strategies are shown in Figure 1.4. The first is referred to as the "dumbest" strategy. You simply ask *direct* questions: "Is the dart on square 1?" "Is the dart on square 2?"

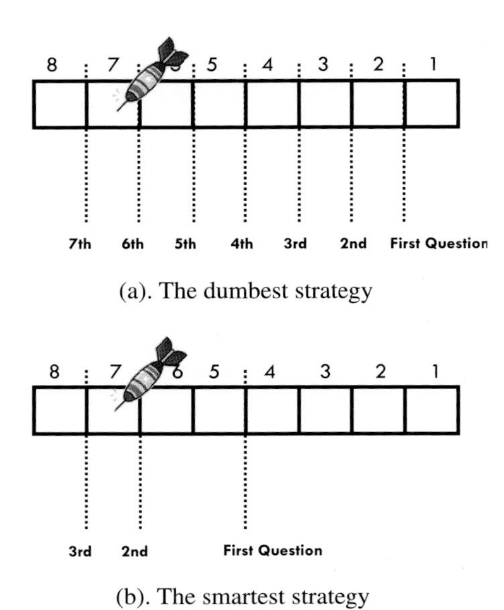

Fig. 1.4 Different strategies of asking binary questions.

And so on until you get a "yes" answer and win the prize. The second extreme strategy is referred to as the "smartest." Here, you divide all the eight squares into two groups of four squares, and ask if the dart is on the left side. If you get a "yes," then remove the four squares on the right, and divide again the remaining four squares on the left into two groups, each containing two squares, and ask, "Is the dart on the right?" If you get a "no," you ignore the two on the right, and you are left with only two squares on the left. Now you ask, "Is it on the right?" Whatever answer you obtain you will know where the dart is after the third answer.

As we have discussed the 20Q game in connection with choosing a person, it is more advantageous to divide all the people into two roughly equal groups — say, men and women, dead or alive, and so on. Of course, in general you cannot divide precisely into two groups with an equal number of people, but in this particular game, as shown in Figure 1.4, this can be done.

If you choose the dumbest strategy and ask, "Is it on square 1?" "Is it on square 2?" etc., you might win after one or two questions. However, in the smartest strategy, you cannot win after either one question or the second question, but you are *guaranteed* to win the prize after the third question.

One can prove mathematically that in any game of equally likely N squares, if you play the game many times you will have to ask on average about $N/2$ questions in the dumbest strategy, but only $\log_2 N$ (base 2 for the logarithm) in the smartest strategy. One can also prove that the dumbest strategy is indeed the dumbest (i.e. on average you will ask the maximum number of questions), and the smartest strategy is indeed the smartest (i.e. on average you will ask the minimum number of questions).

Figure 1.5 shows how the average number of questions varies with the number N. The two curves are very different. To

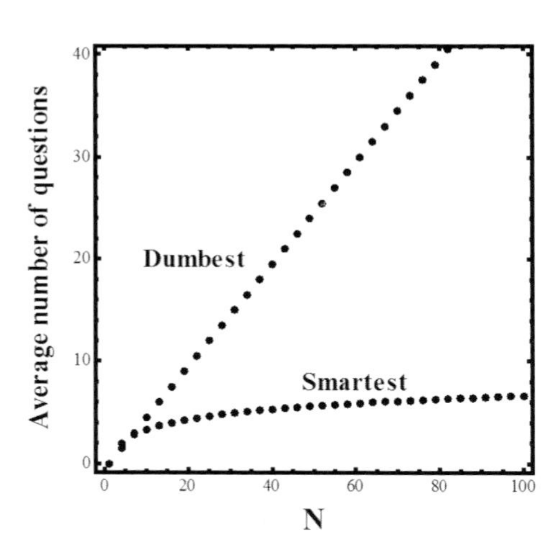

Fig. 1.5 The average number of questions one needs to ask as a function of N using the "dumbest" and the "smartest" strategies.

appreciate the difference, suppose that we play the game with about a million squares. In the dumbest strategy you will ask on average about 500,000 questions. With the smartest strategy you will ask only 20 questions! This is a big difference if you are paying one dollar for each question!

1.2.5 *Choice between uniformly distributed games*

By "*uniformly* distributed game" we mean that the dart is equally likely to hit any one of the N squares. This specification is important when we generalize to nonuniformly distributed games below. Note also that in the conventional 20Q game involving the choice of a person, it is difficult to guarantee that the person I chose was totally random. In fact, the probability of choosing a certain person is never mentioned when we play the parlor game in a party. As we shall see, knowing or even guessing the probabilities of the different people to be chosen might help the player in the planning of asking questions.

In Figure 1.6, we have three games, with $N = 8, 16, 32$. We play the same game as before. I tell you that the dart hit one of the regions on the board. Each region has the same area, and therefore the probability of hitting a point in one of the regions in case A is $\frac{1}{8}$, in B $\frac{1}{16}$ and in C $\frac{1}{32}$. Again, you pay a dollar for each question, and when you find in which region the dart is, you get the prize of five dollars. I assume that you are interested in

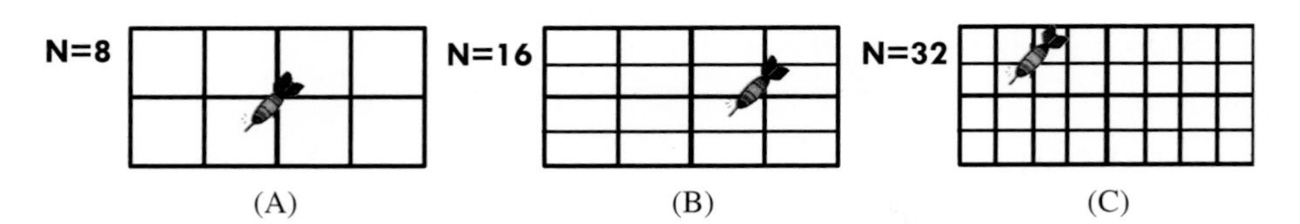

Fig. 1.6 Three different uniformly distributed games.

maximizing your earnings and therefore you will always choose the *smartest* strategy. (This is possible in these games since N is of the form 2^n, and n is an integer. For any other N, it can be shown that you will need to ask about $\log N \pm 1$ questions by using the smartest strategy.[23])

Before we play the 20Q game, let me ask you: Which of the games — A, B or C — will you choose to play under the same rules as described above?

For the choice I gave you in Figure 1.6, your decision is relatively easy (it will be more difficult in the nonuniform games discussed below).

Let us do a quick calculation:

For game A: Three questions are needed; pay three dollars, get five dollars — net earnings two dollars.

For game B: Four questions; net earnings one dollar.

For game C: Five questions; net earnings zero dollars.

Clearly, the best choice is game A. You are guaranteed to earn two dollars in each game you play. (Provided that you are smart enough to choose the smartest strategy. . . .)

The reason I chose these three games was not to excite you about winning the game you chose each time you played it, but to let you observe one important aspect of these games. From A to B we *doubled* the number of *regions*, and we had to add only *one* more question (in the smartest strategy, of course). From B to C we doubled again the number of regions, and we added *one* more question. This is a remarkable result. In these particular cases, the number of questions is $3 = \log_2 8$, $4 = \log_2 16$ and $5 = \log_2 32$ for cases A, B and C, respectively. In general, for N equally probable regions, the number of questions is $\log_2 N \pm 1$. For N

of the form 2^n with n an integer, this is exactly $\log_2 N = n$, but for any N you might need to ask one more or one less question.[24]

In what sense is $\log_2 N$ a measure of information? In this particular game, I *know* where the dart is. I have the *information* on the *location* of the dart. You do not have this information, so you ask binary questions. With each answer you get (for which you pay a dollar), you get some *information* (you first know that it is in the right part, or the upper part of the board, etc.). The more questions you ask the more *information* you get, until you get the *information* on the location of the dart. Thus, the number of questions you ask is related to the *size* of the *information* you are seeking to obtain. The larger the information *contained* in the game, or the larger the amount of information you need to obtain, the larger the number of questions you will have to ask (and the larger the number of dollars you will have to pay to get this information). Thus, we have established a *measure* of the *size* of the *information* contained in each of the games shown in Figure 1.6. The quantity $\log_2 N$ measures the number of questions you must ask in order to obtain the required information.

Note that in the above interpretation of $\log_2 N$ as a measure of the *size* of the *information*, we did not rely directly on the probabilities of the various regions. These probabilities are implicitly given by the assumption of equal probabilities of the outcomes. Hence, the probability of each outcome is $1/N$. (See also the next subsection.) Qualitatively, it is clear that the larger the *uncertainty* you have about the outcome of the experiment (where the dart hit), the more the information you need to obtain in order to remove that uncertainty.

So far the relationship between the *size* of the game and the number of squares has been relatively straightforward. It will become somewhat more difficult to establish a relationship between the number of questions and the *size* of information for nonuniformly distributed games.

One comment before proceeding to nonuniform games. For the particular games with $N = 2^n$ (n is an integer) as in Figure 1.6, in each step you get one bit of information. This is the *maximum* amount of information you can get in one binary question. Therefore, for each game you get the same amount of information by asking the minimum number of questions. This is not possible to do with any N, but we must always plan our strategy in such a way as to extract the maximum possible information at each step, hence asking the minimum number of questions. We shall define the unit of information, the *bit*, and its relation to asking binary questions, in Section 1.3.

1.2.6 *Choice between nonuniformly distributed games*

In Figure 1.7 we show three games. All of these are nonuniformly distributed. This means that each of the board is again divided into N regions, with $N = 4, 8, 16$. The regions have unequal areas, and therefore the chance of hitting a region with a larger area is larger than that of hitting a region with a smaller area. For instance, in game A, the probabilities are $\{\frac{1}{2}, \frac{1}{4}, \frac{1}{8}, \frac{1}{8}\}$.

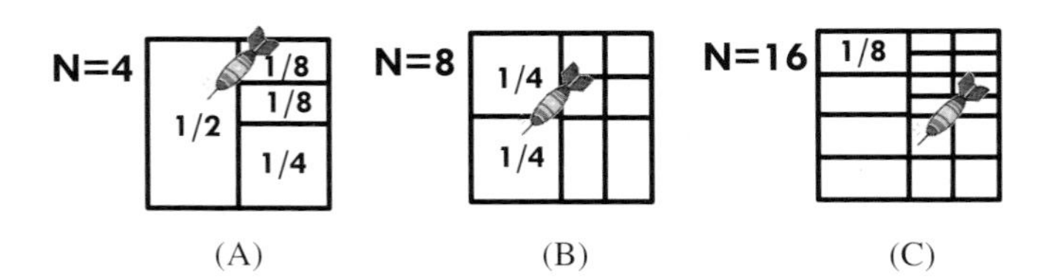

Fig. 1.7 Three different nonuniformly distributed games.

I offer you the opportunity to play the 20Q game as before, with the same rules (one dollar a question; five dollars when you find out where the dart is). Which game will you choose to play in this case?

For the particular games in Figure 1.7, it is also relatively easy to choose. You might make a quick calculation as follows:

In game A, you can find the information (on where the dart is) by asking three questions. You get a prize of five dollars, so you earn two dollars. Not bad if you have to play hundreds or thousands of times.

In game B, you need four questions, and the prize is five dollars, so that one dollar is the net earnings.

In game C, you need five questions, so you will be spending five dollars to earn the prize of five dollars. You will immediately *reject* game C and choose game A.

The choice of game A is indeed the best one among the three games shown in Figure 1.7. However, in your calculations of the expected earnings in the three games, you were *almost* right about your estimated earnings.

Before explaining why, let me leave you with another offer to ponder on. Suppose that I allow you only game C in Figure 1.7 to play under the same conditions (one dollar per question, and a prize of five dollars when you get the information). I am also asking you to play this game 1000 times. Only game C — take it or leave it. Will you play this game?

Let us go back to the three games in Figure 1.7: A, B, C. Obviously, your choice of A was a smart choice. In this game you are *guaranteed* that you will win each time you play the game. As before, the reason is that in this case you can divide the entire board into two equal areas for the first question, and hence you get maximum information. You can do the same for the second

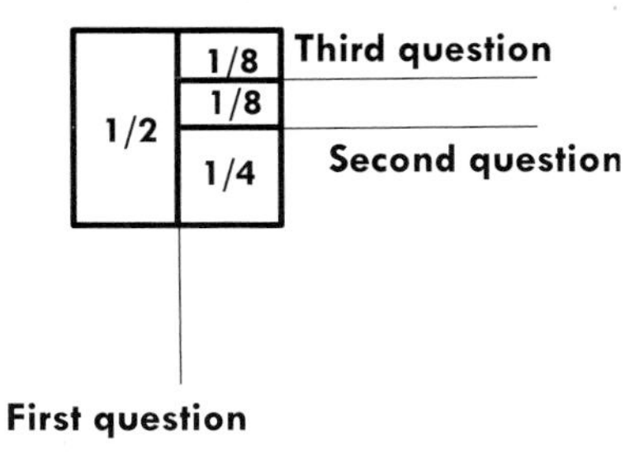

Fig. 1.8 The smartest way of asking questions.

question, and the third question. Thus, for each question you can get the maximum possible information for each dollar you spend. Therefore, you can earn *at least* two dollars each time you play this game. However, you can do better. You can earn more than two dollars per game *on average*, provided that you are "doubly" smart: first, you choose the *smartest* strategy of dividing each time into two equal areas, and second, you have to be even smarter to ask about the *right side*.

Remember the uniform games in Figure 1.6? Once you choose the smartest strategy, you are already guaranteed to win with the minimum number of questions. It does not matter whether you ask "Is it on the right or the left" or "Is it in the lower or the upper part?" In the present game the smartest strategy is to divide as shown in Figure 1.8. It is also true that for each question you can get the maximum information (one bit, as we shall see in Section 1.3). However, for the first question you better ask "Is it on the *left* side?" rather than "Is it on the *right* side?" The reason is that if you ask "Is it on the left side," there is a probability of $\frac{1}{2}$ that you will *win* the prize for *one* question. If you get a "no" answer, then you should ask, after the second question, "Is it in the lower half?" Again, with such a question you can win after the second question. It is only if you get two "no" answers that you must ask the third question.

It is clear that if you choose the smartest strategy *and* always choose that side which has the larger probability of winning for that question, then you can obtain the required information (as well as the prize) with fewer questions. It should be stressed that this is only the *average* number of questions, if you play many times. You might ask the right questions but get a "no" answer from time to time.

How many questions do I need to ask on average to maximize my earnings? The answer is Shannon measure's of information.[25] The values of SMI for the three games in Figure 1.7 are:

$$\text{SMI for A} = 1.75;$$
$$\text{SMI for B} = 2.75;$$
$$\text{SMI for C} = 3.75.$$

You see that you can do much better in game A compared to what you have calculated earlier. Instead of three questions you can on average ask only 1.75 questions (and earn on average $5 - 1.75 = 3.25$ dollars). Similarly, you need to ask on average 2.75 questions in game B, and 3.75 questions in game C.

Thus, SMI is a measure of the amount of information you need to get in a given game. It is also equal to the *average* number of questions you need to ask in order to get this information.

We emphasize the word "average" because if you play game A many times, then there is a high probability ($\frac{1}{2}$) that the dart will hit the left area. In such a case you will win after one question. There is a relatively high probability that it will hit the lower part on the right side. In this case you will win after two questions. And if you get two "no" answers, you will win after three questions.

Therefore, you will on average need fewer than three questions and thanks to Shannon's measure you can calculate the average number of questions, you will ask (if you are smart, of course).

This brings me to the question I asked you about the proposal of playing game C in Figure 1.7. Should you take it or leave it?

The answer is: take it. The SMI for this game is only 3.75. The number of questions is not five, as you might have estimated given the large number of 16 regions. This means that if you play this game many times, you might on *average* earn money ($5 - 3.75 = 1.25$ dollars per game). This is only on average. In some cases, if the dart hits one of the smallest regions in Figure 1.8, you will need to ask five questions. But if it hits one of the larger regions, then you can win with fewer questions. Thus, SMI gives you an estimate of the number of questions you need to ask (if you are smart, of course) only if you play the game many times and each time the dart is thrown anew onto the same board with the same divisions into 16 regions.

1.2.7 *Choice between a uniform (U) and a nonuniform (NU) game*

In Figure 1.9 we compare two games each with four regions. One is uniformly distributed and the other nonuniformly distributed.

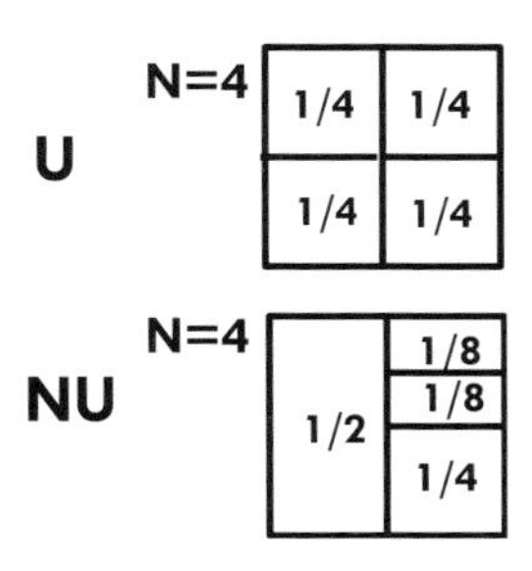

Fig. 1.9 Comparing a uniform and a nonuniform game.

As we have seen, the SMI of the uniform (**U**) game is 2 whereas the SMI of the nonuniform (**NU**) game with the same number of regions is less than 2 (1.75). Recall that from Figure 1.6 we concluded that whenever we have a uniformly distributed game, the larger the number of questions, the more the information you have to gain by asking binary questions.

However, when you have the same number of regions — say, $N = 4$, in Figure 1.9, then it is always true that the SMI of the **NU** game is *smaller* than the SMI of the **U** game. This means that on average we need to ask fewer questions in the **U** case than in the **NU** case.

Note carefully that the information we have in a colloquial sense is larger for the **NU** game than for the **U** game. If we have to describe the game and send this information by a telegram, we should say:

In game **U** there are four regions, each with probability $\frac{1}{4}$.
In game **NU** there are four regions, one with probability $\frac{1}{2}$, one with probability $\frac{1}{4}$ and two with probability $\frac{1}{8}$.

Clearly, the second message describing the **NU** game is *longer* — more words are needed to *describe* the distribution in the **NU** case than in the **U** case.

The fact that the SMI in the **NU** case is *smaller* than that in the **U** case seems paradoxical. But remember that SMI is not the length of the message required to *describe* the game but the amount of information contained in or associated with the game. More generally, if we have an experiment with N possible outcomes, and we know the distribution of the outcomes $p_1, \ldots, p_N$, then we can say that SMI is a measure of the information contained in or associated with this distribution. It is a "measure of information" in the sense of the average

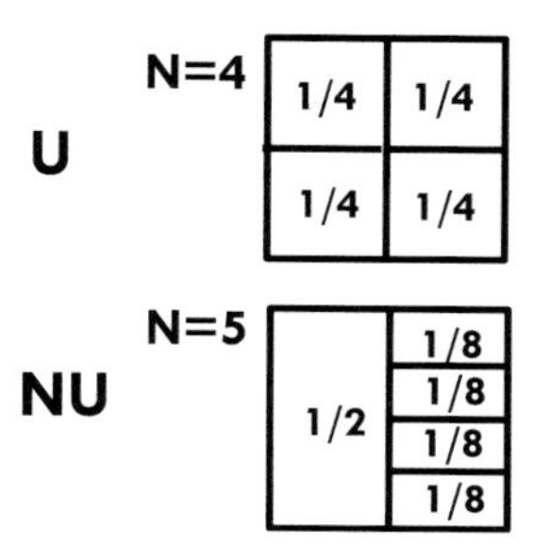

Fig. 1.10 A uniform and a nonuniform game with different N, but having the same SMI$(= 2)$.

number of binary questions we need to ask in order to obtain this information. The larger the SMI, the larger the number of questions. In the example of Figure 1.9 we also see that for a given N the maximum information contained in the game (or the maximum information you have to gain by asking questions) is obtained for the uniform distribution.

Figure 1.10 shows an example of **U** and **NU** games with an unequal number of areas. In this particular case SMI equals 2 for both games. However, in the **U** game, SMI $= 2$ means that you have to ask *two* questions to obtain the information (on where the dart is). On the other hand, in the case **NU** in this figure, 2 is only the *average* number of questions you need to ask. It may be 1 if the dart hits the left region, or it may be 3 if the dart hits the right side of the board — but *on average* the number of questions is 2.

Figure 1.11 shows also that, in general, a larger number of regions (or number of outcomes in an experiment) is not necessarily associated with a larger SMI. In this instance, the **NU** case requires only 1.3 questions on average, much smaller than the case of $N = 4$ with the uniform distribution. I suggest that the reader pause and think about these two cases. Try to

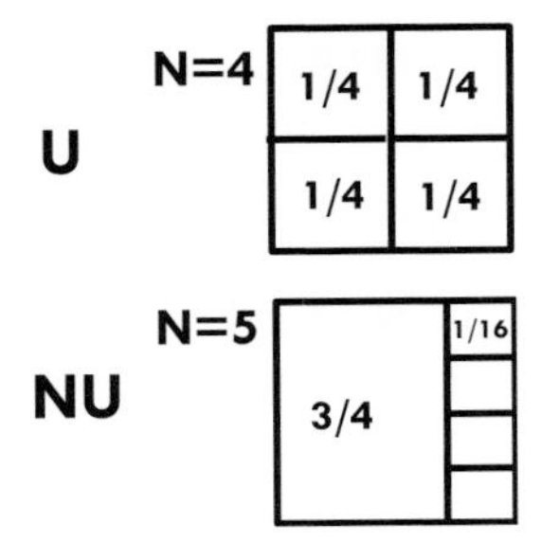

Fig. 1.11 Two games with different N, but the SMI of the $N = 5$ case is smaller than that of the $N = 4$ case.

construct other games having a large number of outcomes but with a small SMI.

1.2.8 *Misinterpretation of SMI as a subjective quantity*

The informational interpretation of SMI has a slippery slope. We have used the 20Q game as an aid for the interpretation of SMI. In this game we refer to the person who *knows* the information and to the person who does not know the information. For instance, in the game of Figure 1.4 we describe the game in which *I know* the information on where the dart is, and *you do not know* that information. In this description of the game, it sounds as if SMI were a subjective quantity. This is a very common misconception, in particular when one is interpreting entropy in terms of SMI.

Indeed, when we play the 20Q game there is information which I know and you do not know. This part of the game is indeed subjective. However, once we defined the SMI of the game we removed any traces of subjectivity. SMI itself *belongs* to the game or is contained in the game — independently of whether we actually play it or not. Think of the 20Q games of

Figure 1.6. We have three games with different values: $N = 8$, 16, 32. These *numbers* are part of the description of the games (or of the experiments in general), and these numbers are independent of whoever is playing the game. You can think of playing this game against a computer, or even one computer playing against another computer.[23]

Thus, the number N in the games of Figure 1.6, or the probability distribution in the games in Figure 1.8, are quantities that *belong* to the game, as much as the number of outcomes in throwing a dice ($N = 6$) belongs to the dice, and is independent of who plays the game, and who knows or does not know the particular outcome of the game.

This pitfall in the interpretation of SMI has caused great confusion and vigorous debates in the interpretation of entropy. We shall discuss this aspect of entropy in Chapter 2. We shall see that the interpretation of entropy as SMI is free of any traces of subjectivity. This conclusion should convince those who have rejected the "informational" interpretation of entropy. The subjectivity associated with the concept of "information" has already been eliminated by the very definition of SMI.

1.2.9 *Summary of the interpretations of SMI*

The interpretation of SMI as "information" should be made with extreme caution. First, because SMI is not *information*. Second, because it is a very specific measure of a very specific kind of information. In Subsection 1.2.5 we have explained in what sense SMI is a measure of information.

One should be even more cautious when interpreting entropy as information. The reason is the following. Suppose that one is pondering about the interpretation of entropy as SMI and

SMI as information. Since information might (or might not) be subjective, and since the relation of "interpretability" might not be transitive, one might reach the conclusion that entropy is (or might be) a subjective quantity. Such a logical deduction has caused *great confusion*. Specifically, some thermodynamicists have rejected the informational interpretation of entropy because this would imply that entropy is not an *objective* quantity. Entropy is as objective as the number of pages in this book, or the number of letters on this page. We shall further discuss the interpretation (and misinterpretation) of entropy in Chapter 2.

Thus, in my view, one should use "entropy" only in the context of thermodynamics. Mathematicians who use this term for any SMI are sometimes careful to refer to "thermodynamic entropy" and "Shannon entropy" (for SMI). However, just talking about the entropy of a dice might be ambiguous. A dice can have a well-defined thermodynamic entropy which depends on temperature, or other specifications of the dice. However, the SMI associated with the probability distribution of the outcomes of the dice has nothing to do with the entropy of the dice. In particular, it is independent of temperature (unless the temperature is so high that the dice will be distorted or melt).

The *uncertainty* interpretation is much better. The smaller p_i is, the larger the uncertainty about the occurrence of the event i will be. Therefore, SMI measures the average uncertainty about all the outcomes of an experiment. It should be noted, however, that when $p_i = 1$, the uncertainty about the occurrence of i is zero. When $p_i = 0$, it is *certain* that i will not occur. It is somewhat awkward to say that in this case the uncertainty (of its occurrence) is minimal, and at the same time that the certainty (of its nonoccurrence) is maximal. Fortunately, this slight awkwardness is "washed away" when we take the product

$p_i \log p_i$, which is zero when $p_i = 0$. Hence, events with a zero probability of occurrence do not contribute to SMI.

The interpretation of SMI as an average *unlikelihood* is in my opinion the best one. Here, the larger the p_i, the greater the likelihood of the occurrence of i. This is true for any value of p_i between zero and one. When $p_i = 0$ the likelihood is zero; when $p_i = 1$ the likelihood is one. Therefore, $\sum p_i \log p_i$ is a genuine measure of the *average* likelihood of occurrence of the outcomes of the experiment. And SMI $(-\sum p_i \log p_i)$ is the average unlikelihood.

To summarize, we saw that the uncertainty and unlikelihood interpretations of SMI are derived from the *meaning* of the probabilities. The informational interpretation is not derived directly from the meaning of probability.

It is appropriate at this point to pause and think about the formal "structure" of SMI. Recall that for any experiment, the outcomes of which are numbers, one can define an average by the expression

$$\langle A \rangle = \sum_{i=1}^{n} p_i A_i.$$

Here, A_i is the numerical value of the outcome i (when the outcomes of the experiments are not numbers, one must first define a random variable and then the average quantity).

SMI has the same "structure" as an average quantity. However, this is a very special average. Here, A_i are not the outcomes of the experiments, but $-\log p_i$, which are related to the probability of the outcome i.

Thus, SMI does not measure the *average information* contained in the message. It is in some sense only a measure of the "size" of the message. It may be *interpreted* as a "measure of information," but that measure has nothing to do with the information itself. It is related to the distribution of letters in an alphabet of a specific language.

Another common misconception is about bits and information. Bits are *units* of SMI. They can be said to be units of information in the sense described in Section 1.3. However, one cannot use bits as a measure of *any* information, simply because not every piece of information, is measurable. The brain receives information through the ears consisting of sound waves, or by specific molecules binding to some specific receptors on our tongues which the brain interprets as different tastes. All these, and many others, are *information* in a colloquial sense, but they are not SMI. Certainly not entropy, as some recent popular-science books claim.

1.3 The Case of an Experiment with Two Outcomes; Definition of the Bit

The case of a two-outcome experiment is important, for several reasons:

(1) The association between SMI and the number of questions is slightly different from the general case of N outcomes.
(2) Understanding the two-outcome case is important for understanding the best strategy of asking questions.
(3) The two-outcome case is the basis for defining the *unit of information*: the bit.

Consider the case $N = 2$. We have only two boxes. I hid a coin in one box. You have to find out where the coin is by asking binary questions. Clearly, by asking one question you will *know* where the coin is. This seems to be in accordance with the conclusion of the previous sections, namely that $\log_2 N = \log_2 2 = 1$, i.e. one question to be asked. Note, however, that in this particular case we did not say anything regarding the probabilities of placing the coin in the first or the second box. Whatever the probabilities are, we must ask one question in order to know where the coin is (we exclude for the moment the case of probability 0 for one box, and probability 1 for the other).

However, we also have the intuitive feeling that if the probability distribution is not symmetric [i.e. not $(\frac{1}{2}, \frac{1}{2})$ for the two boxes], we have more *information* than when the probability distribution is symmetric. To clarify this point, consider a different experiment.

We throw a dart toward a board having a total area A. The board is divided into two parts of areas A_1 and A_2, respectively, such that $A = A_1 + A_2$ (Figure 1.12). You are told that the dart hit the board at some point but not in which of the two areas A_1 and A_2, and that the dart was thrown at random, meaning that all the points on the board are equivalent. Therefore, the probability of hitting the first area is $p_1 = A_1/A$, and that of hitting the second area is $p_2 = A_2/A$, with $p_1 + p_2 = 1$. You are given the probability distribution (p_1, p_2), and your task is to find out in which of the two areas the dart is.

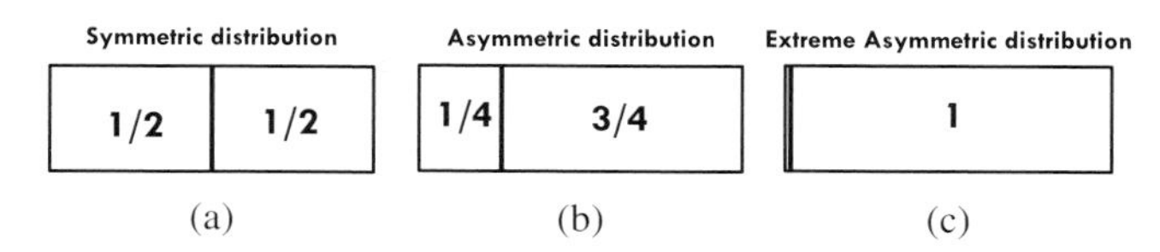

Fig. 1.12 A board with total area A is divided into two regions: (a) $\frac{1}{2}A$ and $\frac{1}{2}A$, (b) $\frac{1}{4}A$ and $\frac{3}{4}A$, (c) 0 and A.

Consider the following three cases:

(a) Symmetric distribution: $(\frac{1}{2}, \frac{1}{2})$
(b) Asymmetric distribution: $(p, 1-p)$, $(0 < p < \frac{1}{2})$
(c) Extreme asymmetric distribution: $(0, 1)$

These three cases are shown in Figure 1.12. In the first case (a), you must ask one question in order to know where the dart hit. In the third case (c), you do not have to ask any question; given that the distribution $(0, 1)$ is equivalent to having the knowledge that the dart hit the *right hand* area, i.e. you *know* where the dart is. In any intermediate case (b) of $0 < p < \frac{1}{2}$, you feel that you "know" more than in the first case (a) but less than in the third case (c). On the other hand, in all cases except the third, one must ask *one* question in order to find out where the dart is. Clearly, we cannot use $\log 2 = 1$ for the "number of questions" to quantify the amount of "information" we have in the general case of $(p, 1-p)$. To do so we modify the rules of the game, or the task you are facing.

Again, you know that the dart hit the board, and you are given the distribution $(p, 1-p)$, i.e. you know the areas A_1 and A_2. You choose one area and ask, "Is the dart in that area?" If the answer is "yes"; you win, and if the answer is "no," you ask, "Is the dart in the second area?" The answer you will certainly get is "yes."

The difference between the rules of this game and the previous ones is simple. When you have only two possibilities, you will *know* where the dart is after the first question, whatever the answer is. However, in the modified game, if the answer is "no," we add one more "verification" step, for which the answer is always "yes." In other words, the game ends when you get a "yes" answer to your question.

Now let us calculate the average number of questions in this modified game.

For the symmetric case (a), the probability of obtaining a "yes" answer for the first step is $\frac{1}{2}$, and that of obtaining a "no" answer is also $\frac{1}{2}$. Therefore, the *average* number of questions in the symmetric distribution case is

$$\frac{1}{2} \times 1 + \frac{1}{2} \times 2 = 1\frac{1}{2}.$$

Thus, on average you need 1.5 questions.

For the extreme asymmetric distribution (c), you will ask exactly *one* question, i.e. "Is the dart on the right hand side?", and the answer will be "yes." In this case you need only *one* question.

In any intermediate asymmetric case (b), we have $(p, 1 - p)$, with $0 < p < \frac{1}{2}$. So the first question should be "Is the dart on the right hand side?" You will obtain a "no" answer with probability p, and a "yes" answer with probability $1 - p$. Therefore, the average number of questions in this case is

$$(1 - p) \times 1 + p \times 2 = 1 + p. \qquad \text{(I)}$$

This result means that, for $0 < p < \frac{1}{2}$, the average number of questions is between 1 and 1.5, provided that we always start by asking whether the dart is in the area having the higher probability (i.e. the one with probability $1 - p$, if $0 < p < \frac{1}{2}$). If we decide to start by asking about the area of the lower probability, the average number of questions will be

$$p \times 1 + (1 - p) \times 2 = 2 - p. \qquad \text{(II)}$$

Of course, if we are interested in asking the minimal number of questions, we should always choose to ask first about the

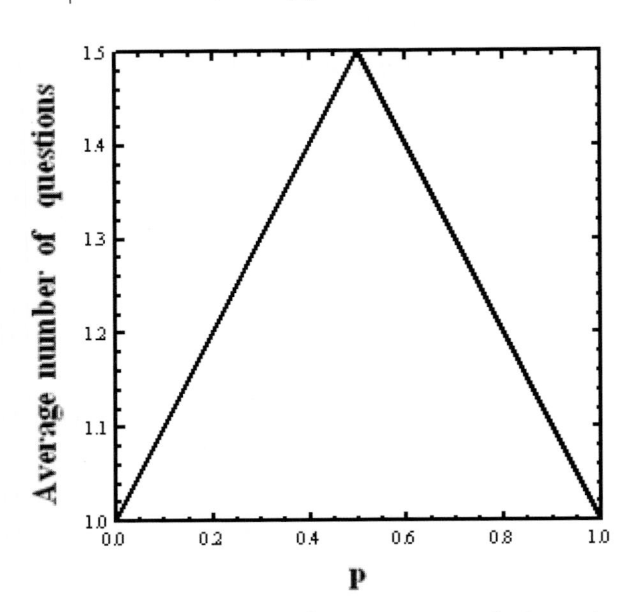

Fig. 1.13 The average number of questions, including the verification question for the games described in Figure 1.12. Note that for $0 \leq p \leq \frac{1}{2}$, the equation (I) applies. For $\frac{1}{2} \leq p \leq 1$, the equation (II) applies. In both ranges we ask first about the region having the *higher* probability.

area having the *larger* probability. With this choice the average number of questions as a function of p is shown in Figure 1.13. This function conforms to our expectations in the following sense:

When p is either zero or one, we *know* where the dart is. Therefore, according to the rules of the previous section, we do not need to ask any questions. According to the modified rules of the game, we must ask only one question: the question of verification. On the other hand, when the distribution is symmetric we have no clue on the location of the dart. In this case we must ask more questions: one to find out where the dart is, and if you get a "no" answer you have to ask a verification question. Thus, you will on average need $1\frac{1}{2}$ questions to gain the information and verify it. For any other distribution $(p, 1-p)$ we have an intermediate value between 1 and $1\frac{1}{2}$. This result conforms to our intuitive feeling that in this particular game the knowledge of the symmetric distribution does not provide any *information* to help us in locating the dart, and therefore we must make more "efforts," i.e. ask more questions to find out where the dart is. The more asymmetric the distribution, the more *information* we have, and hence the less the "effort" needed to locate the dart. In the extreme distribution case, either $(0, 1)$ or $(1, 0)$, we have all the information we need to locate the dart, and therefore we have to ask the minimum number of questions; here, only one — the verification question.

All that we have said above is very qualitative. We could have invented many functions which have a maximum at $p = \frac{1}{2}$, and a minimum at both $p = 0$ and $p = 1$. All of these functions will conform to our expectation of a measure of the amount of missing information or about the result of the experiment. The more asymmetric the distribution, the more certain (or the less uncertain) we are about the outcome of the experiment. Alternatively, we can say that the more symmetric the distribution, the more the *missing information*, or the smaller the amount of information we have on the outcome of the experiment.

Shannon has defined a more general measure of information which is valid for any N outcomes and for any distribution of the outcomes. We shall discuss the general definition in Section 1.4. Here, we introduce Shannon's measure for the particular case of two outcomes. This is defined by

$$\mathrm{SMI}(p) = -p \log_2 p - (1-p) \log_2 (1-p).$$

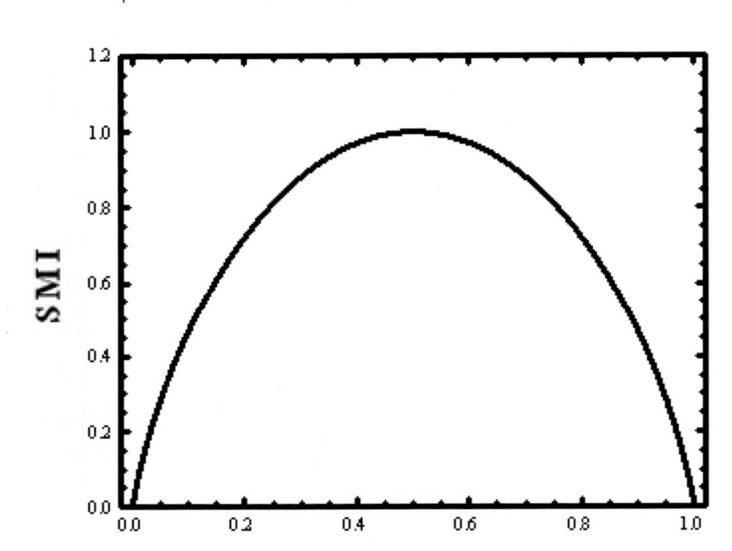

Fig. 1.14 The SMI for the two-outcome experiment.

The form of this function is shown in Figure 1.14. It has the property of having a maximum at $p = \frac{1}{2}$ and a minimum at both $p = 0$ and $p = 1$. We shall further discuss the function SMI for the general case in the next section. Here, we note that if we choose the base 2 for the logarithm, then the maximal value of SMI(p) is

$$\text{SMI}\left(p = \frac{1}{2}\right) = -\frac{1}{2}\log_2\frac{1}{2} - \frac{1}{2}\log_2\frac{1}{2} = 1.$$

This value is used as the *unit of information* called the *bit* (short for "*b*inary dig*it*"). Note carefully that the bit is the unit of information. It measures the amount of information one gets for a "yes"/"no" question, when the probabilities of the two outcomes are equal. If the probabilities are not equal, then the SMI can have any value between zero and one. Note also that this measure of information, for this particular case, is not equal to the average number of questions. (See also Section 1.5.5.)

1.3.1 *Misconception about the bit*

In some recent popular science books one finds statements that claim that a bit is the *atom* of the information. It is not!

An atom is the "atom" of a chemical element. A single molecule is the "atom" of a compound, and perhaps one can say that a "gene" is the "atom" of heredity, and perhaps also loosely that an "idea" is the "atom" of information. But the bit is a *unit* of information like the centimeter is a unit of length, and the gram a unit of mass.

The length of a wire is measured in units of centimeters; a centimeter is not the "atom" of the wire. The mass of a block of iron is measured in grams, but a gram is not the "atom" of iron. In the same way, a special kind of information is measured in units of bits, but the bit is not the "atom" of information.

In the introduction to Lloyd's book (2006), we find the following:

> This is the story of the bit and the universe. The universe is the biggest thing there is and the bit is the smallest possible chunk of information. The universe is made of bits. Every molecule, atom and elementary particle registered bits of information.

I can agree with only the first half of the second quoted sentence. The universe is, by definition, the *biggest* thing. But the bit is a *unit* of information, not *information*, and certainly not "the smallest possible chunk of information."

The universe is made of bits? Why not centimeters or grams?

The claim that "the universe is made of bits" is very common. This arises from the famous "It from bit" idea of Wheeler, to which we shall revert in Section 1.13. Here, suffice it to say that the bit is the *largest* amount of information one can get from a "yes"/"no" question. (See Section 1.3.) The *smallest* amount of information one can get from a "yes"/"no" question is *zero*!

Another common misconception relates to the value of the bit. You might read statements like:

An answer to a "yes"/"no" question gives you one bit of information.
A sequence of *n zeroes* and *ones* consists of *n* bits of information.

Such statements are very common in the literature, and are often made by professional scientists. Unfortunately, they are in general not true.

If we throw a fair coin, whose possible outcomes are *equally likely* to appear, then we can say that we have the maximum uncertainty about the outcome, or we have *minimum* information about the outcomes. By asking a binary question, "Is it H?" we obtain in this case the *maximum information*, which is by definition of the unit of information, *one bit*. For any unfair coin, with probability distribution $(p, 1-p)$, the amount of information obtained by asking a binary question is *less* than one bit; its actual value may be calculated by the SMI associated with this distribution, i.e. $-p\log_2 p - (1-p)\log_2(1-p)$.

In the extreme case of a coin which is infinitely heavier on one side — say, the T side — and therefore will always fall showing the H outcome, we have *minimum* uncertainty about the outcome (or maximum certainty). In this case, by asking a binary question about the outcome of this coin we get zero bits of information.

Thus, for any binary question the answer provides the amount of information between zero and one bit, calculated by the corresponding SMI.

A sequence of n zeroes and ones may be viewed as a message. The amount of information carried by this message is n bits, *provided that the probabilities of occurrence of a zero and a one are equal.*

Not every piece of information is measurable. When we hear a melody, our brain receives *information* from the ears. When we touch a hot object, our brain receives *information* from our fingers. All these and many others are *information*. They are not measured by *bits*!

There are other pieces of information which are not measurable, like love, beauty and any abstract idea. We shall further discuss this aspect of information in Section 1.13.

Pause and think

In a section titled "Coin Flips," Lloyd (2008) discusses the two states of polarization of a photon.

On the first state, he writes:

> Regarding the photon as a bit, we call this state "0."

On the second state, he writes:

> Regarding the photon as a bit, we call this state "1."

It is fine to refer to one state as "0" and another state as "1." But the photon is not a bit!

In another section, the author claims:

> At bottom all that a PC or Mac is doing is flipping bits.

I know what it means to flip a coin or to flip the polarization of a photon, or the orientation of a magnet. But "flipping a bit" is as meaningful as flipping a centimeter or flipping a gram!

1.4 The General Case

For an experiment having *n* possible outcomes, we can say that the amount of uncertainty in the experiment is the SMI defined *on the distribution* of the outcomes. In general, it is not true that the amount of information is equal to the number of "yes"/"no" questions asked to find out which outcome has occurred. The total *amount of information* obtained by asking binary questions is fixed by the experiment, having a given distribution. One can obtain this information by asking binary questions. However, the *number of questions* one asks depends on how one divides the *n* outcomes into two groups, with probabilities $(p, 1-p)$. In some cases it is possible to divide the entire number of outcomes into two groups having equal probabilities. In this case, we obtain the *maximum information* for *each binary question* we ask. Since the total SMI associated with the distribution of the *n* outcomes is fixed, it follows that if one can divide at each stage into two groups having equal probabilities, then the number of questions asked will be *minimal*, and equal to the SMI.

Of course, in general, such a division into two equally probable groups of outcomes is not possible. However, whatever the distribution of the outcome is, and whatever the strategy you choose to ask binary questions, the total amount of information you will obtain is *fixed* by the value of the SMI associated with that distribution.

If you are interested in asking fewer questions, you should try to divide, at each stage of the process of asking questions, the

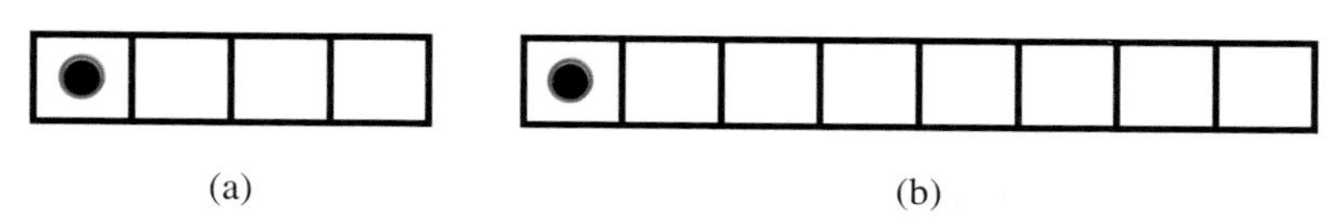

(a) (b)

Fig. 1.15 (a) A coin hidden in cell 1 in a game with four cells; (b) a coin hidden in cell 1 in a game with eight cells.

total number of possibilities into groups having probabilities as close as possible to being equal. Choosing this strategy ensures that *on average* you will get the same total information, which is the SMI of the experiment, by a minimum number of questions. We will present a detailed example in Subsection 1.5.5 (more details in Ben-Naim, 2008).

The following example should be studied carefully in order to remove any confusion between *information* (or the missing information), SMI and the number of bits.

Consider the two 20Q games shown in Figure 1.15. In (a), we hid a coin in cell No. 1 in a game with four cells. In (b), we hid a coin in cell No. 1 in a game with eight cells.

Now, answer the following questions:

(1) Where is the coin hidden in game (a), and in game (b)?
(2) What is the *size* of the missing information in these two games?
(3) What is the SMI of these two games?

The answer to question (1) is straightforward: The coin is hidden in cell No. 1.

This is exactly the same answer one should give to question (1), which is referred to either game (a) or game (b). Thus, the information (in its colloquial sense) on where the coin is, is literally the same for the two games in Figure 1.15.

Regarding the *size* of the missing information, the answer to question (2) depends on how we *define* the size of the information. We may choose to define the size of the information by the number of words or the number of letters in the sentence which carries that information. In this definition, the *number of words* is 8, and the number of letters is 30. And again, this size is the same when one is referring to game (a) or game (b) in Figure 1.15.

Now, we consider the more difficult question (3). Before we answer this question we must be careful to distinguish between the SMI of the *message* carrying the information which is "the coin is hidden in cell No. 1," and the SMI of the *game*. In the first case, the SMI of the message would depend on the number of letters and the frequencies of the letters in the English language (see Section 1.7). In this case, we have the same SMI when the information refers to game (a) or game (b).

In the second case, we are interested in the SMI of the two *games*. Here, we are interested in the *size* of the missing information we want to obtain (on where the coin is) by asking binary questions. In this case, the SMIs of the two games (a) and (b) are different. If the two games are uniformly distributed, then the corresponding SMIs are $\log 4 = 2$ bits and $\log 8 = 3$ bits, respectively.

1.5 Some Elementary Properties of the Function *H*

In this section we present some of the most important properties of SMI without proofs. [More details may be found in Ben-Naim (2008, 2011).] It should be noted that some of these properties were postulated by Shannon when he sought a *measure of information*. He found the quantity H which fulfills these properties. However, one should realize that these properties are not necessarily properties of *any* information. They apply to a very special class of information.

1.5.1 *Continuity of H*

Since the logarithm function is continuous, the function H is also a continuous function of all the variables $p_1, \ldots, p_n$. It is also a differentiable function of its variable.

1.5.2 *Concavity and the maximum of H*

The concavity of the function H can be easily proven for the general case. We seek the condition for the maximum of $H = -\sum_{i=1}^{n} p_i \log p_i$ subject to the condition $\sum_{i=1}^{n} p_i = 1$.

The result is[26]

$$p_i^* = \frac{1}{n} \quad \text{(for all } i\text{)}.$$

Thus, we found that H has a maximum value when all the p_i are equal, i.e. $p_i^* = 1/n$, and that the value of H at this maximum is $H = \log n$.[26] In this case, H is a monotonic function of n. This was the second property required by Shannon.

1.5.3 *The consistency property of H*

The third condition postulated by Shannon is sometimes referred to as the *consistency* property of the function H, or the independence of H on the grouping of the events. This requirement is less obvious intuitively. It states that the amount of information in a given distribution $(p_1, \ldots, p_n)$ is independent of the path or the number of steps we choose to acquire this information. In other words, the same *amount* of information is obtained regardless of the way or the number of steps one chooses to acquire this information. Here, we shall not need the most general formulation of this property. Instead, we shall discuss only the case of placing all the outcomes in two groups. Such grouping is essentially equivalent to playing the 20Q game.

If we divide all the outcomes into two groups, the consistency property of H is very simple. Suppose that we have n events $A_1, A_2, \ldots, A_n$. We need to find out which event has occurred, knowing that one of these events has occurred. Suppose that you group all the n events into two groups — say,

$$(A_1, A_2, A_3, A_4) \quad \text{and} \quad (A_5, A_6, \ldots, A_n).$$

Now call the first group G_1 and the second group G_2. The consistency requirement means that the information associated with the original game is the same as the information associated with the new game of G_1 and G_2, plus the average information associated with the groups G_1 and G_2. In terms of the 20Q game this is equivalent to first finding out which event — G_1 or G_2 — has occurred, and then finding out which event has occurred within either G_1 or G_2. The consistency principle is formulated more generally, and applies to any grouping of the original events into two or more groups. [For more details, see Ben-Naim (2008).] We shall further discuss this meaning of the consistency with an example in Section 1.5.5.

1.5.4 *The case of an infinite number of outcomes*

The case of discrete infinite possibilities is straightforward. First, we note that for a finite and uniform distribution, we have $H = \log n$, where n is the number of possibilities. Taking the limit

$n \to \infty$, we get

$$H = \lim_{n \to \infty} \log n = \infty.$$

This is clear. If we have to find out one out of infinite possibilities, we need to ask an infinite number of questions.

For a nonuniform distribution, the quantity H might or might not exist, depending on whether the quantity

$$H = -\sum_{i=1}^{\infty} p_i \log p_i$$

converges or diverges.

The case of a continuous distribution is more problematic. If we start from the discrete case and proceed to the continuous limit, we get into some difficulties. We will not discuss this problem here. Instead, we will follow Shannon's treatment for a continuous distribution for which a probability density function $f(x)$ exists. In analogy with the definition of the function H for a discrete probability distribution, we define the quantity H for a continuous distribution as follows. Let $f(x)$ be the density distribution, i.e. $f(x)dx$ is the probability of finding the random variable between x and $x + dx$. We defined the function H as

$$H = -\int_{-\infty}^{\infty} f(x) \log f(x) dx.$$

A similar definition applies to an n-dimensional density distribution function $f(x_1, \ldots, x_n)$.

In Chapter 2, we will use this definition of SMI to derive the entropy of an ideal gas. It should be noted that in any application of H in connection with thermodynamics, either we divide the continuous space into a finite number of "cells," or we take *differences* between two values of H. In both cases the problem of divergence of the integral is removed.

1.5.5 *SMI and the average number of binary questions*

We consider here a detailed example where we first calculate the SMI, and then we ask binary questions to acquire the missing information. In this game, the number of questions depends on the strategy of asking questions, but the *total* SMI of the game is fixed, independent of the strategy one chooses to acquire that information.

We have four equally probable events — say, a coin hidden in one out of four boxes, each having probability $\frac{1}{4}$. Let us identify the boxes by the letters a, b, c, d (Figure 1.16). We ask binary

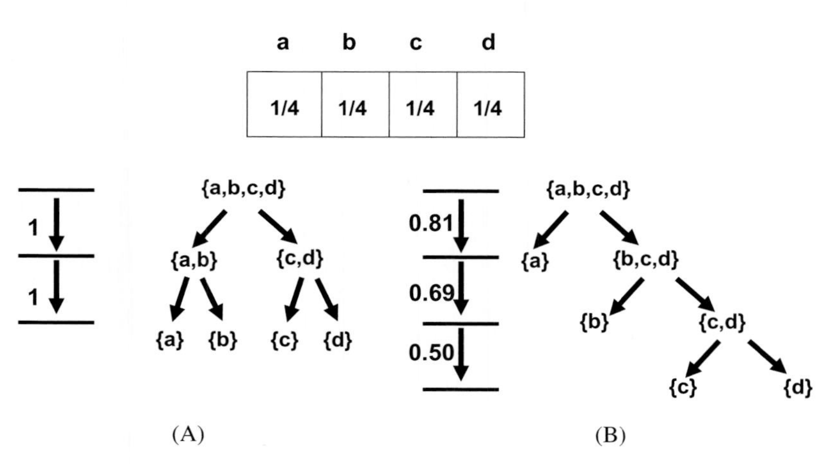

Fig. 1.16 A uniform 20Q game with four possibilities (a, b, c, d). A. The "smartest" strategy of asking questions. Here, we gain one bit at each step. The total number of steps is two, and we get two bits. B. The "dumbest" strategy of asking questions. Here, we gain 0.81 bits at the first step, 0.69 at the second, and 0.5 at the third. Altogether, we gain two bits. The average number of steps in this case is $2\frac{1}{4}$ (see Notes 27 and 28).

questions to locate the coin. The first method, the smartest one, is shown on the left hand side of the figure. Split the four boxes into two halves, then into two half-halves. The diagram in Figure 1.16(A) shows the method of splitting the events into groups, and the amount of information acquired at each step.

The total SMI for this case is

$$H = -\sum_{i=1}^{4} \frac{1}{4} \log_2 \frac{1}{4} = 2.$$

We see that the SMI for this game is equal to the number of questions one needs to ask using the smartest strategy.[27]

The second method, the dumbest one, is shown on the right hand side of Figure 1.16. We ask questions such as "Is the coin in box a," "Is it in box b," etc. The diagram in Figure 1.16(B) shows how we split the events in this instance.

In the case above, we gain less than the maximum information at each step. Therefore, we need to ask more questions.[28] Thus, we see that the SMI in the game is fixed (equal to 2). In the smartest strategy [Figure 1.16(A)] we gain more information at each step. Therefore, we need to ask fewer questions (here 2). In the dumbest strategy [Figure 1.16(B)], we gain less than one bit per question, and hence we need to ask more questions to get the total amount of information.

1.6 Application of the Maximum Uncertainty Principle

Up to this point, we have assumed that the distribution $p_1, \ldots, p_n$ is *given*, and we calculated the SMI *defined* on this distribution. We now turn to the "inverse" problem. We are given some information on some averages of the outcomes, and we want to find out the "best" distribution which is consistent with the available information.

Clearly, having the average value of the outcomes of an experiment (or a random variable) does not uniquely determine the probability distribution. There are an infinite number of distributions that are consistent with the available average quantity.

As an example, suppose that we are interested in assigning probabilities to the outcomes of throwing a single dice without having any information on the dice. What is the best guess of a probability distribution if the only requirement it must satisfy is $\sum_{i=1}^{6} p_i = 1$? There are many possible choices. Here are some possible distributions [Figure 1.17(a)]:

$$1, 0, 0, 0, 0, 0;$$
$$\frac{1}{2}, \frac{1}{2}, 0, 0, 0, 0;$$
$$\frac{1}{4}, \frac{1}{4}, \frac{1}{4}, \frac{1}{4}, 0, 0$$
$$\frac{1}{6}, \frac{1}{6}, \frac{1}{6}, \frac{1}{6}, \frac{1}{6}, \frac{1}{6}.$$

The question now is: Which is the "best" or the "fairest" distribution which is consistent with the available information but introduces no other information?

Clearly, the first choice is highly biased. There is nothing in the available information that singles out one outcome to be deemed the certain event. The second choice is somewhat less biased, but still there is nothing in the given information which indicates that only two of the outcomes are possible, and the other four are impossible. Similarly, the third choice is less biased than the second, but again there is nothing in the

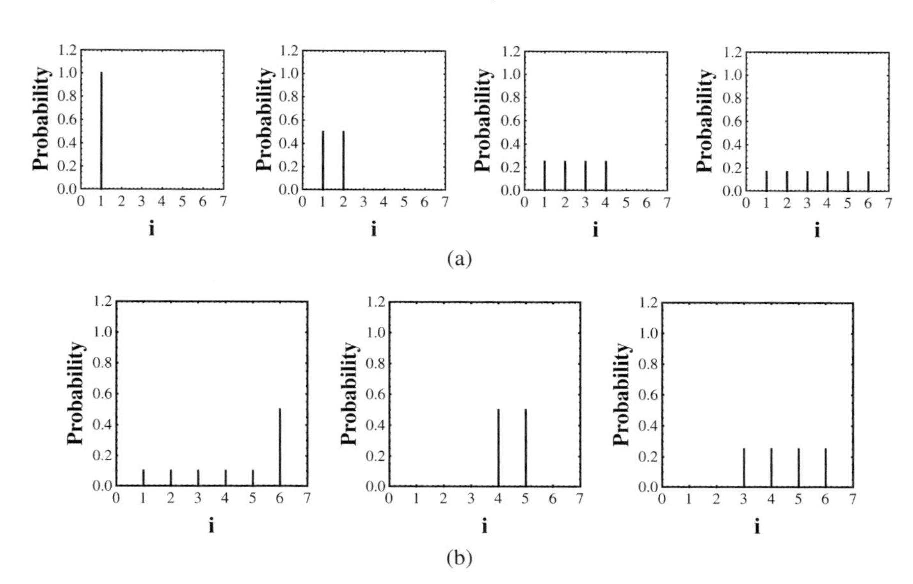

Fig. 1.17 (a) Four possible distributions for a dice; (b) three possible distributions for a dice having an average result of 4.5.

given information which states that two of the outcomes are impossible. Thus, what is left in this list is the fourth choice (of course, there are more possibilities which we did not list). This is clearly the least biased choice. According to Jaynes (1957),[25] the best choice is obtained by maximizing the uncertainty H subject to the available information. We have already done this in Subsection 1.5.2 and found that the "best" guess for this case is the uniform distribution, i.e. the fourth choice in the list above.

Suppose that we are told that the measurements have been done and the average outcome was 4.5. In this case, the distribution must satisfy the two conditions

$$\sum_{i=1}^{n} p_i = 1, \tag{1}$$

$$\sum_{i=1}^{n} i\, p_i = 4.5. \tag{2}$$

Again, it is clear that we have only two equations and six unknown quantities. Thus, there are many solutions to this problem. Some possible solutions are [Fig. 1.17(b)]

$$p^{(1)} = \{0.1, 0.1, 0.1, 0.1, 0.1, 0.5\},$$

$$p^{(2)} = \left\{0, 0, 0, \frac{1}{2}, \frac{1}{2}, 0\right\},$$

$$p^{(3)} = \left\{0, 0, \frac{1}{4}, \frac{1}{4}, \frac{1}{4}, \frac{1}{4}\right\},$$

and many more. Clearly, all these distributions are consistent with the given information.

The general procedure as advocated by Jaynes (1957) is[29]:

> In making inferences on the basis of partial information we must use the probability distribution which has the maximum entropy, subject to whatever is known.

The solution to the problem above is obtained by maximizing the SMI subject to the two constraints (1) and (2) as shown in Figure 1.18. The probability distribution that maximizes the SMI is[24]

$$p_i = 0.037 \times 1.44925^i.$$

This solution satisfies both of the equations (1) and (2). In Chapter 2 we will see another argument that justifies this procedure for choosing the distribution that maximizes the SMI. This occurs when there are many possible probability distributions, and we are interested in the most *probable* probability distribution. In this case, the most *probable* probability distribution is referred to as the *equilibrium* distribution.

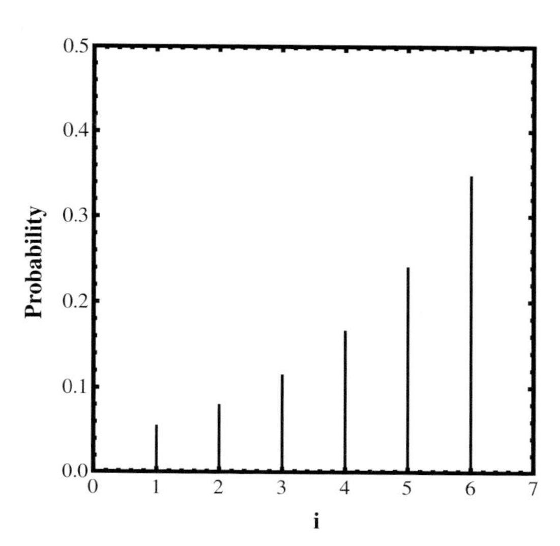

Fig. 1.18 The solution for the distribution for a dice with an average result of 4.5.

The application of this principle to statistical mechanics was suggested by Jaynes (1957). It has given a new view for both the fundamental postulate and other fundamental probability distributions of statistical mechanics. It did not solve any unsolved problem. As Katz (1967) wrote[30]:

> Information theory approach is not a miracle device to arrive at a solution of any statistical problem. It is only a general way to formulate the problem itself consistently. It most often happens that the "best guess" coincided with the educated guess of those who practice guessing as an art.

This "best guess," according to Katz, is the one that is consistent with the "truth, the whole truth and nothing but the truth." In my view, a better way of describing what the "best guess" means is to say that it is consistent with the knowledge, the whole knowledge and nothing but the knowledge that we have on the system under study. This is closer to Jaynes' original description of the merits of the "maximum entropy principle."[29]

1.7 Application of SMI to the English Language

In the previous section we discussed one application of SMI for calculating the most unbiased distribution. There are many applications of SMI itself. Here, we present only one example: the value of the SMI for the English language.

Table 1.3 shows the relative frequencies of letters in English. From this table, one can estimate the value of the SMI per letter for the English language to be about[31]

$$\text{SMI (per letter in English)} \approx 4.03.$$

Note that if all the letters (including the space between words) were equally probable, the value would be

$$\text{Max SMI (of 27 symbols)} = \log_2 27 = 4.76 \text{ bits.}$$

This means that we need at most five binary questions to find out one out of 27 symbols. In fact, if we know the distribution of letters as shown in the table, we can reduce the number of questions. The following example shows how to do it.

Suppose that I am thinking of a four-letter word. You have to find out what the word is by asking binary questions. You know that it is an English word, and you know English. You also have some idea about the relative frequencies of the various letters in the English language. (Table 1.3 and Figure 1.19.)

There are 26 possible letters, and thus there are 26^4 possible "words" of 4 letters. This is quite a large number of words — 456,976 (or about $\log_2 456976 \approx 19$). This means that if we know nothing about the English language we need to ask about

Table 1.3 The relative frequencies of letters in English, in decreasing order (rounded values).

Letter	Probability (frequency)
Space	0.182
E	0.107
T	0.086
A	0.067
O	0.065
N	0.058
R	0.056
i	0.053
S	0.050
H	0.043
D	0.031
L	0.028
C	0.023
F, U	0.022
M	0.021
P	0.017
Y, W	0.016
G	0.011
B	0.012
V	0.008
K	0.003
X	0.002
J, Q, Z	0.001

five questions ($\log_2 26 \approx 5$) to find out the first letter (1 letter in 26 boxes), another five for the second letter, five for the third, and five for the fourth. Altogether, you will need about 20 questions. (It is not exactly 20 questions because you cannot

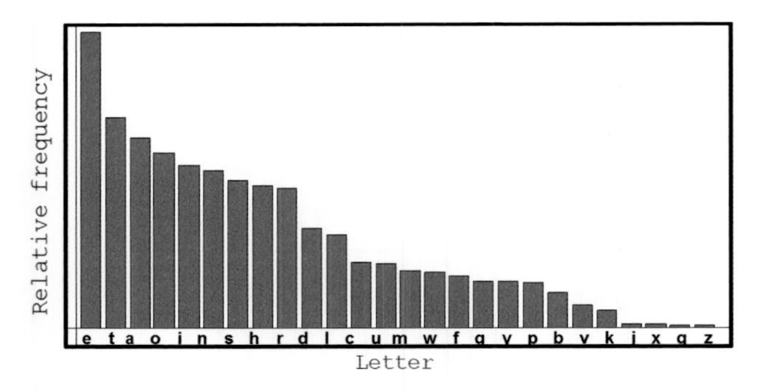

Fig. 1.19 Relative frequencies of letters in the English language.

divide all the letters into two halves having equal probabilities at each time.)

However, if you know something about the distribution of letters in English and you know English, you can do much better. Instead of dividing the total number of letters into two halves, you can divide into two parts of equal probabilities; for instance, the first 6 letters in Table 1.3 have the total probability of about 0.5, and the remaining 20 letters also about 0.5. The idea is that we give more weight to letters that appear more frequently than others. Based on the knowledge of the distribution of letters as provided in Table 1.3, we can calculate that the *SMI* is about 4. This is certainly better than 5, when you do not know the frequencies.

Once you find out the first letter — say, the letter *T* — you can proceed to the next letter. Again, if you know the frequencies of the letters you can find out the second one by asking about four questions. However, if you know something about the frequencies of *pairs* of letters, you might use this information to reduce the number of questions needed to find out the second letter. For instance, the combination *th* is more frequent than the combination *tl* and certainly more frequent than the combination *tq*. Therefore, you can use this additional information to reduce the number of questions — let us say to about three (or perhaps fewer).

Suppose that you have found the first two letters to be *th*. Next, you need to find out the third letter. That is relatively easy. It is most likely that the third letter after *th* is a vowel: *a*, *e*, *i*, *o* or *u*. Therefore, you have to find out one out of five letters, for which, on average, you will need about two questions (note that *e* and *a* are more frequent than *i*, *o* or *u*). If you found that the vowel is *i*, you already have three letters, *thi*. Now you do not need any more frequencies. If you know the English language you know that there are only two four-letter words which start with *thi*. These are *this* and *thin*. So, with one more question you will find out the word I have chosen.

We see that by knowing the distribution of letters, the distribution of pairs of letters, etc., we can actually reduce the number of questions. The difference between the maximum and the actual SMI is referred to as *redundancy* in the language.

We will discuss redundancy in the English language in Section 1.9. Here, we note that the strategy of asking questions as discussed in the previous section also indicates a method of assigning a short code for a given message. The most familiar code which assigns shorter code words to the more frequent letters, and longer code words to the less frequent letters, is the Morse code (Figure 1.20). The Morse code assigns a series of dots and dashes to each letter of the alphabet.

Suppose that it costs you a cent to transmit a dot, and two cents to transmit a dash. Clearly, you would want to assign the shortest code (i.e. the cheapest to transmit) for the most frequently used letter, E in English, and a longer code for the less frequent letters — say, Q or Z. In this way, you can minimize your cost for the transmission of any information.

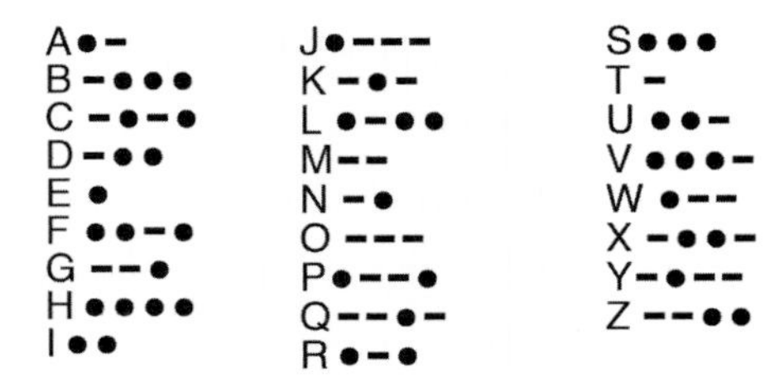

Fig. 1.20 The Morse code.

Note, however, that the size of the encoded message, or the total cost of the transmission of information, has nothing to do with the *meaning* of the transmitted information.

1.8 Conditional Information and Mutual Information

In Subsection 1.1.8 we discussed the qualitative idea about dependence and independence between pieces of information. Here, we will make these notions more precise and more specific. We restrict ourselves to any experiment having N outcomes with probability distribution $p_1, \ldots, p_N$. For simplicity we use the 20Q game we have discussed in the previous sections. Here, the experiment is the random throwing of a dart and the outcomes are the regions in which the dart hit.

Now suppose that we have two boards: one divided into N, and the other into M regions. The two corresponding distributions are $p_1, \ldots, p_N$ and $q_1, \ldots, q_M$, respectively. We throw one dart onto one board and another dart onto the other board.

It is clear that if we are interested in the locations of the two darts [i.e. where the blue (B) dart hit the left board and where the red (R) dart hit the right board], we have to ask SMI(B) questions for the blue dart and SMI(R) questions for the red

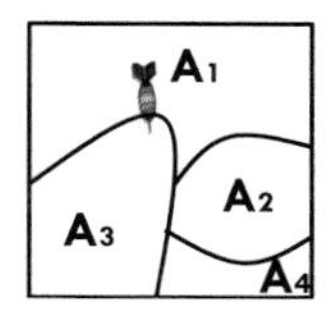

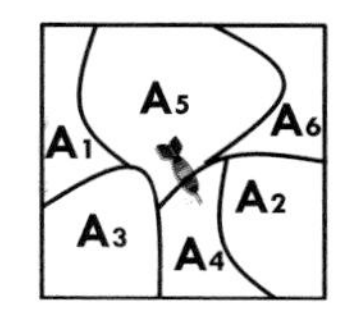

Fig. 1.21 Two independent experiments. Two darts are thrown onto different boards. The SMI of the combined experiments is the sum of the SMIs of the two experiments.

dart. The total information is simply the sum of the information for the location of the blue dart and the location of the red dart (Figure 1.21).

This property of additivity is valid provided that the two experiments are independent. Here, we employ the concept of independence in the sense used in probability theory. Two experiments are said to be independent if the occurrence of one outcome in one experiment does not affect the probabilities of the outcomes in the other experiment.

Intuitively, it is clear that if the boards are far from each other, and two different people throw two darts while blindfolded, and the darts do not interact with each other, we expect the two experiments to be independent. We say that outcome i (say, the blue dart hit region i of one board) and outcome j (say, the red dart hit region j of the other board) are independent if and only if their joint probability $\Pr(i,j)$ is the product of the probabilities $\Pr(i)$ and $\Pr(j)$. Thus, when the experiments are independent the SMI of the two combined experiments is the *sum* of the SMI of the two experiments[32]:

$$\text{SMI}(B \text{ and } R) = \text{SMI}(B) + \text{SMI}(R).$$

In order to discuss dependence, and also to be able to visualize the extent of dependence, we use the same board with the same

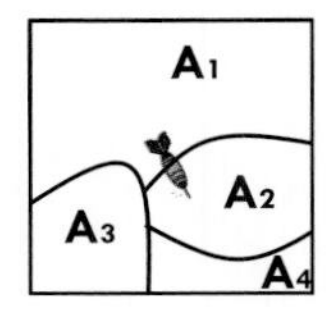

Fig. 1.22 A single board onto which two darts are thrown consecutively.

division into regions such that the distribution of outcomes for the blue dart is $p_1, \ldots, p_4$ (Figure 1.22), with the same distribution for the red dart. If we throw the blue dart first, then remove it from the board, and then throw the red dart at the same board, the SMI of the two experiments is the sum of the SMI of each dart.

Next, we throw the two darts onto the same board simultaneously. In general, we cannot assume independence of the outcomes for the two darts. If the blue dart hits say, region A_1 in Figure 1.22, then the second dart cannot hit the same *point*, or a point near the first dart. If the board is very big, we can assume that the probability of hitting the same or nearly the same point is an extremely rare event, and therefore we shall assume that the outcomes of the two throws are effectively independent.

Now we introduce dependence between the two experiments as follows. We attach a string of length r to the two darts (Figure 1.23). This means that when we throw the two darts onto the same board simultaneously, the outcomes are correlated. Clearly, if the blue dart hits a point x on the board, the red dart must be somewhere in a circle with a radius of about r around the point x (excluding the immediate vicinity of x itself). Remember that the probability distribution for each dart separately is $p_1, \ldots, p_4$. Given that the blue dart hit a point x — say, in region A_1 of Figure 1.23 — modifies the probabilities of

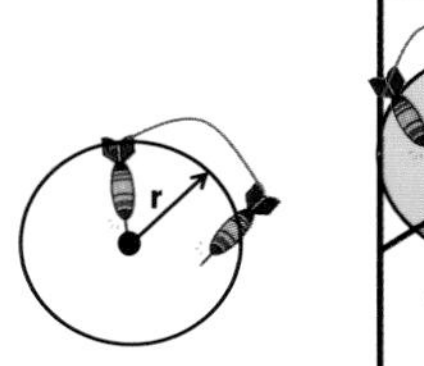

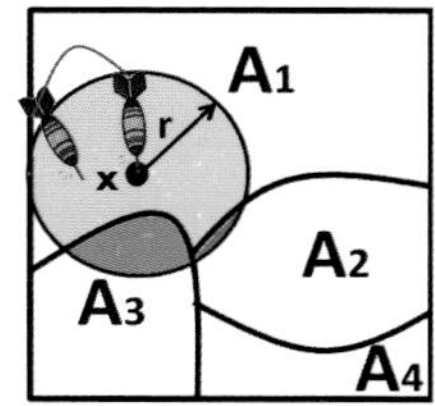

Fig. 1.23 A single board onto which two connected darts are thrown simultaneously.

the locations of the red dart. We can define a set of conditional probabilities for these two experiments. For instance, if the blue dart hits x within A_1, then the conditional probability of the red dart being in A_1 is different from the original $\Pr(R \text{ at } A_1)$. We write this as $\Pr(R \text{ at } A_1|B \text{ at } x)$. We can also define the conditional probability of R hitting A_2 given that B hit anywhere in A_1; denote this by $\Pr(R \text{ at } A_1|B \text{ at } A_2)$.

With these probability distributions we can define the corresponding *conditional* SMI.

This is done in two steps. First, define the SMI of the red dart given a *specific* outcome of the blue dart, and then take the average over all possible outcomes of the blue dart. This is referred to as the *conditional* SMI, which may be denoted by $\text{SMI}(R|B)$. It turns out that $\text{SMI}(R|B)$ is always *smaller* than the unconditional SMI,[33] i.e.

$$\text{SMI}(R|B) \leq \text{SMI}(R).$$

Qualitatively, this means that if we know the outcome of one experiment (any experiment or a game), then the amount of information in the second experiment is always smaller than the amount of information prior to the given information. The equality sign holds if and only if the two experiments are independent. Loosely speaking, this inequality means that we cannot lose information on one experiment, by being given some information on another experiment.

There is another example where information given on one thing enhances the information one has on another thing. This is the famous Monty–Hall dilemma, or its more dramatic version known as the "three prisoners problem." In her book, vos Savant discusses at great length the story of the Monty–Hall problem, her answer to the problem, and the reaction of thousands of readers who claimed that her answer to the problem was wrong. [For a complete solution to the problem based on Bayes' theorem, see Ben-Naim (2008, 2014).]

In terms of the 20Q game, we can say that if we know the outcome of one experiment, the number of questions we need to ask in order to obtain the information on the outcome of the second experiment is always smaller than if we do not know the outcome of the first. This is quite a remarkable result. Note that the *conditional probability* can be either supportive or unsupportive. For instance, given the result "even" in throwing a dice, the probability of the outcome "4" is larger than the unconditional probability, i.e. $\Pr(\text{"4"}|\text{"even"}) - \Pr(\text{"4"}) > 0$. We say that there is a *positive correlation* between the two events. On the other hand, given an "even" result, the conditional probability of "5" is zero, i.e. $\Pr(\text{"5"}|\text{"even"}) - \Pr(\text{"5"}) = -\frac{1}{6} < 0$. We say that there is a *negative correlation* between the two events.

The conditional information is always positive; given the result of one experiment — say, X — *always reduces* the SMI for the other experiment Y, and we always have the inequality $\text{SMI}(Y|X) - \text{SMI}(Y) \geq 0$, as well as $\text{SMI}(X|Y) - \text{SMI}(X) \geq 0$. (The equality sign holds if and only if the two experiments are independent.[33])

A quantity which measures the "average" correlation between all the outcomes of one experiment, and the outcomes of another experiment, is called the *mutual information.*

This quantity is defined by

$$\begin{aligned}\mathrm{I}(X;Y) &= H(X) - H(Y) - H(X,Y)\\ &= \mathrm{SMI}(X) - \mathrm{SMI}(Y) - \mathrm{SMI}(X,Y).\end{aligned}$$

If $P_X(i)$ is the probability of the outcome i in the experiment X, and $P_Y(j)$ is the probability of the outcome j in the experiment Y, then we define the *correlation* between these two outcomes by

$$g_{XY}(i,j) = \frac{P_{XY}(i,j)}{P_X(i)P_Y(j)}.$$

A positive correlation corresponds to $g > 1$, a negative correlation corresponds to $g < 1$, and no correlation corresponds to $g = 1$. Thus, the quantity g measures the extent of *correlation* between two single events.

In terms of the correlation function as defined above, the mutual information is

$$I(X;Y) = \sum_{i,j} P_{XY}(i,j) \log\left[g_{XY}(i,j)\right] \geq 0.$$

Note carefully that $\log[g_{XY}(i,j)]$ can be positive or negative for a specific pair of outcomes i and j. However, the *average* of this quantity must always be *positive* (it is zero when the two experiments are independent).

Figure 1.24 shows the relationship between the various quantities defined above. Note that $I(X;Y)$ is defined symmetrically with respect to the two experiments X and Y. It is always positive (except for independent experiments), and it is a measure of the

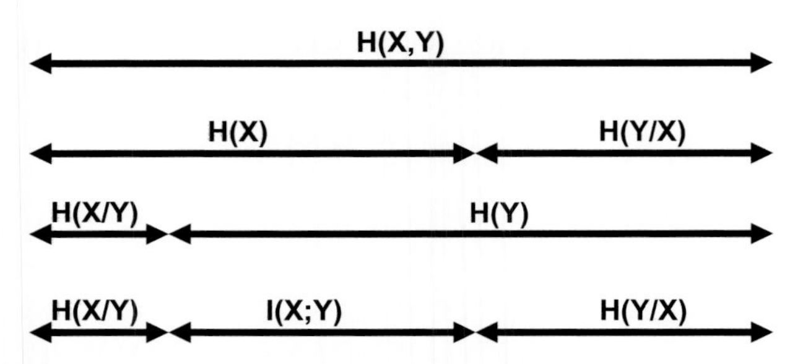

Fig. 1.24 The relationship between the quantities $H(X), H(Y), H(X, Y)$ and $I(X; Y)$.

extent of the dependence between the two experiments, or how much information is given on X knowing Y, and vice versa.

Yaglom and Yaglom (1983) refer to the mutual information $I(X; Y)$ simply as *information*. This interpretation is potentially misleading. Information in its colloquial sense is always *about* something. For instance, information about a dice could include its color, size, shape, etc. Suppose that we have two *dependent* experiments with two different dice. In this case, knowing the result of one dice does not provide *information* about the color, size, shape, etc. of the other dice. It provides only information on the change in the value of the SMI of the other dice — or, equivalently, on the change in the extent of uncertainty about the outcomes of the other dice.

1.9 Redundancy

The term "redundancy" used colloquially means that something is excessive, superfluous, repetitious, etc. We have seen some examples in Subsection 1.1.9. In information theory this term was distilled from the general term and is defined in a more precise manner. For any experiment with a given distribution of

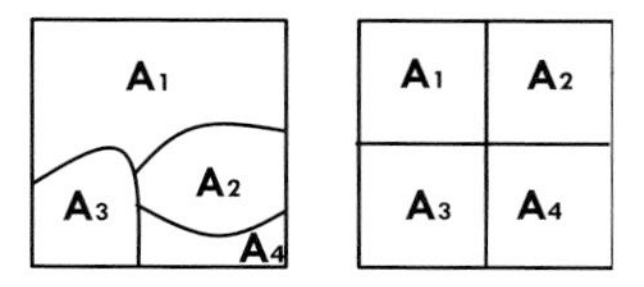

Fig. 1.25 Two boards with the same number of regions but different probability distributions. The dart game is easier to play with the board on the left than on the right.

the outcomes $p_1, \ldots, p_N$, we can define the fraction:

$$\frac{\text{SMI}(p_1, \ldots, p_N)}{\text{Max[SMI]}} = \frac{\text{SMI}}{\log_2 N}.$$

We know that the maximum SMI for an experiment with a given N is obtained when all the outcomes are equally likely, i.e. their probabilities are $p_i = 1/N$ (for $i = 1, 2, \ldots, N$). Therefore, the maximum value of the SMI is simply $\log_2 N$.

Figure 1.25 shows two boards, each divided into four regions. On the left we have regions of unequal areas, and on the right we have four equal-area regions. If we are playing the dart game, and we can choose between the two boards, it will *always* be advantageous to play the game on the left. We can say that the game on the right does not provide any *information* we can use to our advantage (i.e. reducing the number of questions).

In Figure 1.26 we show three games, each with four outcomes. The corresponding SMIs are $\text{SMI}(A) = 2$, $\text{SMI}(B) = 1.79$ and $\text{SMI}(C) = 1.21$. These values indicate that it is "easier" to play game C, in the sense that the distribution in C provides more information, and hence you will need to ask fewer questions. The *redundancy* as defined in information theory is a measure of how much a game is easier to play relative to the one

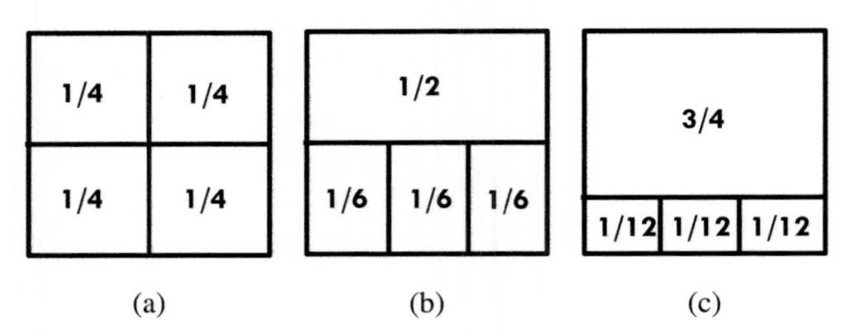

Fig. 1.26 Three boards with the same number of regions but different probability distributions. The redundancy increases from the left to the right game.

with the same number of outcomes but with equal probabilities. The formal definition is

$$R = \frac{\log_2 N - \text{SMI}}{\log_2 N} = 1 - \frac{\text{SMI}}{\log_2 N}.$$

For the three games in Figure 1.26 we have $R(A) = 0$, $R(B) = 0.104$ and $R(C) = 0.39$. This means that the distribution in C is the most redundant. This is vaguely reminiscent of the redundancy in spoken language (see Section 1.19 on redundancy in the Jacob and Laban story).

The concept of redundancy is useful in the application of information theory to transmission of information. The redundancy in this case is a measure of how easy it is to guess a letter from an alphabet when we know the frequency of its occurrence in that particular language.

The following data are taken from Yaglom and Yaglom (1983):

For the English language, the maximum SMI is

$$H_0 = \log 27 \approx 4.75 \text{ bits}$$

(H_0 refers to the case where all symbols are equally probable).

Taking the real frequencies of letters in the English language, one gets

$$H_1 = -\sum p_i \log p_i \approx 4.03 \text{ bits},$$

where p_i is the relative frequency of the letter i.

If we know the distribution of pairs of letters, we can define the *conditional* SMI of a letter i followed by a letter j:

$$H_2 = -\sum p(i|j) \log [p(i|j)].$$

One can also define the conditional information of a letter k followed by two letters i and j (in the order ij), by

$$H_3 = -\sum p(k|i,j) \log [p(k|i,j)].$$

Table 1.4 shows the values of the SMI and the conditional information for single letters in different languages.

As expected, we always have $H_0 \geq H_1$, and the value of the conditional information decreases from H_1 to H_2 to H_3, etc.

It is interesting to note that in English (as well as in other European languages) the vowels are more frequent than the consonants. In modern Hebrew, the vowels are not used in

Table 1.4 Values of H_0, H_1, H_2 and H_3 for different languages.

Language	English	French	Italian	Portuguese	Russian
H_0	4.75	4.75	4.39	4.52	5.00
H_1	4.03	3.95	3.90	3.91	4.35
H_2	3.32	3.17	3.32	3.35	3.52
H_3	3.10	2.83	2.76	3.20	3.01

written text. This fact dramatically changes the redundancy in the Hebrew language. As an example, Bluhme [quoted in Yaglom and Yaglom (1983)] compared the statistical characteristic of a collection of three-letter words in Hebrew and in English, and found the following results:

$$H_3^{(\mathrm{Heb})} \approx 3.73(\mathrm{bit/letter}),$$

$$R_3^{(\mathrm{Heb})} = 1 - \frac{H_3}{H_0} \approx 0.16,$$

$$H_3^{(\mathrm{Eng})} \approx 0.83(\mathrm{bit/letter})$$

$$R_3^{(\mathrm{Eng})} \approx 0.82.$$

Note that the redundancy (for the three-letter words) in Hebrew is significantly *smaller* than in English.

Similar analysis of H_0, H_1 and H_2 was done in music, in the arts, and in many other fields. For an interesting discussion on the SMI of different languages, as well as of different books by the same author (e.g. Shakespeare), see Landsberg (1961).

We have discussed the term "redundancy" in its colloquial sense and its definition within information theory. As we have seen, these two concepts of redundancy are very different: one is qualitative and the other is quantitative.

Confusing between the two is very common in popular-science literature. Seife (2007) starts his book with a chapter on "Redundancy." On page 6, we find: "World War II was the first information war." I do not know what "information war" is, but from the content of the chapter it is clear that the author refers to problems of coding and decoding of information. I believe that these kinds of problems are part of any war, from the biblical story on the Samson enigma to the present wars around the world. In the same chapter, Seife (2007) writes:

> The idea that something as seemingly abstract as information is actually measurable — and tangible — is one of the central tenets of information theory ... right after World War II, when mathematicians laid out a set of rules that defined information....

Not *any* "abstract information" is measurable and mathematicians did not "set rules that *defined* information," and all this has nothing to do with the title of the chapter, "Redundancy."

The same chapter on redundancy ends with the following statement:

> ... this hard-to-define concept of information holds the key to understanding the nature of the physical world.

Lamentably, I am not aware of any understanding of the physical world, which was brought about by the *concept* of information, and certainly not by the concept of "redundancy" defined in information theory.

I would like to conclude this section with a very interesting example, which I believe is related to the redundancy in the English language.

Start by reading the following paragraph as quickly as you can:

> "You arne't ginog to blveiee that you can aulactly uesdnatnrd what I am wirtnig. Beuacse of the phaonmneal pweor of the hmuan mnid, aoccdrnig to a rscheearch at Cmabrigde Uinervtisy, it deosn't mttaer in what order the ltteers in a wrod are, the olny iprmoatnt tihng is taht the frist and lsat ltteer be in the rghit pclae. The rset can be a taotl mses and you can sitll raed it wouthit a porbelm. Tihs is

bcuseae the huamn mnid deos not raed ervey lteter by istlef, but the word as a wlohe. Amzanig huh? Yaeh and you awlyas tghuhot slpeling was ipmorantt!"

This quotation was taken from Patel (2008), based on Rawlinson's thesis (1976) on "The Significance of Letter Position in Word Recognition." Patel uses this example in connection with the "language" of the proteins written in amino acids (see Chapter 3). The explanation as to why one can read and understand the paragraph in spite of the many "spelling mistakes" is that we tend to read *whole words*, and not letter by letter. (Note that from the point of information theory this paragraph has the same SMI as the original, correct paragraph.)

In my opinion, the main reason for recognizing the words in spite of the jumbled letters is the redundancy in the English language. Although I cannot offer a proof of my contention, I can offer a plausible argument. As noted above, the Hebrew language is far less redundant than the English language (this is due mainly to the absence of most vowels in written Hebrew).

If my conjecture is correct, then I would expect that the less the redundancy is in the language, the more difficult it would be to read text with jumbled letters. Indeed, I tried several examples and found to my surprise that even when I *knew* the original text of the paragraph (because I had *invented* it), I had a hard time reading the jumbled letters. Here is an example of two sentences in Hebrew. If you know Hebrew, try to read this as quickly as you can (the original sentences are given at the end of this section).

Hebrew text:

בירעבת כובתה ברדך כלל חורסת הועתונת

למיפעם זה משקה על הויהגי הוכנן של הילמים

I also notice that because of the lack of vowels, many words in the jumbled letters have a meaning in Hebrew, differing from the original words.

Another explanation for this phenomenon was offered by Velan and Frost (2007). They presented sentences in both English and Hebrew to bilingual students. It was found that word recognition was much more difficult for jumbled Hebrew letters than for English. These authors pointed out that although the English reader tends to read whole words, the Hebrew reader first tries to identify the *triconsonantal roots* in the Hebrew words. This identification is sometimes difficult, because the same three consonants usually have different meanings when written in a different order (example: שחל,שלח,חלש,חשל,לחש,לשח).

This fact makes it very difficult to read quickly the jumbled letters text in Hebrew. Hence, the authors concluded:

> It seems that **research at Hebrew University** may produce quite different results than **rsheearch at Cmabrigde Uinervtisy**, where visual word recognition is concerned.

Here are the correct sentences in Hebrew presented above:

בעברית כתובה בדרך כלל חסרות התנועות

לפעמים זה מקשה על ההיגוי הנכון של המילים

1.10 Some Simple Examples of SMI

The following examples are presented here as illustrations. All these will be relevant to the calculation and interpretation of the entropy in Chapter 2. I suggest to the reader to pause at each example, and try to grasp the qualitative meaning of the results for each of these examples.

(a) One marble in M boxes

Suppose that we have a system of M boxes and we know that a marble is placed in one of the boxes, and that the boxes are equivalent. The marble can be in any specific box with probability $1/M$. The SMI for this case is

$$\mathrm{SMI}_a = -\sum_{i=1}^{M} p_i \log p_i = -\frac{M}{M} \log \frac{1}{M} = \log M.$$

Clearly, the larger M is, the larger will be the uncertainty about the location of the marble (i.e. in which box it is). Equivalently, we may say that we shall need a larger number of binary questions to find out where the marble is.

(b) Two different marbles in M boxes

We assume that each box can contain at most one marble. We also assume that the boxes are equivalent. In this case the probability of finding the first marble — say, the blue one — is as in case (a): $1/M$. However, the conditional probability of finding the second marble — say, the red one — in one of the remaining $M - 1$ boxes is now $1/(M - 1)$, which differs from $1/M$. Clearly, in this case there is *dependence* between the two events: the location of the blue marble and the location of the red marble. For instance, for $M = 4$, we have probability $\frac{1}{4}$ for the location of the first marble, and a conditional probability of $\frac{1}{3}$ for the location of the second marble, given that the first marble is in some specific box.

We can define a correlation function to account for the dependence between the two events. For this particular example the correlation is

$$g(1,2) = \frac{\Pr(1 \text{ in } j \text{ and } 2 \text{ in } k)}{\Pr(1 \text{ in } j) \times \Pr(2 \text{ in } k)}$$

$$= \frac{\frac{1}{M(M-1)}}{\frac{1}{M} \times \frac{1}{M}} = \frac{M^2}{M(M-1)} > 0.$$

Thus, in this case, the correlation is *larger* than 1, which means that the conditional probability of finding the second marble in box $k (k \neq j)$ is *larger* than the probability of finding the first marble in box j.

We can also define the mutual information for this case as

$$\begin{aligned} I(1;2) &= \mathrm{SMI}(1) + \mathrm{SMI}(2) - \mathrm{SMI}(1,2) \\ &= \log M + \log M - \log[M(M-1)] \\ &= \log\left[\frac{M^2}{M(M-1)}\right] = \log[g(1,2)] \\ &= -\log\left(1 - \frac{1}{M}\right), \end{aligned}$$

where SMI(1) is the SMI for marble 1 in M boxes, SMI(2) is the SMI for marble 2 in M boxes, and SMI(1, 2) is the SMI for the joint events, i.e. both marble 1 and marble 2 in M boxes, with the condition that no two marbles can be in the same box at the same time.

Note that the mutual information $I(1; 2)$ is positive. This result is much more general, and it is independent of whether the correlation is positive or negative.[33]

The reason for this correlation is obvious. The fact that a marble occupies one box reduces the number of boxes available to the second marble. Therefore, the conditional probability of finding the second marble in any of the available (empty) boxes is different from the probability of finding the first marble in one of the M boxes.

Before we continue to the next case, where we shall encounter a new type of correlation, let us assume from now on that M is a very large number. If we have N marbles distributed in M boxes, we shall always assume that $M \gg N$. Therefore, within this assumption, $M - 1 \approx M$, and hence for the example above $g(1, 2) \cong 1$.

(c) Two indistinguishable marbles in M boxes

The next example is important for understanding the entropy of a system of atoms and molecules. Here, however, we continue to use the language of marbles in boxes. As in case (b), we have M boxes which are equivalent. We place two indistinguishable marbles in the boxes with the condition that no more than one marble can be placed in a single box. Also, we assume that M is very large. In this case the probability of finding a marble in box j is $1/M$. The probability of the joint event, one marble in j *and* one marble in $k(j \neq k)$, is denoted by Pr(1 in j and 2 in k). In this case the probability is different from the one we calculated in the previous example (b). Here, the total number of configurations is not $M(M - 1)$ as in case (b), but $M(M - 1)/2$ (Figure 1.27). The reason is that different configurations which are obtained by exchanging two different particles on their locations become

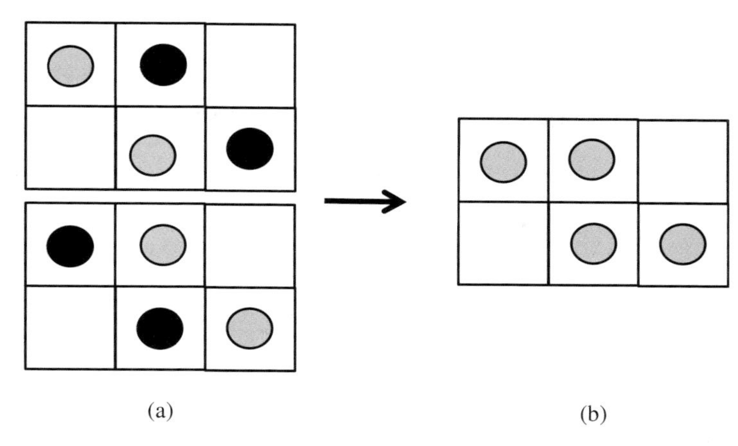

(a) (b)

Fig. 1.27 Two two configurations on the left (a), became one configuration when the particles are identical (b).

identical when the particles are indistinguishable. For very large M, we can neglect the difference between M and $M - 1$. The number of distinguishable configurations is $M^2/2$, and hence the probability of a specific configuration is

$$\Pr(\text{one in } j \text{ and one in } k) = \frac{2}{M^2}.$$

The correlation function in this case is

$$g(1,2) = \frac{\Pr(\text{one in } j \text{ and one in } k)}{\Pr(\text{one in } j) \times \Pr(\text{one in } k)} = \frac{\frac{2}{M^2}}{(\frac{1}{M})^2} = 2.$$

We see that in this case the correlation is positive. The reason for this correlation is the indistinguishability of the marbles. As in case (b), the *singlet* probability of finding one marble in j is the same as before $(1/M)$. However, the pair distribution is different. Since we now have half as many configurations as in case (b), the joint probability is twice as large as in case (b). Hence, we have a correlation which is positive. This kind of correlation cannot be removed as we let M be very large. The corresponding mutual information is

$$I(1;2) = \text{SMI}(1) + \text{SMI}(2) - \text{SMI}(1,2)$$
$$= \log M + \log M - \log \frac{M^2}{2} = \log 2 > 0.$$

(d) Unlabeling the marbles

We combine the results of the two last examples to reach an important conclusion. We start with a system of M boxes in which N *different* (or labeled) marbles are distributed. We also assume that $M \gg N$.

The number of configurations for such a system is

$$W(D) = M(M-1)(M-2)(M-N+1) \cong M^N.$$

The reason is simple. The first particle can be placed at one of the M boxes. The second particle can be placed in one of the $M-1$ boxes, but since $N \ll M$ we can replace $M-1$ by M, and so on until we place the Nth particle in the $M-N+1$ boxes, or approximately M boxes.

Now we unlabel the marbles, i.e. we remove the colors, or the numbers or any other labels that make the marbles distinguishable. We now have a system of N *indistinguishable* marbles (ID). The corresponding number of configurations is

$$W(\text{ID}) \cong \frac{M^N}{N!}.$$

The number of configurations is reduced by the factor $N!$, which is the number of permutations of the N marbles.

The change in the SMI in the process of unlabeling the marbles is thus

$$\text{SMI(ID)} - \text{SMI(D)} = \log \frac{M^N}{N!} - \log M^N = -\log N!$$

For the case of $N = 2$ and $M = 4$ we have the result

$$\text{SMI(ID)} - \text{SMI(D)} = \log\frac{(4\times 3)}{2} - \log(4\times 3) = -\log 2.$$

We see that the process of unlabeling *decreases* the SMI of the system. This result seems to be counterintuitive. We perceive the process of unlabeling as a process in which we *lose* information; we initially had a label which identified each marble and this label is now *lost*. We "erased" the identity of the marbles, but the uncertainty, as measured by the SMI, had *decreased*. The reason is that the SMI computed above is not about the identity of the marbles, but about the *configurations* of marbles in the boxes. The number of configurations in the process of unlabeling has *decreased*, and hence the corresponding SMI has decreased too. In Chapter 2, we shall see that this process of unlabeling is important for calculating the entropy of real systems.

1.11 The Change in SMI for Some Simple Processes

The processes described in this section will involve N particles distributed in M cells, with the condition $M \gg N$. All of these processes are the analogs of real processes involving atoms or molecules in an ideal gas phase, which we shall discuss in Chapter 2. Instead of marbles in the boxes we used previously, we shall use the language of *particles* in cells. The assumptions will be the same as in the examples discussed in the previous section. The number of cells M will always be very large compared with the number of particles. We also assume that each cell contains at most one particle at the time. This restriction will be removed in Chapter 2, when we discuss *real* particles in cells. However, the assumption $M \gg N$ makes the event "two or more particles in a single cell" very rare and therefore can be neglected.

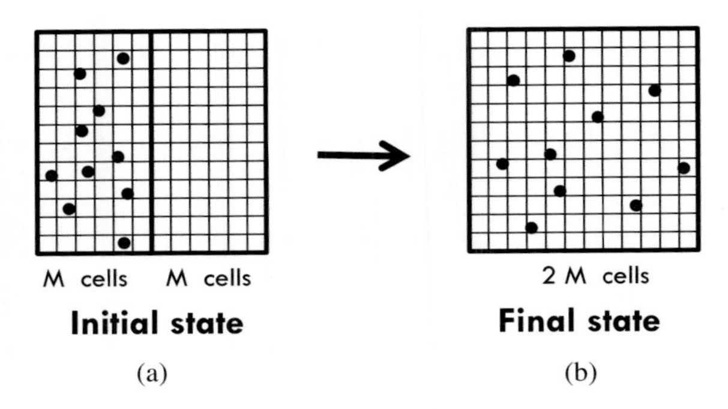

Fig. 1.28 Expansion of N particles from M cells to $2M$ cells.

1.11.1 *The expansion process*

Figure 1.28(a) shows a box with $2M$ cells divided into two equal compartments, each containing M cells. Initially, we distribute N indistinguishable particles in one compartment. There are altogether $\binom{M}{N}$ configurations, and all these are presumed to be equally probable. Therefore, the corresponding SMI is

$$\text{SMI(initial)} = \log \binom{M}{N}.$$

After the removal of the barrier between the two compartments [Figure 1.28(b)], we have N particles distributed in $2M$ cells, and hence the SMI is

$$\text{SMI(final)} = \log \binom{2M}{N},$$

$$\text{SMI}(f) - \text{SMI}(i) = \log \binom{2M}{N} - \log \binom{M}{N}.$$

For large N and M, but ($M \gg N$), one can easily show that in this process the change in the SMI is[33]

$$\text{SMI}(f) - \text{SMI}(i) \cong N \log \left[\frac{2M}{M} \right] = N \log 2 = N.$$

For $N = 2$ and $M \gg 2$ we have

$$\mathrm{SMI}(f) - \mathrm{SMI}(i) = \log \binom{2M}{2} - \log \binom{M}{2} = 2 \log 2 = 2.$$

This result can be interpreted in terms of the loss of *locational* information of one bit per particle. Initially, each particle could be found in one of M cells. After we remove the barrier, each particle can be found in one of $2M$ cells. Thus, there is a loss of one bit per particle. Note that this result is independent of the particles being labeled or unlabeled. In both cases, the *expansion* of the available "volume" per particle has increased by a factor of 2.

1.11.2 *The pure mixing process*

We start with N_A particles of type A, distributed in M cells, and N_B particles of type B, distributed in M cells. We bring *all* the particles — N_A A particles and N_B B particles — into M cells (Figure 1.29). The number of configurations in the initial state

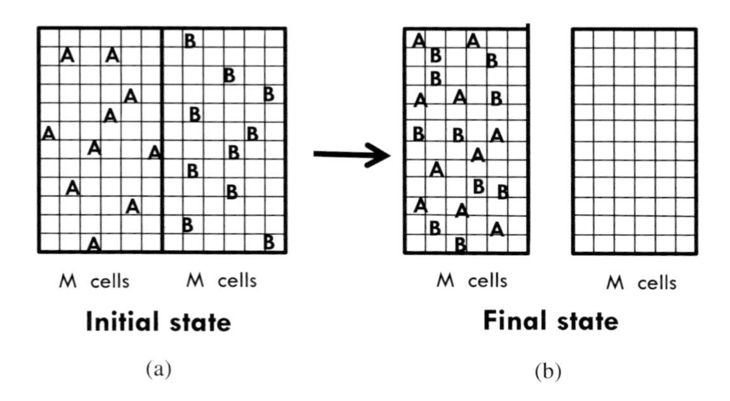

Fig. 1.29 A process of "pure mixing."

(i) is

$$W(i) = \binom{M}{N_A}\binom{M}{N_B},$$

and in the final state (f)

$$W(f) = \binom{M}{N_A}\binom{M - N_A}{N_B}.$$

It is easy to see that when $M \gg N_A, N_B$, the change in the SMI for this process is zero,[33] i.e.

$$\text{SMI}(f) - \text{SMI}(i) = \log \frac{W(f)}{W(i)} = 0.$$

The reason is simple. In both the initial and the final state each particle can be found in one of the M sites. Thus, there is no change in the *locational information* about the particles in this process. This process is referred to as *pure mixing*, since the only change that we observe is the *mixing* of the particles and there is no change in the locational information (see also Subsection 2.7.2).

1.11.3 *The pure assimilation process*

In this process we start with N particles in M cells and another N particles in M cells, and we bring all the $2N$ particles into M cells. This process is the same as in Figure 1.29, but now the N particles in each system are of the same kind (Figure 1.30). In this case the number of configurations in the initial and in the

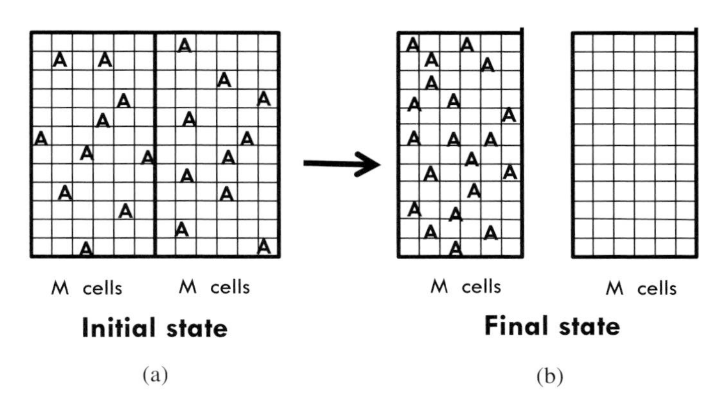

Fig. 1.30 A process of "pure" assimilation.

final state are

$$W(i) = \binom{M}{N}\binom{M}{N},$$
$$W(f) = \binom{M}{2N}.$$

We refer to this process as *pure assimilation*. Note that the locational information does not change in this process. The corresponding change in the SMI is

$$\mathrm{SMI}(f) - \mathrm{SMI}(i) = \log\frac{W(f)}{W(i)} \cong \log\frac{(N!)^2}{(2N)!} < 0.$$

This quantity is always negative.[33]

For $N \gg 1$ we can use the Stirling approximation to obtain

$$\mathrm{SMI}(f) - \mathrm{SMI}(i) \xrightarrow{N\gg 1} -2N\log 2 < 0.$$

Thus, we conclude that in the pure assimilation process the change in the SMI is always *negative*.

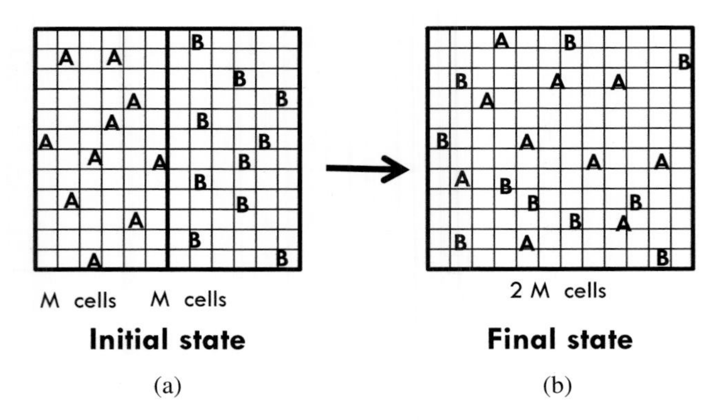

Fig. 1.31 A process of mixing of A and B particles and expansion.

1.11.4 *Mixing and expansion*

We refer to the process depicted in Figure 1.31 as *mixing and expansion*. In this process, we observe mixing, and in addition each particle, A or B, expands from M to $2M$ cells. The number of configurations in the initial and in the final state are

$$W(i) = \binom{M}{N_A}\binom{M}{N_B},$$

$$W(f) = \binom{2M}{N_A}\binom{2M - N_A}{N_B}.$$

The corresponding change in the SMI is

$$\text{SMI}(f) - \text{SMI}(i) = \log\left[\frac{W(f)}{W(i)}\right]$$

$$\xrightarrow{M \gg N_A, N_B} N_A \log 2 + N_B \log 2.$$

In the particular case where $N_A = N_B = N$ (but the particles in each compartment are different), we have

$$\text{SMI}(f) - \text{SMI}(i) = 2N \log 2.$$

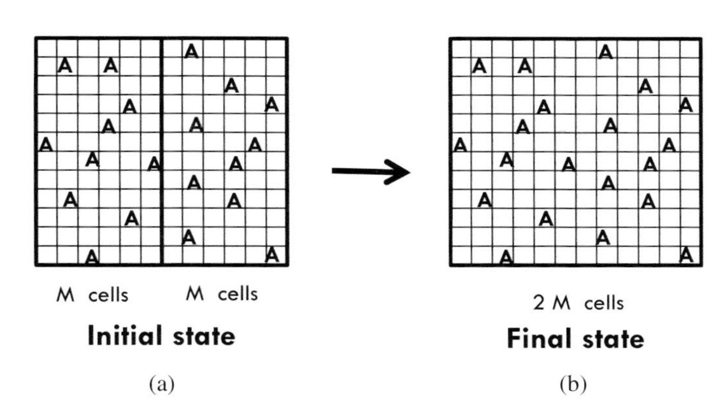

Fig. 1.32 A process of assimilation and expansion.

We can conclude that in this process the change in the SMI is due to the change in the *locational information*, $\log 2 = 1$ i.e. one bit per particle, exactly as in the expansion process. Clearly, we observe *mixing* in this process but the change in the SMI is due only to the *expansion* of each particle from M to $2M$. The mixing in this process does not affect the value of the SMI.

1.11.5 *Assimilation and expansion*

We refer to the process depicted in Figure 1.32 as *assimilation and expansion*. The "expansion" part is obvious. Each particle was initially confined to the "volume" M. Upon removing the partition it can access the larger "volume." Therefore, if we have N particles in each compartment, the change in the SMI due to the expansion process is exactly as in the previous example, namely $2N \log 2$. However, unlike the process in Figure 1.31, in this process there is a change in the *number* of *indistinguishable particles*. Initially, we had two groups of N indistinguishable particles, but each group was distinguished from the other group. In the final state, the N particles of one compartment are *assimilated* with the N particles of the second compartment, i.e. we have $2N$ indistinguishable particles.

The total number of configurations in the initial and in the final state are (*cf.* Figure 1.31)

$$W(i) = \binom{M}{N}\binom{M}{N},$$
$$W(f) = \binom{2M}{2N}.$$

The corresponding change in the SMI is

$$\begin{aligned}\text{SMI}(f) - \text{SMI}(i) &= \log\left[\frac{W(f)}{W(i)}\right] \\ &= 2N\log 2 - \log\left[\frac{(2N)!}{(N!)^2}\right] > 0.\end{aligned}$$

It is easy to show that this is always a positive quantity.[34]

Table 1.5 Values of $R(\text{SMI}) = [2N\log 2 - \log[\frac{(2N)!}{(N!)^2}]/\,2N\log 2]$ for different N.

N	$R(\text{SMI})$
10,000	3.7×10^{-4}
20,000	1.99×10^{-4}
30,000	13.7×10^{-4}
70,000	6.34×10^{-5}
80,000	5.60×10^{-5}
100,000	4.56×10^{-5}
1,000,000	5.39×10^{-6}

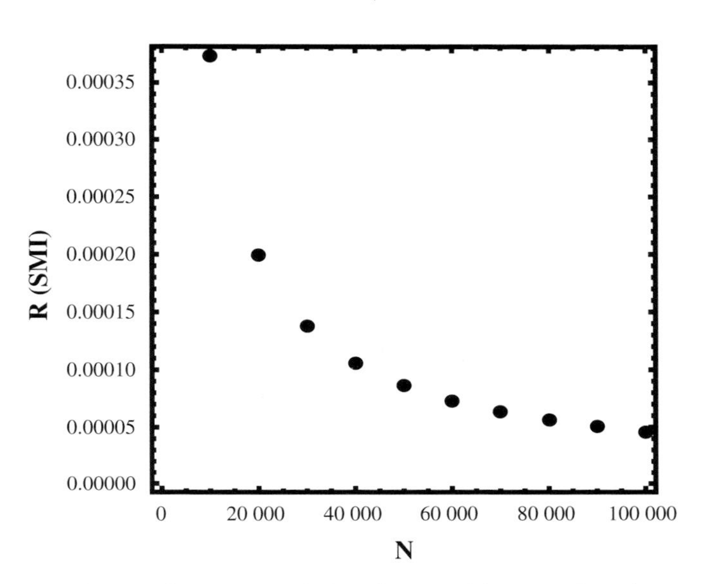

Fig. 1.33 Values of R(SMI) defined in Table 1.5 for the process of assimilation and expansion.

Table 1.5 and Figure 1.33 show a few values of the relative change in the SMI, for some values of N. As can be seen, these numbers are always positive. We shall see in Chapter 2 that for a thermodynamic system this change becomes negligible compared with the value of the SMI.

1.11.6 *Expansion into infinite volume*

Figure 1.34 is an extension of Figure 1.28. In the expansion process shown in the latter, we calculated the change in the SMI to be

$$\mathrm{SMI}(f) - \mathrm{SMI}(i) = N \log \left[\frac{2M}{M}\right] = N \log 2.$$

This is for the expansion from M cells to $2M$ cells. Now, we repeat the same experiment of expansion, but from M cells to

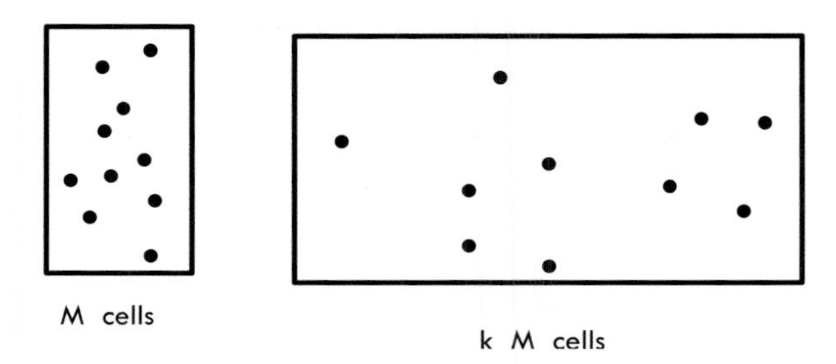

Fig. 1.34 Expansion of N particles from M cells to kM cells.

$k \times M$ cells, where k is any integer (the cells are not shown in Figure 1.34). The change in the SMI is now

$$\text{SMI}(f) - \text{SMI}(i) = N \log\left[\frac{kM}{M}\right] = N \log k.$$

We interpret this result in terms of loss of *locational information* of $\log k$ bits per particle. Now suppose that we increase k to infinity. In other words, we let the particles expand to an *infinite* number of cells (i.e. $k \to \infty$). What is the change in the SMI for this process? Clearly, when $k \to \infty$ also $\log k \to \infty$, and therefore we conclude that in this case the change in the SMI is infinite. In Subsection 2.7.9 we shall discuss a similar process with ideal gases.

1.12 Can SMI Flow, or Be Created or Destroyed?

In Subsection 1.1.5 we asked whether information can flow, or be created or destroyed. The answer was affirmative. As we have emphasized several times, SMI is not *information* but a measure of information (in a very restricted sense). Therefore, while it is meaningful to say that *information flows* from one place to another, it would be awkward to say that the *measure of information* flows from one place to another. When I tell you,

"Tomorrow it will rain," information flows from me to you. However, the *number of letters*, or words, or the SMI associated with this message, is not transferred from me to you.

When I throw ten marbles into a box, the *marbles are flown* into the box, but the *number* of marbles is not. Similarly, we can talk about creating or deleting information, but the *measure* of the information is neither created nor destroyed.[35]

I urge the reader to think of examples of various measures of objects, and examine the adequacy of talking about the flow of these measures, their creation and their deletion. This mental exercise will be important when we discuss in Chapter 2 the flow and creation of entropy. Fortunately, we will not need to talk about *destroying* entropy, though some do say that entropy *destroys* . . . whatever. (See Chapters 2–4.)

In Chapter 2, we will discuss the flow of entropy. We will see that this is an accepted way of talking about entropy mainly for historical reasons. Clausius defined $dS = dQ/T$. Here, whenever *heat flows* into or out of the system we can also speak of entropy flow. However, as we will see in Chapter 2, accepting the SMI interpretation of entropy make the phrase "entropy flows" somewhat awkward. The SMI, unlike the entropy, is not associated with any substantial quantity. Therefore, we cannot say that it *flows* from A to B, much as you cannot say that the *number* of letters in this book flows from A to B.

The situation is different when *information* is understood colloquially. We do say that information flows from person to person, or from the book into the person who reads it, and so on. However, unlike a flow of matter or energy where *flow* means that it is physically moved from A to B (i.e. it is at A and *then* it is at B), this is not the case for flow of information in its colloquial sense.

Information is *not physical* (contrary to many statements in popular-science books; see also Subsection 1.13). When you read a book containing information on thermodynamics or history, you say that the information *flows* from the book into your brain. But that information still remains in the book, as well as in all other books with the same title.

When I tell you that it is raining today in Jerusalem, you get that information from me. The information does not pass physically from me to you; it passes conceptually. A "copy" of the same information is retained in me, as well as in any other person who heard that information.

Does the information multiply when I tell it to you? Colloquially, you can say that the information has been multiplied, and there is one copy in me, one copy in you, one copy in each book where it is written, etc. This is what Brillouin calls the "distributed information."[35] However, you can also choose to view information as an abstract concept. There is a piece of *specific information* in all the books containing it, or all the people who know it. That is perfectly valid for the colloquial information. However, it will be absurd if one views information as a physical entity.

As I will show throughout the book, there is a huge confusion between the *concept* of information, the measure of information, the specific SMI, and entropy. The ultimate manifestation of this confusion is expressed by Seife, where a new law is pronounced[36]:

> And while the concept of entropy leads to the second law of thermodynamics, the idea of decoherence is related to what may be an even more powerful law, a new law:

Information can neither be created nor destroyed

> This is a law that encapsulates the laws of thermodynamics, and the explanation for the weirdness of quantum mechanics and relativity. It describes the way physical objects interact with one another and the way scientists can acquire understanding about the natural world. It is the new law.

In my view, this "new law" is *meaningless* if one does not define what one means by "information." Colloquially speaking, some information can be created, some can be destroyed, and some may be said to be conserved. However, if by "information" the author means the SMI, it is at best awkward to say that the SMI is created, destroyed or conserved — as much as it will be awkward to claim that the number of letters in this book is created, destroyed or conserved.

Whether Seife uses the term "information" in its colloquial or its SMI sense, it has *nothing* to do with *thermodynamics*, nor is it governed by any other laws of physics. Besides, this "new law" does not *explain* the weirdness of either quantum mechanics or relativity! This "law" does not "describe" the way physical objects interact with one another. It certainly has nothing to do with the way "scientists can acquire understanding about the natural world." More generally, this "law" explains *nothing*!

C-information is an abstract concept. It is, in general, not measurable. The facial expression of a crying child conveys information which is not measurable. In fact, some facial expressions can convey different information to different people. The melody played by an orchestra conveys some information to the listener. It might, however, convey different information to different listeners. A grain of salt on your tongue conveys some information to the brain. It might convey different information to different people. All of these are nonmeasurable in information theory. One can assign various measures of information to the facial expression, to the musical notes or to the taste of salt, but these measures are different from SMI. Certainly, they have nothing to do with entropy or thermodynamics.

There are many examples of nonverbal communication of information in the animal kingdom. Bees, for example, buzz naturally as they fly, but they buzz aggressively when their nests are being disturbed, or if they are being attacked or defending themselves. Through a special "choreographed" dance, they inform the other members of the colony on the direction of a food source. Monkeys inform the other members of the barrel of an impending danger by emitting a distinct screech.

It goes without saying that we humans also communicate through nonverbal means. Such information might be ambiguous, or might be understood differently by different people.

If you see someone waving his or her hand, he or she might be "saying" to you (or to any other person), "Hi, here I am," or "Goodbye, I am leaving." The *meaning* depends on the context in which it is "said" or "done."

The other day my wife and I were swimming in the pool. I saw my wife leave the pool, waving her hand, presumably saying, "I am leaving — bye." Then she pointed her finger at the wall clock. I "answered" by raising my two hands and exposing all ten fingers. For us, it was a clear exchange of messages: by pointing at the clock, she asked me how much more time I would spend swimming, and by showing my ten fingers I answered, "Ten minutes."

Clearly, this nonverbal communication is not measurable by SMI. Its meaning also depends on the context in which it was "said."

Here is another nonverbal message, which I found in the train in Stuttgart:

Non verbal messages in a train in Stuttgart

Regarding the conservation of information, consider the following example. A new drug has been developed and manufactured by a pharmaceutical company. It is sold with a small piece of literature providing *information* on the drug's chemical composition, administration, storage, etc.

Clearly, colloquially speaking, this information was *created* by the inventors of the drug. No one knew or possessed the information contained in the literature before the drug was invented (including the inventors themselves). This is accepted colloquially. On the other hand, one can claim that the same information had always existed — somewhere in an abstract space of concepts — and the inventor did not *create* that information, but only discovered it. (This is the same as the great debate in mathematics: Is it discovered or invented?)

Suppose that there are a million copies of the same literature, each containing the same information on the newly discovered drug. By printing millions of copies of the literature, have we multiplied the information contained in one piece of literature by a million? Again, you can answer with either a yes or a no, depending on how you define the concept of information.

Now, suppose that one of these pieces of literature is burned. Has that information been destroyed? Again, if you refer to the specific information in that specific literature, you may say that that particular information contained in the literature has been destroyed. However, the information about the drug is still contained in all the other pieces of literature, and in this sense it has not been destroyed.

Next, consider the case where all the millions of copies of the literature have been burned, and all the inventors, as well as all the people who have used the drug and read the literature, have died. Not a single copy of the literature has survived, and no one knows that information, i.e. there exists absolutely no record of that information. Would you say that the information on the drug was destroyed, or perhaps it is stored somewhere in some abstract space of concepts? You can choose to answer this question either way.

The situation is very different when we talk about the SMI associated with the drug which is contained in the literature. In this case the SMI of the information on the drug is a well-defined quantity. It might have different values (different numbers of bits) for the same information written in different languages. And if there are millions of copies of the same literature, the SMI of the million copies is *a million times* the SMI of a single booklet. This is so because the SMI of any text in a specific language depends on the *total number* of letters in that text. No matter what the content of the text (or the sequence of symbols) is, and no matter how many copies there are of the same information, the SMI is related to the *total number* of letters in the entire text.

If you translate the same text into a different language, then the SMI of the two texts containing the *same information* might be different in the two languages. If you read the text of the literature, remember its content and destroy it afterward, you can say that the information in the specific literature was destroyed, but you can still have a "copy" of that information, and therefore the information itself is still stored in your brain. Regarding the SMI of the information stored in the brain, it might or might not have the same value as the SMI in the booklet. It depends on which "language" or what sequence of symbols the information in the brain is stored in.

It should be stressed again that SMI applies whenever a probability distribution is *given*. The SMI belongs to, or is associated with, the given distribution. It would be awkward to talk about either creating or destroying the SMI associated with a given distribution. The distribution could be of the letters in a given language, the musical notes of a melody, the bases in DNA or the amino acids of a protein. All these sequences can be assigned an SMI. This SMI has nothing to do with the *meaning* of information in the sequence of letters, the melody of the musical notes, the instructions contained in the DNA or the function of the specific protein.

It would be meaningless to talk about either creating or destroying the SMI of an unspecified piece of information. And, of course, neither the information nor the SMI obeys any physical law.

Finally, consider the following:

While I am writing this very sentence I can say that I am creating information, and the information is transmitted to you while you read this sentence.

I have noticed that while writing this book, I am adding more and more letters. The number of letters will increase until I finish writing the book and send it to the printer. Would you say that I *created* the *number* of letters in this book? Would you say that the SMI associated with this book was created by me? What is the SMI of this book if it is printed in 1000 copies?

What is the SMI of the same book when it is translated into Chinese?

1.13 Is Information Physical?

Colloquially, we say that information flows from person A to person B, or from the transmitter to a receiver. However, if information is "physical" as some people claim, then the following question might be raised: Is information physically transferred from A to B? If this is true, does it mean that it was initially in A (or at A) and now it is in B (or at B)? Or perhaps it is both in A and in B?

The answer to this question is important when one is talking about *conservation* of the total information in the universe. Does this mean that if I tell a thousand people that "today is Monday," there is a "copy" of this information in each of those people, or is there only one copy? Is this information quantifiable at all? What about a person who writes a fiction book — does he or she create information, or did the information exist somewhere before?

All these questions must be answered when one claims that information is *physical*, i.e. it is *conserved* or obeys some physical laws.

Recently, in some popular-science books, one has been able to read many statements about *information* being *physical*, that *information* obeys some physical laws, and that everything is *information*. The most famous statement was made by Wheeler: "It from bit," which is supposed to convey the idea that all physical things are informational-theoretical in origin.

I cannot agree with such statements. If you kick a stone, either it will fly away or your foot will hurt. You cannot kick the information on today's weather, or the information contained in this book. Not everything is considered information, and not everything can be bits. A bit is a unit of information and not a physical entity. You cannot say that this book is a centimeter, or a gram — and certainly not a bit!

Here is a quotation regarding information as being "physical" [Wheeler (1990)]:

> It from bit. Otherwise put, every "it" — every particle, every field of force, even the space–time continuum itself — derives its function, its meaning, its very existence entirely — even if in some contexts indirectly — from the apparatus-elicited answers to yes-or-no questions, binary choices, bits. "It from bit" symbolizes the idea that every item of the physical world has at bottom — a very deep bottom, in most instances — an immaterial source and explanation; that which we call reality arises in the last analysis from the posing of yes-or-no questions and the registering of equipment-evoked responses; in short, that all things physical are information-theoretic in origin and that this is a participatory universe.

I definitely disagree with the "it from bit" doctrine, as well as the idea "that all things physical are information-theoretical in origin."

I think such statements have misled many scientists, as well as popular-science writers, into claiming that "information theory" is a revolution in *physics*, equal to or perhaps greater than the two great revolutions in physics: relativity and quantum mechanics. I personally maintain that information theory is not a theory of information in the first place. Secondly, information theory is certainly not a theory of physics. Therefore, the creation of information theory was not, and is not, a revolution in physics.

Thus, the apparent information revolution is only an illusion. Information is an abstract concept and it is not subjected to any (existing) physical law, or any new (nonexisting) law of information. In fact, even the well-defined SMI is not a physical quantity unless one maintains that probabilities are *physical* quantities. Most of the pieces of information stored in our brains (including feelings, the sense of redness or blueness, sweetness or bitterness, and many more) are not registered in a digital format. They are probably more like analog, and are not measured in bits.

We may have information about the location or velocity of an atom, or we may have information about the energy level in which a molecule is, but whether any atom or molecule *is* information, or even is a carrier of information, is in my opinion untenable.

Following the discussion of "it from bit," Gleick (2011) asks rhetorically:

> Why does nature appear quantized? And his answer: "Because information is quantized. The bit is the ultimate unsplittable particle.

Of course, *information*, in its colloquial sense, is "quantized" — but not in the sense used in quantum mechanics. Any C-information is expressed with a *finite number* of discrete words, letters, symbols, etc. It is ironic that SMI, which is one specific *measure* of information, is *not* quantized, and the bit is not an "unsplittable particle." The bit is *not* a particle in the first place, and SMI is, by its very definition, a continuous function of the probabilities $p_1, \ldots, p_N$ (see Section 1.3). SMI is a well-defined quantity (provided that we agree on the meaning of a probability distribution), and its units are bits.

Today is Monday.

This information is *true* today! Tomorrow, it will be false. A week from today, it will be true again. Does the truthfulness of information change with time? Yes, some information can change their meaning with time. On the other hand, the SMI of a given message does not change with time. Of course, the *same* information — say, "the color of oranges is orange" — can have different SMIs if expressed in different languages. Perhaps this *information* might not have an SMI at all if no one has ever uttered its words or written it in any language. Yet, this is information nonetheless. In this chapter, I hope that I have made the distinction between *information*, the abstract concept, and its measure (SMI) clear. It is unfortunate that the term "SMI," and also "entropy," are very often misused, and perhaps also abused in some instances.

Another famous quotation from Wheeler [quoted by Jacob Bekenstein (2003)]:

> . . . regard the physical world as made of information, with energy and matter as incidentals.

In my earlier book (Ben-Naim, 2008) I quoted this statement by Wheeler in connection with the informational interpretation of entropy. Since then I have read Wheeler's book (1994) and other statements made by him. My views have changed diametrically. I have commented on the "it from bit" doctrine above. Here, I want to add a short comment regarding this particular statement. What does it mean when one says that "the world is made of information?" How can one say that matter and energy are incidentals? This quotation *provides information* on the relative importance of "information," "energy" and "matter." I do not believe that energy and matter are incidentals.

Such a statement does not convey information on physics, on reality or on information theory.

Siegfried (2000) has a whole chapter titled "Information is Physical." On page 2 the author writes:

> Wheeler observes that a black hole keeps a record of the information it engulfs. The more information swallowed, the bigger the black hole is — and thus the more space on the black hole's surface to accommodate boxes depicting bits . . . the black hole converts all sorts of real things into information. Somehow, Wheeler concludes, information has some connection to existence, a view he advertises with the slogan "It from bit."

Of course, no one knows what goes on in a black hole. Certainly, one cannot claim that "the black hole converts all sorts of real things into information." Such statements are meaningless at best! They arise from confusing entropy with SMI, then confusing SMI with information, and then confusing information with bits.

Pause and think

Suppose that you send a rocket with a copy of this book toward a black hole. There is no doubt that the rocket and the book will be crushed by the immense gravity of the black hole. Can you figure out into what kind of information this book will be converted?

Is it information about the locations and velocities of the atoms of the book?

Is it information *contained* in the book?

Is it the SMI associated with the text of this book?

I do not think anyone can give an answer to these questions.

On page 9 of Siegfried (2000):

> Many scientists may still regard talk about the "reality" of information to be silly. Yet the information approach already animates diverse fields of scientific research. It is being put to profitable use in investigating physics, life, and existence itself, revealing unforeseen secrets of space and time.

This is a wild and unfounded exaggeration! Of course information is *real*! The question whether information is *physical* or has any *substantial reality* is not silly, but more likely meaningless. I am not aware of anyone who thinks that the "reality" of information is silly, but I think that talk of the *physical nature* of information is silly, indeed.

It is true that information theory has found applications in diverse fields of science. To the best of my knowledge, information theory *did not* reveal any unforeseen secrets of space and time, nor did it explain anything about "life and existence itself." (See also Chapter 3.) Perhaps I missed some revelations of information theory. I would appreciate it very much if the reader could reveal to me some of these secrets.

One more quotation from Siegfried (page 197):

> Physicists don't worry about losing information when a bomb blows up in a library. In that case information isn't really lost. Subtle patterns remain in the atoms and photons that fly away. It wouldn't be easy, but in principle all atoms and photons could be tracked down and measured. Their current conditions would permit a reconstruction of the precise initial conditions before the explosion.

My views are diametrically different. If a bomb blows up a library I will be *very worried* about those books that contain

Fig. 1.35 A burning book. Some of the molecules that are released into the air, e.g. carbon dioxide, are assimilated by the same molecules in the air.

information that cannot be found in any other book, or any other storage device. If this is the case, then *the information contained in those books will be lost forever.* There is no way, *in principle*, as well as in *practice*, to retrieve this information from the atoms and photons that fly away. Once the atoms and molecules that fly away are assimilated by other atoms and molecules, there is no way *in principle* to trace their origin, or to reconstruct the lost information: see Fig. 1.35 and also Section 1.1.5 and Note 10. Therefore, I believe that such statements are highly misleading: not only misleading, but every one should be worried if a bomb explodes in a library, not only physicists!

I would be delighted to be proven wrong.

In his autobiography Wheeler writes:

> I have been led to think of analogies between the way a computer works and the way the universe works. The computer is built on Yes–No logic. So, perhaps is the universe . . . . The universe and

> all that it contains (it) may arise from myriad Yes–No choices of measurement (the "bits").

Perhaps? More probably not! I do not believe there is any evidence to support the claim that "The universe and all that it contains (it) may arise from myriad Yes–No choices of measurement (the "bits")." As I said before, I cannot agree with such statements either, whether by "bit" Wheeler meant *information* or the *units* of information.

Wheeler's slogan "it from bit" embodies the widespread confusion between the three concepts: information, SMI and the bit. From what he has written (and some is quoted above), it is clear that he meant to say, "Everything is information." Personally, I do not share that view. However, what Wheeler actually says is quite different from what he meant to say. Even if one accepts the view that every physical object (an atom, a molecule or a table) is information, one cannot say that every physical object is SMI — and certainly not a bit!

To clear up the common confusion between information, SMI and the bit, consider the following triplet of concepts: the sense of heat or cold, the temperature of a body, and the degree centigrade.

The analogy between information and the sense of heat or cold, between SMI and the temperature, and between the bit and the degree is not perfect. It is raised here only to express the big conceptual difference between the three.

The sense of heat or cold, as information, can be either subjective or objective. Place one hand on a hot body (say, 50°C), and the other hand on a very cold body (say, ice at 0°C). After some time, immerse the two hands in water at 25°C. One hand will feel *cold*, while the other will feel *hot*. We also use the words "hot" and "cold" or "cool" and "warm" to describe a human being's personality, or an event, or other things without any reference to their temperature.

The *temperature* of a body is an objective *measure*, and the units of *temperature* are degrees. These two are the analogs of SMI and bits. Note that not all objects may be assigned a temperature and measured in degrees. Similarly, not all information can be assigned an SMI and measured in bits.[37]

Clearly, one cannot claim that any physical object is information, or that any information is physical. Seife (2007; page 2) made the following statement:

> Information theory is so powerful because information is physical. Information is not an abstract concept, and it is not just facts or figures, dates and names. It is a concrete property of matter and energy that is quantifiable and measurable . . . And everything in the universe must obey the laws of information because everything in the universe is shaped by the information it contains.

Clearly, such statements embody the same confusion between information, the measure of information, and physical objects. Information *is* an abstract concept, and information *is not* a concrete property of matter and energy. The book you are holding in your hand was not "shaped by the information it contains" (whatever "shaped" might mean).

In Gleick's book *Information*, Wheeler is quoted as saying:
What we call the past is built on bits.

> What we call the *past* is the past! Not bits, not centimeters and not grams. Not even seconds or minutes, or any other units of time. Of course, you *can say* that the past is built on bits, the future is built on bits and the present is built on bits. No one can prove you wrong or right.

1.14 Evolving Games

Until now we have discussed an experiment having N possible outcomes with a given probability distribution $p_1, \ldots, p_N$ on which we have defined the corresponding SMI. We will see in Chapter 2 that the understanding of the concept of SMI is crucial to the understanding of *what* entropy is.

However, understanding *what* entropy is does not give us any hint about the question of *why* it always increases (under certain conditions). In other words, SMI in itself does not provide any explanation for the Second Law of Thermodynamics. We will discuss this aspect of the Second Law in Chapter 2. For now we will continue with the 20Q game, adding one new complication which never (as far as I know) occurs when we actually play the 20Q game in a party. However, in nature there are many examples where the same system can have different distributions of outcomes at different times or under different conditions. If we measure the distribution of people's heights or weights in different cities or different countries, we might find that these distributions change with time. The corresponding SMI will also change with time.

We will now formalize this change in the SMI as the system's distribution evolves. Then we will discuss a particular experiment (or a game which evolves "naturally" when we remove a constraint on the system). The study of such systems will be relevant to the Second Law discussed in Chapter 2.

Note that I am using the term "evolving game." The evolution could occur either artificially or naturally. This is different from "natural evolution" of species, which has been applied to "ideas" or to "information" in its colloquial sense.

Consider again the dart game with four regions in Figure 1.36A. Now, while we play this game the borderlines between

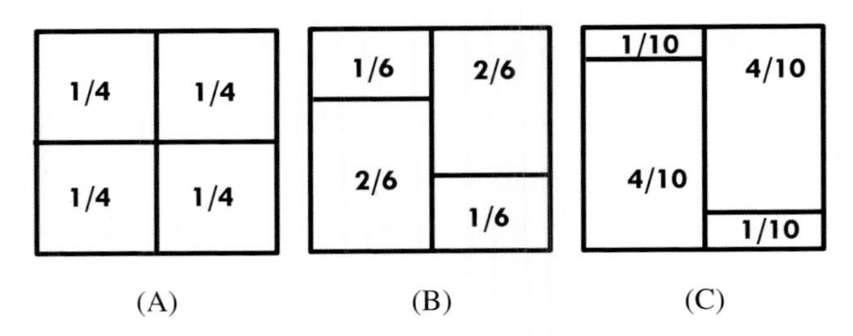

Fig. 1.36 An evolving game from A to B to C.

the regions start to move. For each move we get a different distribution and therefore a different SMI. For instance, if the horizontal borderlines change as in Figure 1.36B we have a new SMI for this game [in these particular examples $\text{SMI}(A) = 2$ and $\text{SMI}(B) = 1.92$]. We can further decrease the SMI in moving from B to C in Figure 1.36. Of course, we can think of other kinds of changes of the borders between the regions. We already know that whatever the changes are, if we start with game A in Figure 1.36, the game will always be *easier* to play, i.e. fewer questions need to be asked to find the location of the dart. Game B is easier than A, and game C is easier than B. Note that in these particular changes in the game, we can go either from A to B to C, or in the other direction, from C to B to A. The "evolution" in this series of games is purely artificial.

Next, we describe a different "evolution" of the game. Here, the total number of outcomes is kept fixed; only the distribution will change with time. We will see that whatever the state of the initial game is, the game will evolve (in most cases) in such a way that it will be *more difficult to play*. We will also see that we can choose games such that instead of "most cases" we can claim that it will become more difficult to play "almost always," and in Chapter 2 we will drop the "almost" and leave the "always" when discussing a thermodynamic system and the Second Law.

The game we will play is again a variation of the 20Q game. However, this game is designed in such a way that it evolves in time in one direction.

Consider a two-dimensional tray having M compartments. In these compartments we distribute N marbles, such that N_i marbles are in compartment i, and $\sum_{i=1}^{M} N_i = N$. We assume that the marbles are very small and there is no limit on the number of marbles in each compartment. There are also no interactions between the marbles, so that the probability of a marble occupying any of the compartments is independent of the number of marbles which are already in that compartment.

The distribution of marbles $N_1, N_2, \ldots, N_M$ also defines a probability distribution $p_1, p_2, \ldots, p_M$, where $p_i = N_i/N$. Each of the p_i may be interpreted as the probability of finding a *specific* marble in compartment i, with $\sum_{i=1}^{M} p_i = 1$.

We can play the 20Q game on this system as follows. I choose a *specific* marble — say, numbered "6" or colored red. You have to find out in which compartment the marble is. You are given the total number of compartments, and you are given the probability distribution $p_1, p_2, \ldots, p_M$. This game is equivalent to the game with a dart hitting a board divided into M regions, such that the fraction of area of the ith region is $p_i = A_i/A$. As in the dart game, the more asymmetric the distribution, the easier the game. Two extreme cases are shown in Figure 1.37.

In Figure 1.37A all the ten marbles are in one compartment. Hence, the distribution of marbles is $(10, 0, 0, \ldots, 0)$. In this case you *know* where the specific marble I chose is. This distribution provides the maximum *information* on the location of the marble I chose. On the other hand, in Figure 1.37B, we have ten marbles distributed uniformly, one in each compartment. In this

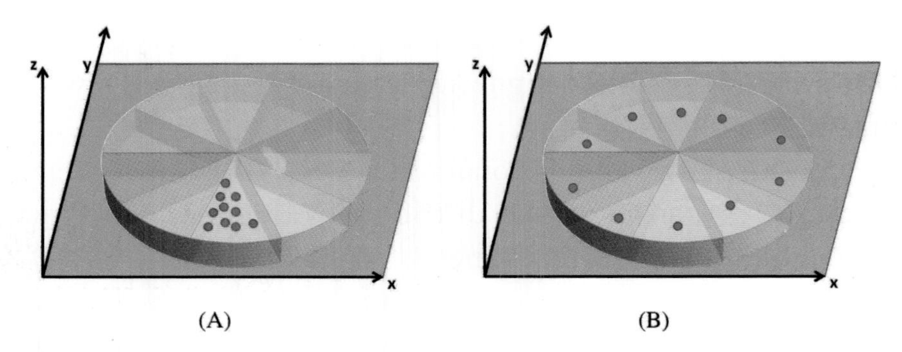

Fig. 1.37 Two 2Q games: the easier (A) and the more difficult (B).

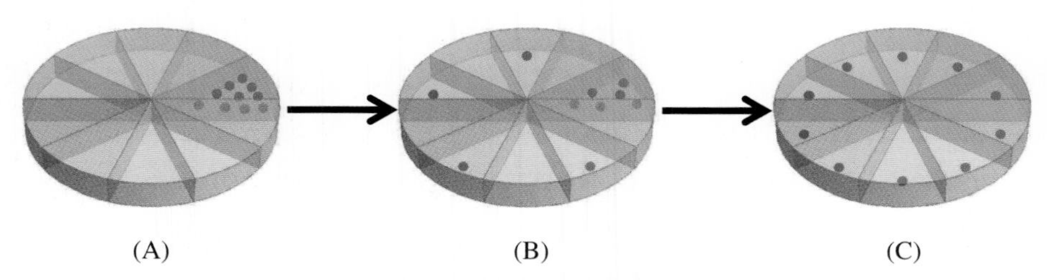

Fig. 1.38 An evolving 20Q game. By shaking the tray the game will change from A to B to C.

case the game is the *most difficult* to play. On average you have to ask $\log_2 10 \approx 3.3$, or between 3 and 4 questions. We can say that this distribution provides the *minimum information* on the location of the specific marble. For any other distribution the number of (smart!) questions you need to ask is between 0 and 3.3. For each distribution the SMI provides the average number of questions you need to ask.

We now change the game. For simplicity, suppose that we start with the (easiest) game in Figure 1.38A. We shake the tray in the *xy* plane, and from time to time also in the *z* direction so that marbles can jump from one compartment to another. The shaking is done at random, which means that each *specific* marble has the same probability of crossing to either the right or the left compartment. If the shaking is very slow and mild, no marble can cross from one compartment to another. If the shaking is very violent, then marbles can cross one, two or more borderlines between the compartments. (We assume that all the marbles stay in the tray.) The randomness of the shaking means that whatever the probability of a "specific" marble to cross from left to right is, it is the same as for crossing from right to left. Note the emphasis on the word "specific." Note also that in a circular tray all the compartments are equivalent. In a real three-dimensional box divided into compartments, the compartments are not equivalent.

Clearly, the probability of jumping from one compartment to another depends on the vigor of the shaking. For a given shaking we denote by q the probability of a *specific* marble to cross either from i to $i+1$ or from i to $i-1$, and this is the same probability for any i.

On the other hand, the probability of *any* marble in compartment i to cross to either $i+1$ or $i-1$ depends on the *number* of marbles in that compartment. Below, we shall not be interested in these probabilities. Instead, we shall be interested in the following probabilities. First, recall that p_i is the probability of finding a specific marble in compartment i. The second probability is the following: while we shake the tray, the *distribution* of the marbles in the compartments changes. Figure 1.38B shows a distribution of marbles after a brief shaking. We now ask: What is the probability of finding a particular distribution of marbles, $N_1, \ldots, N_M$? We shall refer to this as the *superprobability*.

The total number of specific configurations is M^N. (The first can be in one of M compartments, the second can be in one of M compartments, etc.). We assume that all these configurations are equally probable, and hence

$$\Pr(\text{specific configuration}) = \frac{1}{M^N}. \tag{1}$$

The probability of obtaining a distribution $(N_1, \ldots, N_M)$, i.e. N_1 in 1, N_2 in 2, etc. (independently of which *specific* marble is in compartment i or j), is

$$\Pr(\text{of a distribution } N_1, \ldots, N_\mathrm{M}) = \left(\frac{1}{M}\right)^N \frac{N!}{\prod_{i=1}^{M} N_i!}. \tag{2}$$

The factor $N!/\prod_{i=1}^{M} N_i!$ is the number of *specific* configurations which are consistent with the distribution $N_1, N_2, \ldots, N_M$.

While we are shaking the tray the probability distribution of marbles in the compartment changes. For each distribution we can define the corresponding SMI:

$$\mathrm{SMI}(p_1, \ldots, p_M) = -\sum_{i=1}^{M} p_i \log p_i. \tag{3}$$

We can also define the *superprobability* as the probability of finding this probability distribution by

$$\Pr(p_1, \ldots, p_M) = \left(\frac{1}{M}\right)^N \frac{N!}{(p_i N)!}, \tag{4}$$

where $p_i = N_i/N$. The two quantities SMI and Pr are related to each other. For a very large number N, and when all the N_i are very large, we can use the Stirling approximation for the factorial:

$$\ln N_i! = N_i \ln N_i - N_i. \qquad (5)$$

Here, ln is the natural logarithm. With this approximation we can rewrite the probability Pr in (4) as

$$\ln \Pr = -N \ln M - N \sum_{i=1}^{M} p_i \ln p_i. \qquad (6)$$

Dividing the equation (6) by $\log_2 e$ and rearranging, the equation (6) can be written as

$$\Pr(p_1, \ldots, p_M) = \left(\frac{1}{M}\right)^N 2^{N \times \mathrm{SMI}(p_1, \ldots, p_M)}. \qquad (7)$$

This is a remarkable relationship. The *superprobability* of obtaining a probability distribution $p_1, \ldots, p_M$ (assuming that all of the M^N configurations are equally probable) is related to the SMI associated with that distribution. In the language of the 20Q game, we can say that while we are shaking the tray, the probability distribution of the marbles changes. Each probability distribution defines both an SMI, equation (3), and a *superprobability*, equation (4). Because of the relationship (7) between Pr and the SMI, we can conclude that the *larger* the probability of obtaining a specific game, the larger its SMI, or the more difficult it *will be* to play that game.[38] We already know that the distribution that maximizes the SMI is the uniform distribution. Therefore, we conclude that the uniform distribution is also the distribution which maximizes the probability Pr. Figure 1.38(C) shows the distribution having maximal probability.

From equation (7) we can conclude that for any given N and M, the distribution which maximizes the SMI is also the distribution which maximizes the probability function Pr. The relevance of this conclusion to the Second Law of Thermodynamics will be discussed in Chapter 2. Here, the conclusion is strictly valid for any SMI and for any game in which the distribution changes with time.

The connection between the maximal SMI and the maximum Pr justifies the reference to the distribution that maximizes the SMI (and Pr) as the *equilibrium* distribution. In the present example we saw that the equilibrium distribution is the uniform distribution. In the language of the 20Q game we can say that starting with any initial game and shaking the tray, the game will evolve to a more difficult one to play (more questions to ask). [For some simulated games of this kind, the reader is referred to Ben-Naim (2010).]

1.15 Summary of Chapter 1

In this chapter we have made a distinction between the general, colloquial term "information" (C-information) and the specific measure of information as used in information theory. The central quantity in information theory is Shannon's measure of information (SMI). It is most important to realize that SMI is a specific measure of the size of the message carrying the information. It is not concerned with the meaning, value, importance, etc. of the information itself.

Confusing C-information with SMI is commonplace in the literature. This confusion is prevalent especially when *information* is used in connection with *entropy*, life and the universe. We shall discuss a few examples in the following chapters.

It should be clear by now that what is referred to as information theory is *not* a *theory of information*. It is basically a theory that measures (in a restricted sense) the size of information, and also deals with questions of efficiency of transmitting information, etc. Information theory does not provide explanation of any phenomena, or an answer to problems in physics, chemistry or biology. This conclusion is in stark contrast to the subtitle of Seife's book:

> How the New Science of Information is Explaining Everything in the Cosmos, from our Brains to Black Holes.[39]

This subtitle is extremely misleading. Whatever Seife means by the "new science of information," it does not explain anything. Certainly not the "brains" or the "black holes."

When we discuss C-information we may take the liberty to say anything that comes to our mind about information: a "hot dog" is not a dog, and "butterflies in the stomach," does not literally mean that there are butterflies fluttering their wings inside our stomach. Yet, both phrases convey some information. Also, the same information can mean different things to different people.

Colloquially, we also say that information flows, and is created or destroyed. All these are acceptable when we talk about information as an abstract concept. One cannot say the same thing about SMI. SMI does not flow; nor is it created or destroyed.

We have seen that the slogan "it from bit" embodies the confusion between the three concepts of information, SMI and the bit. We will see in the next chapter that one more concept — entropy — is added to this muddy whirl of confusion.

We hear quite often about the *information revolution* — and that this revolution is in physics! Of course, it is true that in the past decades there has been some kind of revolution in terms of the *amount* of information which is accessible to us. This is certainly not a revolution in physics, and cannot be compared to the two great revolutions that occurred in physics during the early 20th century.

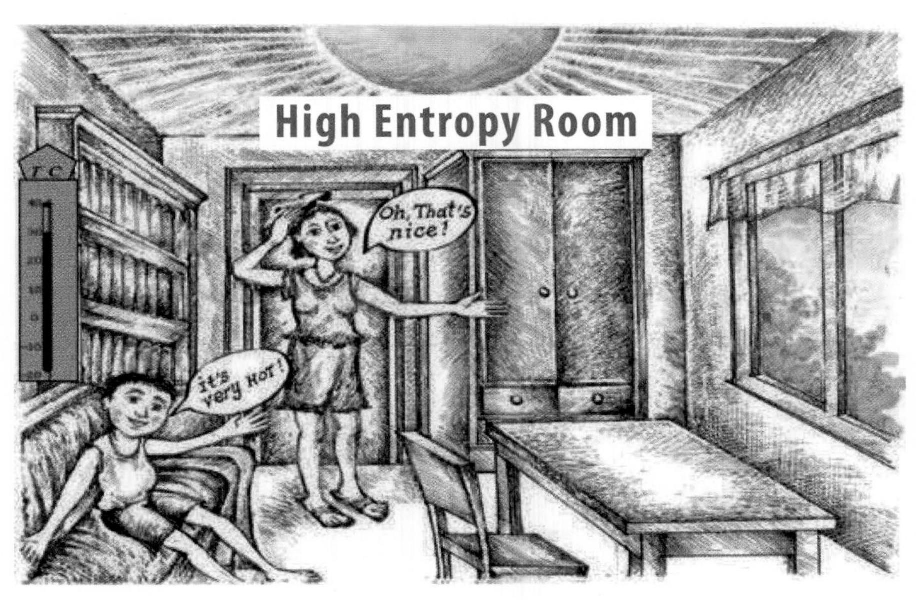
High Entropy Room
Oh, That's nice!
it's very HOT!

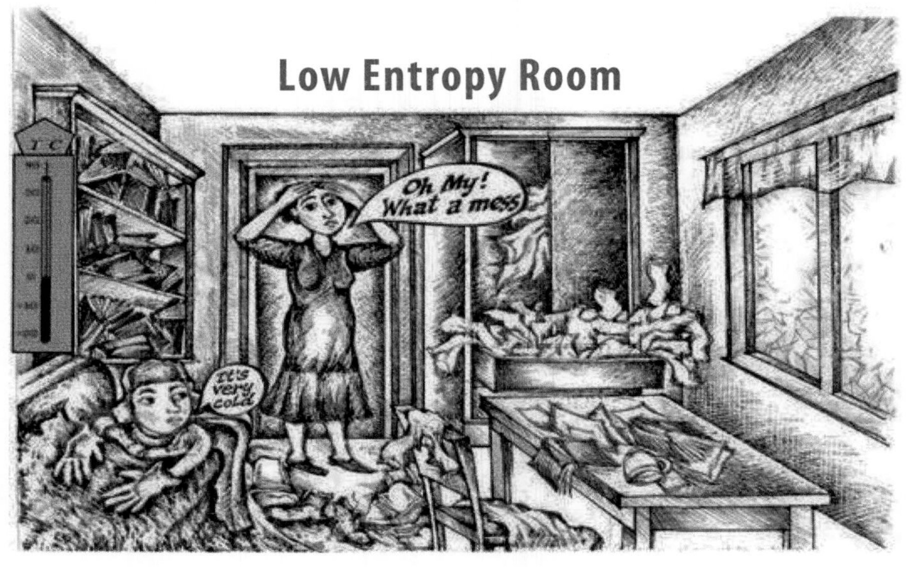
Low Entropy Room
Oh My! What a mess
It's very cold

Chapter 2

Entropy

In this chapter we will briefly discuss the concept of entropy — first as it was defined within thermodynamics, second as it was defined by Boltzmann, and finally as it can be derived from Shannon's measure of information. The last is of particular interest in connection with the *meaning* of the concept of entropy. The fact that thermodynamic entropy may be derived from SMI provides a solid and well-established interpretation of entropy. Once we have derived the explicit entropy function, we will discuss the Second Law of Thermodynamics (referred to as the Second Law or 2nd Law). This part answers the following question: Why does entropy always increase? We will also see that the seeming connection between the Second Law and the *arrow of time* is only an illusion.

The next question we will examine is whether the so-called Maxwell's demon "defeats the entropy." We will see that the whole idea of Maxwell's demon cannot be discussed within thermodynamics. This conclusion is the same as the one we will arrive at in connection with the applicability of entropy and the Second Law to living systems (discussed in Chapter 3).

Examples of some simple processes for which the entropy change may be calculated are presented in Section 2.7 and compared with the corresponding changes in the SMI.

Finally, we emphasize again that in this book we reserve the term "entropy" for thermodynamic entropy. What is commonly referred to as Shannon entropy will be referred to as SMI.

2.1 Nonatomistic Formulation of the Second Law

Traditionally, the birth of the Second Law is associated with Sadi Carnot (1796–1832). Although Carnot himself did not formulate the Second Law, his work laid the foundations on which the Second Law was formulated a few years later by Clausius and Kelvin.

Carnot was interested in heat engines — more specifically, the *efficiency* of heat engines. Before describing an idealized heat engine, let me say very loosely that such an engine converts *heat* into *work*. Take a volume of gas and heat it, and its volume will increase. We can use this expansion of the gas to do some work, such as lifting a brick from the first to the second floor. Once we bring the brick to the second floor, we unload it, and then cool the gas to its original temperature. The sequence of the processes of heating, expansion, cooling and compression can be repeated. Let me describe the simplest of such an engine (Figure 2.1). Suppose that you have a vessel of volume V containing any fluid — a gas or a liquid. The upper part of the vessel is sealed by a movable piston. The vessel is initially in state 1, thermally insulated, and has a temperature T_1 — say, 0°C.[1] In the first step of the operation of this engine (step I), we place a weight — say, 1 kg — on the piston. The gas will be compressed somewhat. The new state is state 2. Next, we attach the vessel to a *heat source*

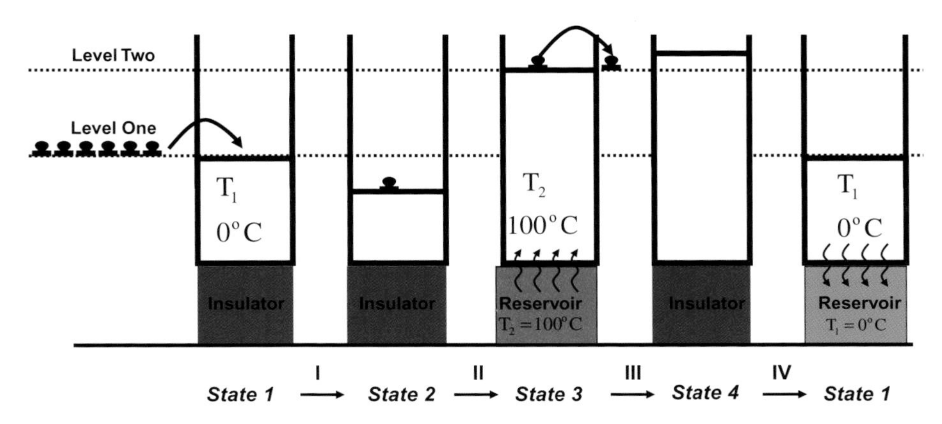

Fig. 2.1 A schematic heat engine operating between two temperatures, $T_1 = 0°\text{C}$ and $T_2 = 100°\text{C}$.

(step II). The heat source is simply a very large body at a constant temperature — say, $T_2 = 100°\text{C}$. When the vessel is attached to the heat source, thermal energy will flow from the heat source to the engine. This flow of heat is depicted by the four wiggly arrows in Figure 2.1. For simplicity, we assume that the heat source is immense compared with the size of the system or the engine. This ensures that after equilibrium is reached, the system will have the same temperature, T_2, as the heat source, and though the heat source has "lost" some energy, its temperature will be nearly unchanged. As the gas (or the liquid) in the engine heats up, it expands, thereby pushing the movable piston upward. In this step, the engine has done some *useful work*: lifting a weight placed on the piston from a lower level to a higher level. The new state is state 3. Up to this point, the engine has *absorbed* some quantity of energy in the form of heat which was transferred from the heat source to the gas, thereby enabling the engine to do some work by lifting the weight. Removing the weight (step III) might cause a further expansion of the gas. The final state is state 4.

If we want to convert this device into an *engine* that repeatedly does useful work, like lifting weights (from level 1 to level 2), we need to operate it in a complete cycle. To do this, we need to bring the system back to its initial state, i.e. cool the engine to its initial temperature, T_1. This can be achieved by attaching the vessel to a *heat sink* at $T_1 = 0°C$ — step IV (again, we assume that the heat sink is much larger compared with our system, such that its temperature is nearly unaffected while it is attached to the engine). In this step heat will flow from the engine to the reservoir. Again, we depict this heat flow by the wiggly arrows in Figure 2.1. As a result, the engine will cool to its initial temperature, T_1, and we will return to the initial state and the cycle can start again.

This is not the so-called Carnot cycle. Nevertheless, it has all the elements of a heat engine, doing work by operating between the two temperatures, T_1 and T_2.

The net effect of the repeated cycles is that heat, or thermal energy, is pumped *into* the engine from a body at a high temperature, $T_2 = 100°C$; work is done by lifting a weight, and another amount of thermal energy is *pumped out* from the engine into a body at a lower temperature, $T_1 = 0°C$. The Carnot heat engine is schematically shown in Figure 2.2. It consists of four steps, two isothermal processes (from A to B and from C to D in the figure) and two adiabatic processes (from B to C and from D to A). We are not interested in these details here.

Carnot was interested in the *efficiency* of such heat engines operating between two temperatures under some ideal conditions (e.g. a massless piston, no friction, no heat loss). Qualitatively, the efficiency of the engine is defined as the amount of work done by the system per unit of thermal energy

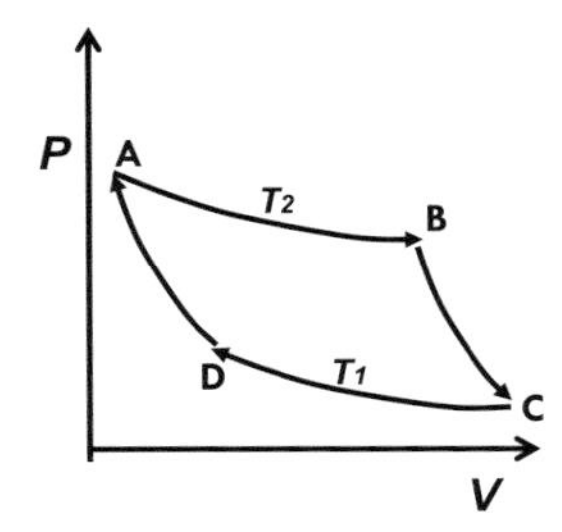

Fig. 2.2 A schematic Carnot cycle. The arrow from A to B and from C to D represent processes at constant temperatures, T_2 and T_1, respectively. The arrow from B to C and from D to A represent adiabatic process, i.e. no heat is exchanged between the engine and its surroundings.

flown into the system.[2]

$$\text{Efficiency} = \frac{\text{net work done}}{\text{thermal energy flown into the system}}.$$

Carnot was mainly interested in the limits on the efficiency of heat engines. He found out that the limiting efficiency depends only on the temperatures between which the engine operates, and not on the substance (i.e. which gas or liquid) that is used in the engine. Later, it was shown that the efficiency of Carnot's idealized engine could not be surpassed by any other engine. This laid the cornerstone for the formulation of the Second Law and paved the way for the appearance of the new term "entropy."

It was William Thomson (1824–1907), later known as Lord Kelvin, who first formulated the Second Law of Thermodynamics. Basically, Kelvin's formulation states:

> There could be no engine, which when operating in cycles, the sole effect of which is pumping energy from a source of heat and completely converting it into work.

Although such an engine would not have contradicted the First Law of Thermodynamics (the law of conservation of the total energy), it did impose a limitation on the amount of work that can be done by operating an engine between two temperatures.

In simple terms, recognizing that heat is a form of energy, the Second Law of Thermodynamics is a statement on the impossibility of converting heat (thermal energy) completely into work (though the other way is possible, i.e. work can be converted completely into heat; for example, stirring a fluid with a magnetic stirrer, or mechanically turning a wheel in a fluid). This impossibility is sometimes stated as "perpetual motion of the second kind is impossible." If such perpetual motion were possible, one could use the huge reservoir of thermal energy of the oceans to propel a ship, leaving a trail of slightly cooler water. Unfortunately, this is impossible.

Another formulation of the Second Law of Thermodynamics was later given by Rudolf Clausius (1822–1888). Basically, Clausius' formulation is what every one of us has observed: heat always flows from a body at a high temperature (and hence is cooled) to a body at a lower temperature (which is heated up). We never observe a spontaneous occurrence of the reverse process. Clausius' formulation states that no process exists such that its net effect is only the transfer of heat from a cold to a hot body. Of course, we can achieve this direction of heat flow by doing work on the fluid (which is how refrigeration is achieved). What Clausius claimed was that the reverse of the process of heat transferred from a hot to a cold body, when brought into contact, can never be observed to occur *spontaneously*. This is shown schematically in Figure 2.3, where two bodies initially isolated are brought into thermal contact.

While the two formulations of Kelvin and Clausius are different, they are in fact equivalent. This is not immediately

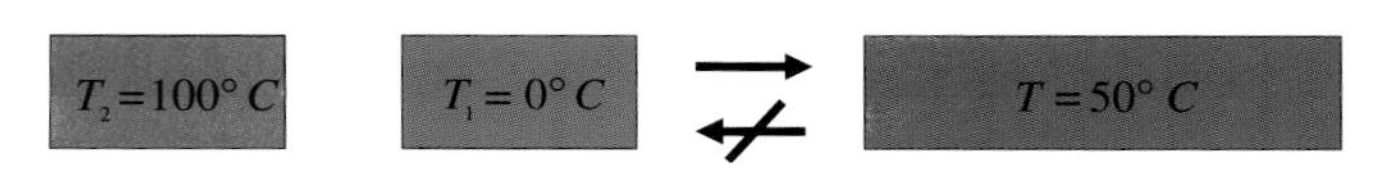

Fig. 2.3 Two bodies at different temperatures (T_1 and T_2) when brought into contact; heat will always flow from the hot to the cold body — never in the reverse direction.

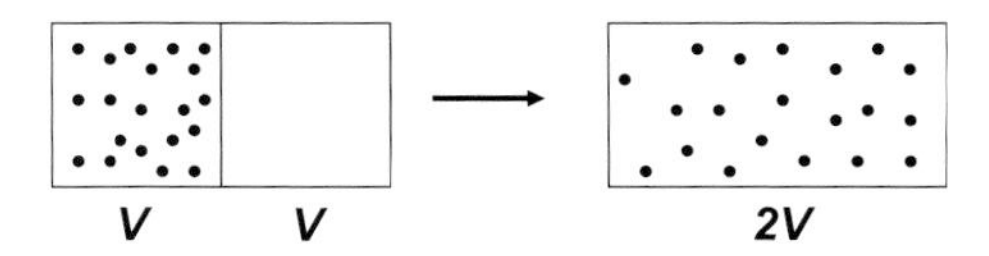

Fig. 2.4 Expansion of gas from volume V to $2V$. Upon the removal of the partition between the two compartments, the gas will always expand to occupy the entire volume, $2V$.

Fig. 2.5 Mixing of two different gases. Upon removal of the partition between the two compartments, the two gases will always mix.

apparent. However, one can prove their equivalency by using a simple argument. The interested reader can find this proof in any book on thermodynamics. We do not need the details here.

There are many other formulations or manifestations of the Second Law of Thermodynamics. For instance, a gas in a confined volume V, if allowed to expand by removing the partition, will always expand to fill the entire new volume — say, $2V$ (Figure 2.4). We never see a spontaneous reversal of this process, i.e. gas occupying volume $2V$ will never spontaneously converge to occupy a smaller volume — say, V. Similarly, two different gases when brought together as shown in Figure 2.5 will always mix. We never see the demixing of two gases occupying spontaneously.[3]

All these processes have one thing in common. They always proceed in one direction, never proceeding *spontaneously* in the reverse direction. But it is far from clear that all these processes are driven by a common law of nature. It was Clausius who saw the general principle that is common to all of these processes. Recall that his formulation of the Second Law is nothing but a statement of what every one of us is familiar with. The greatness of Clausius' achievement was his outstanding prescience that all of these spontaneous processes are governed by one law, and that there is some quantity that determines the direction of the unfolding of events, a quantity that always changes in one direction in a spontaneous process. Clausius introduced the new term *entropy*. In choosing the word "entropy," he writes[4]:

> I prefer going to the ancient languages for the names of important scientific quantities, so that they mean the same thing in all living tongues. I propose, accordingly, to call S the **entropy** of a body, after the Greek word "**transformation**." I have designedly coined the word entropy to be similar to **energy**, for these two quantities are so analogous in their physical significance, that an analogy of denominations seems to me helpful.

With the new concept of entropy one could proclaim the general overarching formulation of the Second Law. *In any spontaneous process occurring in an isolated system, the entropy never decreases.* This formulation, which is very general, embracing many processes, sowed the seeds of the mystery associated with the concept of entropy, the mystery involving a quantity that does not subscribe to a conservation law.

We are used to conservation laws in physics. This makes sense: material is not created from nothing; energy is not given to us free.[5] We tend to conceive of a conservation law as being "understandable" as something that "makes sense." But how can a quantity increase indefinitely and why? What fuels that unrelenting, ever-ascending climb? It is not surprising that the Second Law and entropy were shrouded in mystery. Indeed, the Second Law is unexplainable within the context of the macroscopic theory of matter.

Leon Cooper (1968), after quoting the above passage from Clausius, comments:

> By doing that, rather than extracting a name from the body of the current language (say: lost heat), he succeeded in coining a word that meant the same thing to everybody: *nothing*.

Cooper was right — the term "entropy" has become one of the most mysterious terms in physics.[6]

Notwithstanding the immense generality of the Second Law, some authors extend the application of this law beyond its realm of applicability, such as in processes of life, or of the entire universe. We shall discuss these unwarranted applications of the Second Law in Chapters 3 and 4.

In modern thermodynamics one summarizes the Second Law with the statement that there exists a *state function*, denoted as S and referred to as *entropy*, which in any spontaneous process occurring in an isolated system always increases. A *state function* means that when the *thermodynamic state* is defined — say, by giving the temperature, pressure and composition — the entropy of the system is also defined.

Thermodynamics does not offer an explicit functional dependence of the entropy on the parameters that define the system. Nevertheless, it offers a general procedure for calculating the difference in entropy for any change in the system from state A to another state, B.

This is far from a trivial matter. In Section 2.4 we shall see how easy it is to calculate the change in entropy for some simple processes once we have the entropy *function* itself. In thermodynamics we do not have the luxury of having the entropy function. Moreover, the Clausius definition provides the change in entropy for one specific process; transferring a small amount of heat dQ to a system at a fixed temperature T involves the change in entropy $dS = dQ/T$. In many textbooks one finds the subscript "rev" (in Q_{rev}), which is short for "reversible." Here, we simply assume that the system is at equilibrium at temperature T, and we add a very small amount of heat dQ such that the temperature of the system is almost unchanged. We shall soon discuss the concept of the *quasistatic* process. This term is preferable to "reversible," since the latter is loaded with too many meanings.[7]

It should be clear that Clausius' definition of "entropy" is not a *definition* of "entropy." It is only a change in entropy in a very specific process of heat transfer.

How do we calculate the change in entropy for processes which do not involve heat transfer? We shall demonstrate the way thermodynamics handles this question with a simple example.

2.1.1 *Entropy change in the expansion of an ideal gas*

Consider again the process depicted in Figure 2.4. We start with a system of N particles in a volume V and having a total energy E. We assume that the gas is *ideal*, i.e. we neglect intermolecular interactions. We also assume, for simplicity, that the total energy of the system is the sum of the kinetic energies of all its particles.

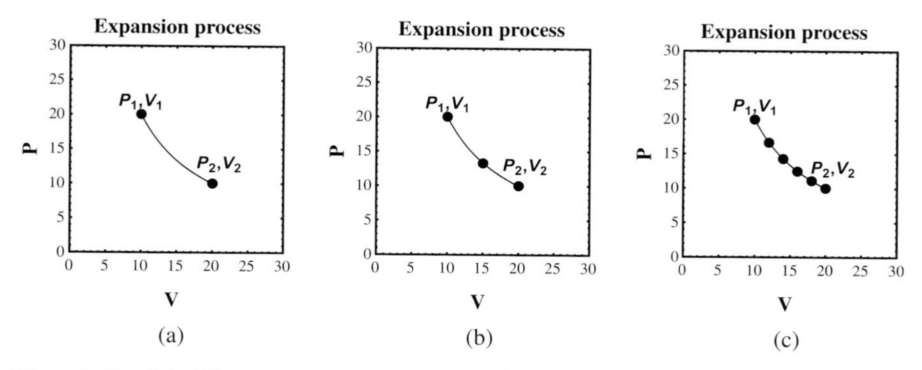

Fig. 2.6 (a) The expansion process from the initial state (P_1, V_1) to the final state (P_2, V_2); (b) the same expansion process with one known intermediate equilibrium state; (c) the same expansion process along a sequence of intermediate equilibrium states.

We remove the partition between the two compartments in the system (Figure 2.4), and we observe expansion of the gas to occupy the entire new volume $2V$.

How do we calculate the change in entropy for this process? Recall that Clausius' definition is $dS = dQ/T$. In the process of expansion, there is no transfer of heat into or out of the system.

The process depicted in Figure 2.4 may be described in a *PV* diagram, shown in Figure 2.6. The initial state is P_1, V_1, T ($V_1 = V$ and $P_1 = Nk_BT/V_1$), and the final state is P_2, V_2, T ($V_2 = 2V$ and $P_2 = Nk_BT/2V = P_1/2$). Note that the temperature, or the energy of the system, does not change in this process. Because the gas is ideal, the doubling of the volume of the system causes a reduction in the pressure, from P_1 to $P_1/2$.

Thus, on the *PV* diagram we have the two points P_1, V_1 and P_2, V_2. After removing the partition between the two compartments, we cannot describe any intermediate points in this diagram.

In order to calculate the change in entropy in this process, we exploit the fact that entropy is a *state function*. This means that if we know the initial and final *states* of the system, the change in entropy is *determined*. In our case we write $\Delta S = S(P_2, V_2, T) - S(P_1, V_1, T)$. This difference depends only on the initial and final states but is independent of the *path* along which the system moves from the initial to the final state.

Thus, we know that ΔS for the process in Figure 2.4 is well-defined. However, we still have to find a way to calculate ΔS for this process. Thermodynamics offers a truly ingenious way of doing this calculation. Devise *any* process leading from the initial to the final process for which we *know* how to calculate ΔS, and there we are! ΔS will be the same for whatever process we can find to get to P_2, V_2 from P_1, V_1.

To do this we start with the First Law of Thermodynamics, which we write as

$$\Delta E = Q + W.$$

ΔE is the change in the internal energy of the system, Q the heat exchanged with the system, and W the work done on the system.[8] This is simply a statement of the conservation of energy.

It should be emphasized that in the particular spontaneous expansion from V to $2V$ in an isolated system, ΔE is zero, and Q and W are also zero. When we choose a path from the initial to the final state, and we perform the process quasistatistically, the changes in all the *state functions* are the same as in the spontaneous process. This is the essence of the meaning of "state function."

However, when we choose different paths the total heat exchanged and the total work might be different for the different paths.

When the process is carried out in very small steps, we can use the Clausius relationship $dQ = TdS$, and if it is only expansion work we have $dW = -PdV$, where P is the pressure, dV the change in volume, and dW the expansion work (sometimes referred to as the PV work).

In order to calculate the change in entropy for the expansion process, we must integrate the equation

$$dE = TdS - PdV.$$

We know that the total change in the internal energy along any path will be zero. Therefore, the change in entropy can be calculated from the total expansion work (carried out quasistatistically):

$$\Delta S = \int \frac{dQ}{T} = \int \frac{P}{T} dV.$$

This equation tells us that if we can find a path along which we can move from the initial to the final state, and for which we can calculate the *work* done by the system or on the system, then we can calculate the change in entropy of the system. It is here that the concept of a *reversible* process is introduced. We prefer to use the concept of a *quasistatic* process instead. The details of the calculations are given in Note 9. The result is

$$\Delta S(\text{expansion}) = Nk_B \ln 2.$$

This is a remarkable result. The process depicted in Figure 2.4 does not involve any transfer of heat into or from the system, yet we could use the Clausius definition along with the fact that the entropy is a state function to calculate the entropy change (ΔS) for this process. We did this by constructing a *path* consisting of a sequence of equilibrium states[9] (Figure 2.6).

It should be stressed that by "reversible" in this process we mean that the process can be *reversed* along the *same* sequence of states along the way from A to B. At each point, *including* the initial and final points, the system is assumed to be at equilibrium. It is only for equilibrium states that the entropy is defined. This is the meaning of entropy being a *state function.* In the literature one finds the notation $dQ_{\text{rev}} = TdS$. It should be noted, however, that "rev" is *not* part of the definition of the entropy or of the Second Law. Once we have the entropy function itself we can calculate the entropy change between any two well-defined states without ever using the quasistatic process (or the term "reversible").

Qualitatively, we *define* an equilibrium state as a state for which all the parameters of the system which *define* the state of the system do not change with time. In practice, we might not be able to determine whether or not a system is at equilibrium.

We note that we do not have a *bona fide* definition of an equilibrium state. As Callen pointed out, any definition of "equilibrium" is circular.[10] An "equilibrium state" is "defined" as a state for which the thermodynamic relationships (in particular, entropy as a function of the thermodynamic variables) apply. On the other hand, the thermodynamic relationships are presumed to apply to equilibrium states.

Thermodynamics does not tell us how exactly the entropy of a system depends on the parameters that characterize the system, such as temperature, pressure or volume. Yet, we can deduce that for an isolated system characterized by fixed energy E, volume V and number of particles N (assuming for simplicity that the system contains only one kind of particles), the function $S = S(E, V, N)$ must be a monotonically increasing function of E, V, N. Furthermore, from the Second Law it follows that

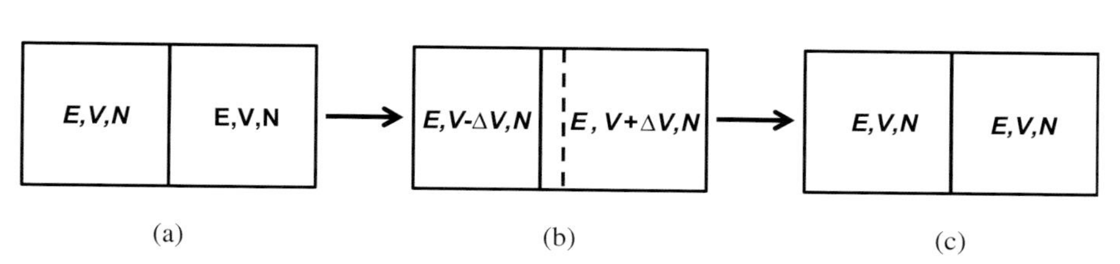

Fig. 2.7 (a) A system consisting of two compartments, each characterized by the variables (E, V, N); (b) the same system as in (a), but the partition between the two compartments is moved leftward; (c) the final equilibrium state obtained after the partition is released.

this function must be concave downward, i.e. the slope of S as a function of either E, V or N must be *positive* and *decreasing* as E, V or N increases.

We demonstrate this behavior for one variable — say, the volume of the system. Suppose that we have an *isolated* system which is divided into two compartments. For simplicity, suppose that each compartment contains an ideal gas with N particles, in a volume V and having an energy E. Thus, the entire system as a whole is characterized by $(2E, 2V, 2N)$. Now, suppose that we move the partition between the two compartments such that the volume of one compartment — say, the left one — is decreased by a small volume ΔV, and the volume of the second compartment — [the right one in Figure 2.7(b)] is increased by the same amount, ΔV. Clearly, if the system was initially at equilibrium the pressure in the left compartment will increase and the pressure in the right will decrease. This difference in the pressures will tend to push the partition in such a way that the initial equilibrium state will be restored, i.e. the volume of the left compartment will increase from $V - \Delta V$ to V, and the volume of the right compartment will decrease from $V + \Delta V$ to V. The final state is shown in Figure 2.7(c). Since this process

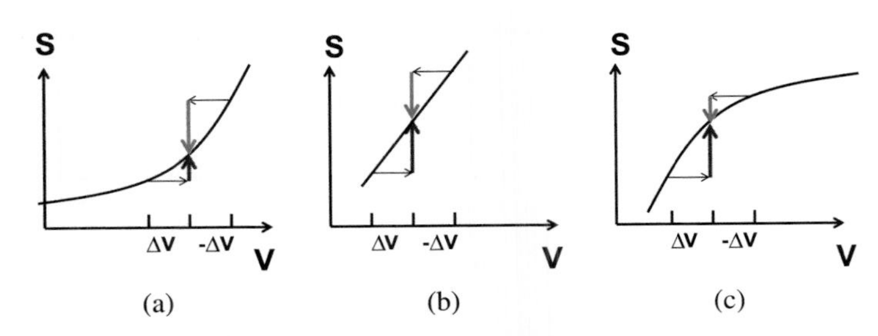

Fig. 2.8 Three possible curves for the entropy as a function of volume.

occurs spontaneously in an isolated system, the total entropy of the system must increase.

Figure 2.8 shows three possible curves for the entropy as a function of V (for fixed values of E and N). From thermodynamics we know that the entropy must increase with V. Before the spontaneous process occurred, the left compartment had a volume $V - \Delta V$ and the corresponding entropy S_L. On the other hand, the right compartment had a volume $V + \Delta V$, and the corresponding entropy S_R.

When the spontaneous process occurs, the total entropy of the system must increase. In each curve in Figure 2.8 we show the increase in the entropy of the left compartment and the decrease in the entropy of the right compartment. Clearly, if we require that the *sum* of the changes in the entropy of the two compartments be *positive*, we must have an *increase* in the entropy of the left compartment which is *larger* than the *decrease* in the entropy of the right compartment. This occurs only when the entropy S, as a function of V, has a negative curvature (i.e. that the slope decreases as V increases), as shown in Figure 2.8(c).

One can use the same argument for the dependence of S on the energy E (at fixed V and N), and for the dependence of S on the number of particles (at fixed V and E).[11] The conclusion is

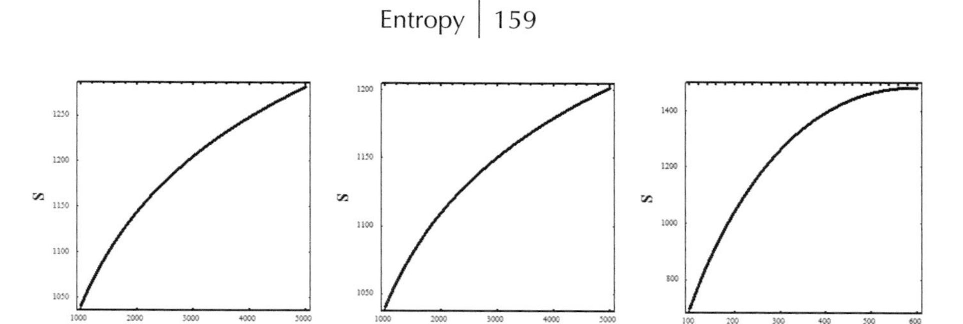

Fig. 2.9 Dependence of S on E, V and N for an ideal gas.

that S must be a monotonically increasing function of E, V and N, and it must have a negative curvature with respect to each of these variables (Figure 2.9). (See also Subsection 2.3.2.)

We can summarize this section as follows. Thermodynamics provides a method for calculating the entropy change between any two *equilibrium states*. It does not tell us the *value* of the entropy of a system at equilibrium. It only states that there exists such a *state function*. When the state of the system is defined by the variables E, V and N, the entropy function must be a monotonically increasing function of each of these variables, and in addition this function must have a negative curvature (or concave downward) with respect to these variables. More generally, the Second Law states that in any process occurring spontaneously, in an isolated system in which the system moves from one equilibrium state to a new equilibrium state, the entropy must increase. Thermodynamics does not reveal to us *what* the meaning of entropy is, or *why* it always changes in one direction only. We shall see in Section 2.3 that entropy may be derived from information theory. This derivation provides answers to both of these questions.

2.2 Atomistic Formulation of the Second Law

Before the development of the kinetic theory of heat (which is based on the recognition of the atomistic theory of matter), thermodynamics was applied without any reference to the atomistic nature of matter — as if matter were a continuum. Thermodynamics did not offer any interpretation of entropy. This, in itself, is not unusual. Any law of physics reaches a dead end when we have to accept it as it is, without any further, "deeper" understanding. In addition, the Second Law was formulated as an absolute law — entropy *always* increases in a spontaneous process in an isolated system. This is not different from any other law; for example, Newton's laws are always obeyed — no exceptions.

A huge stride toward understanding of entropy and of the Second Law of Thermodynamics was made possible following Boltzmann's statistical interpretation of entropy — the famous relationship between entropy and the total number of microstates of a system characterized macroscopically by a given energy, volume, and number of particles. Ludwig Boltzmann (1844–1906), along with Maxwell and many others, developed what is now known as the kinetic theory of gases, or the kinetic theory of heat. This not only led to the identification of temperature, which we can feel with our sense of touch, with the motion of the particles constituting matter, but also to the interpretation of entropy in terms of the number of states that are accessible to the system. This formula is shown in Figure 2.10(a), and it is engraved on Boltzmann's tombstone [Figure 2.10(b)].

Note that W in Boltzmann's equation is the *number* of states or arrangements. This number was interpreted as measuring a degree of disorder in the system. This interpretation is vague, mainly because order and disorder are ill-defined quantities.

(a)

(b)

Fig. 2.10 (a) Boltzmann's equation relating the entropy S to the total number of accessible states W; (b) Boltzmann's tombstone in Vienna.

We shall see some examples in Section 2.7, where this interpretation does not hold. We also note here that this relationship between entropy and the number of states holds only for isolated systems. It is for isolated systems that the first postulate of statistical thermodynamics applies, namely that all the quantum-mechanical states are equally probable.

At that time, the atomistic theory of matter had not yet been established, or universally accepted. Although the idea of the atom was in the minds of scientists for over 2000 years, there was no compelling evidence for its existence. Nevertheless, the kinetic theory of heat did explain the pressure and temperature of the gas in terms of the motion of the atoms and the molecules. But what about entropy, the quantity that Clausius introduced without any reference to the molecular composition of matter?

Boltzmann also introduced *probabilistic* arguments in his formulation of the Second Law. This was quite foreign to classical physics at that time.

Boltzmann's formulation of the Second Law drew criticisms from many of his contemporaries. The gist of the criticisms (known as the reversibility objection or the reversibility paradox) is the seeming conflict between the so-called time reversal or time symmetry of the Newtonian equations of motion and the time asymmetry of the behavior of Boltzmann's entropy. This conflict between the reversibility of the molecular motion and the irreversibility implied by the Second Law was a profound one, and could not be resolved. How can one derive a quantity that distinguishes between the past and the future (i.e. always increasing with time), from equations of motion that are indifferent and do not care about the past and the future? Newton's equations can be used to predict the evolution of the particles into the past as well as into the future.

Boltzmann's response to the reversibility objection was that the Second Law holds most of the time but in very rare cases it can go the other way, i.e. entropy might decrease with time.

This was untenable. The (nonatomistic) Second Law of Thermodynamics, like any other law of physics, was conceived and proclaimed as being absolute — no room for exceptions, not even rare exceptions. No one had ever observed violation of the Second Law. As there are no exceptions to Newton's equations of motion, there should be no exceptions to the Second Law, not even in rare cases. The Second Law must be absolute and inviolable. At this stage, there were two seemingly different views of the Second Law. On the one hand, there was the classical, nonatomistic and *absolute* law as formulated by Clausius and Kelvin, encapsulated in the statement that entropy *never decreases* in an isolated system. On the other hand, there was the atomistic formulation of Boltzmann, which claimed that entropy increases "most of the time" but there are exceptions, albeit very rare ones. Boltzmann proclaimed that entropy could decrease — that it was not an impossibility, but only highly improbable. However, since all observations seem to support the absolute nature of the Second Law, it looked as if Boltzmann had suffered a defeat, and along with that the atomistic view of matter. We shall further discuss Boltzmann's reaction to this criticism in Section 2.4.

There seemed to be a state of stagnation as a result of the two irreconcilable views of the Second Law. It was not until the atomic theory of matter gained full acceptance that the Boltzmann formulation had the upper hand. Unfortunately, this came only after Boltzmann's death in 1906. A year earlier, a seminal theoretical paper published by Einstein on Brownian motion had provided the lead to the victory of the atomistic view of matter. At first sight, this theory seems to have nothing to do with the Second Law.

Brownian motion was observed by the English botanist Robert Brown (1773–1858). The phenomenon is very simple: tiny particles, such as pollen particles, are observed to move in a seemingly random fashion when suspended in water. It was initially believed that this incessant motion was due to some tiny living organisms, propelling themselves in the liquid. However, Brown and others showed later that the same phenomenon occurs with inanimate, inorganic particles, sprinkled into a liquid.

Albert Einstein (1879–1955) was the first to propose a theory for this so-called Brownian motion. He believed in the atomic composition of matter and was also a staunch supporter of Boltzmann. (It is interesting to note that Einstein, who lauded Boltzmann for his probabilistic view of entropy, could not accept the probabilistic interpretation of quantum mechanics.) He maintained that if there is a very large number of atoms or molecules jittering randomly in a liquid, there must also be fluctuations. When tiny particles are immersed in a liquid (tiny compared to macroscopic size, but still large enough compared to the molecular dimensions of the molecules constituting the liquid), they will be "bombarded" randomly by the molecules of the liquid. However, once in a while there will be asymmetries in this bombardment of the suspended particles, as a result of which the tiny particles will be moving one way or the other in a zigzag manner.

In 1905, Einstein published as part of his doctoral dissertation a theory of these random motions. Once his theory was corroborated by experimentalists [notably Jean Perrin (1870–1942)], the acceptance of the atomistic view became inevitable. Classical thermodynamics, based on the continuous nature of matter, does not have room for fluctuations. Indeed, fluctuations in a macroscopic system are extremely small. That is why we do not observe fluctuations in a macroscopic piece of matter. But with the tiny Brownian particles the fluctuations are magnified and rendered observable. With the acceptance of the atomic composition of matter also came the acceptance of Boltzmann's expression for entropy. It should be noted that this formulation of entropy was not affected or modified by the two great revolutions that took place in physics early in the

20th century: quantum mechanics and relativity. The door to understanding entropy was now wide open.

Boltzmann's heuristic relation between entropy and the logarithm of the total number of states did open the door to an understanding of the meaning of entropy. However, there were several reasons for the lingering of the mystery surrounding entropy and the Second Law. I have discussed some of the reasons in my previous books.[12]

2.3 Informational Derivation of Entropy and the Second Law

The third approach to the Second Law is based on Shannon's measure of information (SMI). This may be loosely referred to as the *informational-theoretical* approach. The SMI approach is both elegant and illuminating. It provides an answer to the question of *what* entropy is, as well as to *why* entropy always increases in an isolated system. Personally, I strongly believe that once this approach is accepted and taught as part of a course in thermodynamics, there should be no reason to be mystified by entropy and its behavior. It will also contribute to the clarification of the limits of applicability of the concept of entropy and the Second Law. (These are discussed in Chapters 3 and 4.)

In Section 1.2, we mentioned von Neumann's suggestion to Shannon to call his quantity (H) "entropy," adding that since no one understands what entropy is, you can say anything about it and no one would be able to prove you wrong. This is not true today. Entropy is well understood, and calling SMI "entropy" had caused a great deal of confusion. As we shall see in the next chapters, many statements about life and the universe involve the concept of entropy. Some cannot be *proven* to be wrong, but they can neither be proven to be right. Besides, referring to just "entropy" leaves one wondering whether it means thermodynamic entropy or SMI.

In this section, we present very briefly the steps leading from SMI to entropy. All the mathematical details will be skipped [they are provided elsewhere, in Ben-Naim (2008, 2011)]. As noted in Section 1.2, there are two main uses of SMI. One is to obtain the most plausible probability distribution given whatever information we have on the moments of the distribution (e.g. averages or standard deviations). The second is the evaluation of SMI itself, which provides a *measure of information*, or a measure of uncertainty associated with a probability distribution. We show in this section that both of these applications of SMI are used to derive entropy. SMI provides an explanation for both entropy and the Second Law. This is truly a remarkable achievement, yet it is also a source of great confusion.

2.3.1 *An outline of the derivation of the entropy of an ideal gas*

We start by playing the 20Q game for a particle in a one-dimensional box. Then we play the 20Q game for a particle moving in one dimension. Adding two correction terms due to the uncertainty principle and the indistinguishability of the particles, we can get the SMI for a system of ideal gas consisting of simple particles (i.e. particles having no internal degrees of freedom). Lo and behold! From the 20Q game spawned the entropy function of an ideal gas.

The overall plan for obtaining the entropy of an ideal gas from SMI consists of four steps:

First, we calculate the *locational* SMI, associated with the *equilibrium* distribution of locations of all the particles in the system.

Second, we calculate the *velocity* SMI, associated with the *equilibrium* distribution of velocities (or momenta) of all the particles.

Third, we add a correction term due to the quantum-mechanical *uncertainty* principle.

Fourth, we add a correction term due to the fact that the particles are *indistinguishable*.

Note that in the first two steps, we use SMI to find out the *equilibrium distribution*, and then we use this equilibrium distribution to *evaluate* the corresponding SMI. In the last two steps the corrections can be expressed as *mutual information*. Once we combine the results of the four steps, we get, up to a multiplicative constant, the *entropy* of an ideal gas.

This is a very remarkable achievement. Starting with the familiar 20Q game, we get the entropy of an ideal gas. It should be noted that this approach is superior to both Clausius' definition and the Boltzmann definition of entropy. In contrast to Clausius' definition, which provides only *differences* in entropy, the SMI approach provides the entropy function itself. Unlike the Boltzmann definition, which requires a lengthy calculation of the number of energy levels (W) of an ideal gas, then calculating the entropy, the SMI approach provides the entropy function directly, without going through the calculation of W.

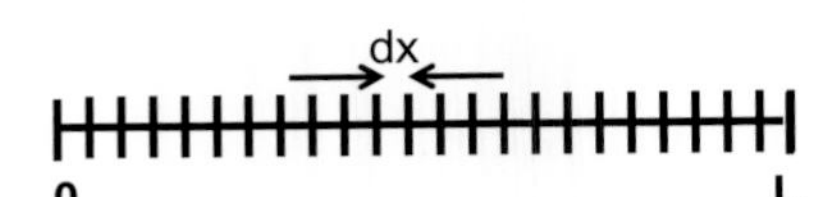

Fig. 2.11 The total length of the box is divided into small intervals, each of the size dx.

(i) The locational SMI of an ideal gas

Consider first the case of one particle which moves freely in a one-*dimensional* "box" of length L. Clearly, there are infinite numbers of points at which the center of the particle may be found. However, we are never interested in the *exact* point at which the particle is, but rather in which interval dx, where dx is very small (Figure 2.11). The formal problem we pose is to find the maximum value of the SMI, subject only to the normalization condition.[13]

The result is

$$f^*(x) = \frac{1}{L}.$$

Note that in this case the probability distribution $f^*(x)$ is constant, independent of x. Also note that the distribution $f^*(x)$, which maximizes the SMI, is also the actual experimental distribution at equilibrium.

The corresponding SMI (at equilibrium) is

$$H_{\max}(\text{location of one particle along the } x \text{ axis}) = \log L.$$

We will refer to the density function $f^*(x)$ as either the function that maximizes H or the *equilibrium* density function. The latter interpretation will be clear when we interpret the final result of H as the entropy of an ideal gas at *equilibrium*.

We note here that $\log L$ is to be understood as the SMI associated with the location of a particle in the 1D system of length L. The larger L is, the larger the SMI will be. One should keep in mind that the SMI is actually a divergent quantity. However, in all the applications of this equation, we either make a *discretization* of the length L or take the *difference* of the SMI for the two values of L. In both cases we remove any problem arising from the divergence of the SMI.[12]

Two comments are now in order. First, "log" stands for the logarithm with base 2. Second, we shall treat L as a pure number. In fact, this number is L/dx, the number of intervals of length dx (Figure 2.11). We will ignore the units of length, as well as any other quantity under the logarithm sign. In the final expression, we will have a pure number under the logarithm.

The generalization of this result to the three-dimensional case is straightforward. Suppose that the particle is confined to a cubic box of edge L and volume $V = L^3$. Clearly, the SMI associated with the y axis and the z axis will be the same as in $f^*(x)$. Furthermore, we assume that the events "being at a location x," "being at a location y" and "being at a location z" are independent events. Therefore, the SMI associated with the location x, y, z within the cube of volume V is the sum of the SMI associated with the three axes. Thus, if we use the shorthand notation $H_{\max}(x)$ for the quantity in the above quotation, we can write

$$\begin{aligned} H_{\max}(x, y, z) &= H_{\max}(x) + H_{\max}(y) + H_{\max}(z) \\ &= 3 \log L = \log V \end{aligned}$$

and the equilibrium density is

$$f^*(x, y, z) = \frac{1}{L} \times \frac{1}{L} \times \frac{1}{L} = \frac{1}{L^3} = \frac{1}{V}.$$

We next extend this result to the case of N *independent* and *distinguishable* (D) particles. We also use the shorthand notation $\boldsymbol{R}_i = (x_i, y_i, z_i)$ for the locational vector of particle i, and $\boldsymbol{R}^N = \boldsymbol{R}_1, \ldots, \boldsymbol{R}_N$ for the locational vector of all the N particles.

Since the particles are *independent* the SMI of the N particles is simply the sum of the SMIs of all the single particles, and since the SMI for a single particle is the same for each particle we have for the N *independent* and *distinguishable* particles.

$$H_{\max}^{D}(\boldsymbol{R}^N) = NH_{\max}(x, y, z) = N \log V.$$

Note that we added the superscript D for *distinguishable* particles. We will soon see that the fact that the particles are *indistinguishable* (*ID*) introduces a *correlation* between the particles which causes a *reduction* in the SMI of N particles. Note also that we retain the subscript "max." This will be important when we identify the maximal value of H with the entropy of the system at equilibrium.

Thus, the quantity $N \log V$ is the locational SMI of N independent and indistinguishable particles, evaluated for the distribution which *maximizes* the SMI. Therefore, we can refer to this as the *equilibrium locational* SMI.

It should be noted that $1/V$ is the probability density for finding a *specific* particle at a point $\boldsymbol{R}$. The corresponding SMI is $\log V$. The probability density for finding *any* particle at $\boldsymbol{R}$ is N/V, where N is the number of particles. The corresponding SMI is not $\log(N/V)$, but $N \log V$, provided that the particles are indistinguishable. Once we introduce indistinguishability, we do obtain $\log(N/V)$ as part of the SMI of a system of N particles. However, the "N" in $\log(N/V)$ does not come from the locational probability density N/V, but from the *indistinguishability* of the particles. For more details, see subsection (v) below and Ben-Naim (2008).

(ii) The momentum SMI of an ideal gas

Again, we start with particles moving along the 1D system of length L. We are interested in the probability density for finding a specific particle with velocity between v_x and $v_x = v_x + dv_x$. We assume that the particles can have any value of v_x from $-\infty$ to $+\infty$, but we require that the *average* kinetic energy of the particles be constant. We will skip the details of the derivation. The solution to this problem is[14]

$$f^*(v_x) = \frac{\exp(-v_x^2/2\sigma^2)}{\sqrt{2\pi\sigma^2}}.$$

We will refer to $f^*(v_x)$ as the equilibrium density distribution of the velocities in one dimension. This distribution is recognized as the normal distribution with the average at $x = 0$ and variance σ. The variance σ^2 may be expressed in terms of the absolute temperature T[15]:

$$\sigma^2 = \frac{k_B T}{m}.$$

With this identification, we rewrite the distribution of velocities in one dimension as

$$f^*(v_x) = \sqrt{\frac{m}{2\pi k_B T}} \exp\left[\frac{-mv_x^2}{2k_B T}\right].$$

Figure 2.12 shows the distribution $f^*(v_x)$ for various values of T (for this illustration we take $m = 1$, $k_B = 1$). We see that the larger the temperature, the larger the *spread* of the distribution of the velocities. Thus, the average "width" of the distribution

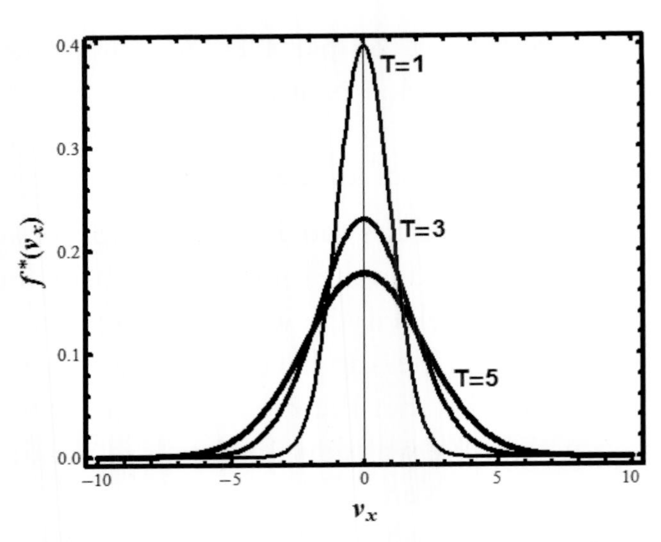

Fig. 2.12 The density distribution of velocities in one dimension (T is the absolute temperature).

may be measured with either the variance σ^2 or the temperature T.

In terms of the temperature, we rewrite the SMI associated with the *equilibrium* distribution $f^*(v_x)$ as

$$H_{\max}(v_x) = \frac{1}{2}\log\left(\frac{2\pi e k_B T}{m}\right).$$

The meaning of this quantity is derived from the meaning of the SMI. Note again that the SMI of any continuous variable has a divergent part.[12] However, in actual application of $H_{\max}$ we either make a *discretization* of the infinite range $(-\infty, \infty)$ into a finite number of small intervals, or take *differences* in the values of H between two states — here, two temperatures. Doing this removes the divergent part of the SMI. The important result we have obtained is that the larger the temperature (or, equivalently, the average kinetic energy of the particles), the larger the SMI or

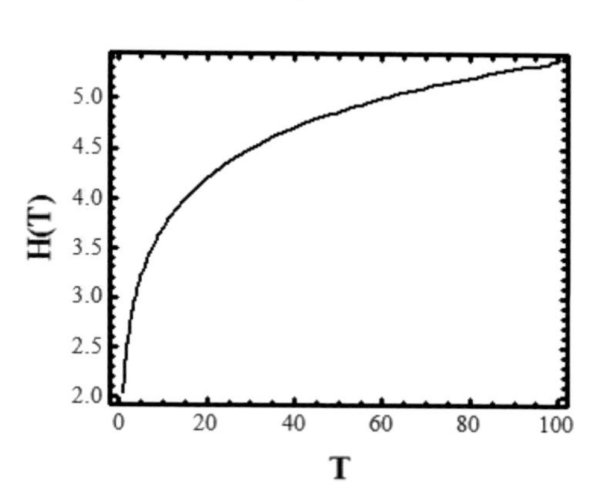

Fig. 2.13 The velocity SMI as a function of the temperature T.

the uncertainty associated with the distribution of the velocities. This is shown in Figure 2.13. Note again that we ignore the units in the expression under the logarithm.

We next assume that the velocities along the three axes v_x, v_y, v_z are independent. Therefore, the SMI for a single particle moving with velocities v_x, v_y, v_z is given by

$$\begin{aligned} H_{\max}(v_x, v_y, v_z) &= H_{\max}(v_x) + H_{\max}(v_y) + H_{\max}(v_z) \\ &= 3H_{\max}(v_x) = \frac{3}{2}\log\left(\frac{2\pi e k_B T}{m}\right). \end{aligned}$$

To summarize, we have calculated the *equilibrium* density distribution of velocities (this is the same as the distribution which maximizes the SMI). Then we calculated the SMI corresponding to this equilibrium distribution of velocities.

For the purpose of constructing the entropy of an ideal gas, we will need the distribution of the momenta. This is simply obtained by the transformation $p_x = mv_x, p_y = mv_y, p_z = mv_x$. The one-dimensional distribution of momenta is given in Note 16. Although we will not need the speed distribution, it

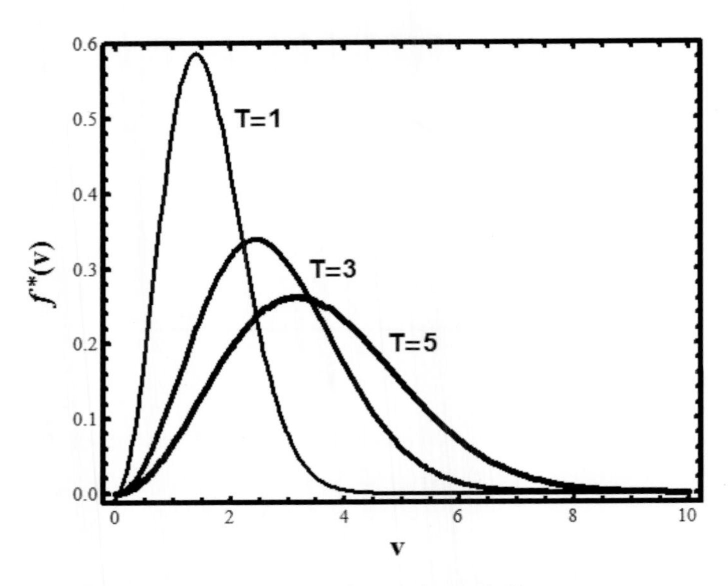

Fig. 2.14 The distribution of speeds at different temperatures.

is helpful to have an idea of the form of this distribution. This distribution is shown in Figure 2.14 and discussed in Note 17.

(iii) The mutual information due to indistinguishability of the particles

We have already seen in Chapter 1 that indistinguishability of the particles introduces correlation, and hence also mutual information.

Consider two particles 1, 2 distributed in M equal cells. The particles are independent and distinguishable (D). The number of possible arrangements of one particle is simply M and the corresponding SMI is

$$H(1) = \log M.$$

The same is true of the second particle, i.e.

$$H(2) = \log M.$$

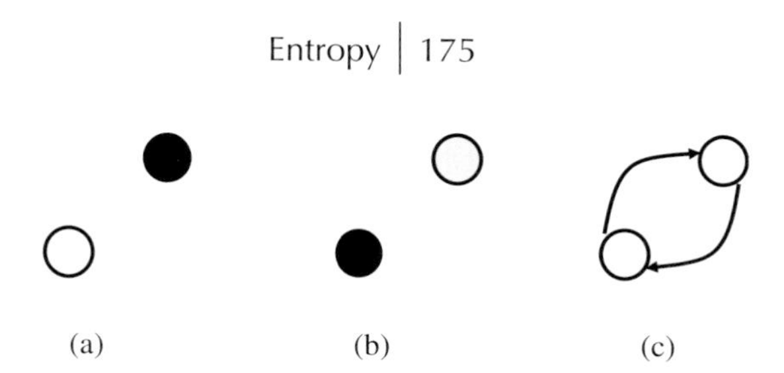

Fig. 2.15 Two different configurations, (a) and (b), become one configuration when we erase the "labels" on the particles (c).

Therefore, the SMI of the two particles, denoted as $H(1,2)$, is the sum of $H(1)$ and $H(2)$:

$$H^D(1,2) = H(1) + H(2) = 2\log M.$$

Thus, the number of possible arrangements of two independent and distinguishable particles in M cells is simply M^2, and the corresponding SMI is $H^D(1,2)$, as given above. Note that the particles do not interact. In this example there is no limit on the number of particles that can occupy the same cell.

If the particles are indistinguishable (*ID*), then the counting of the number of arrangements is different. The reason is quite simple. We do not distinguish between two configurations that are obtained by exchanging the particles. Figure 2.15 shows two possible configurations, (a) and (b), of two particles. Clearly, when the two particles are distinguishable, these two configurations are *different*, and therefore should be counted as two configurations. However, if the two particles are indistinguishable, this amounts to erasing the colors or the labels of the two particles, and hence the two configurations coalesce into one, (c). Thus, in general, the total number of configurations is *reduced* when we "unlabel" the particles.

In the general case of N particles in M cells, the counting of the total number of configurations of N distinguishable particles is simply M^N. The number of configurations of N *indistinguishable* particles is more complicated to count. However, for the case where $N \ll M$ (i.e. when the number of cells is so large that the occurrence of more than one particle in a cell is a rare event), the number of configurations is reduced from M^N to $M^N/N!$ [For more details, the reader is referred to Ben-Naim (2008).]

When the number M is very large, the number of configurations of the system of the two indistinguishable particles is $M/2$, and the corresponding SMI is

$$H^{ID}(1,2) = \log\frac{M^2}{2}.$$

The difference between $H^D(1,2)$ and $H^{ID}(1,2)$ can be interpreted in terms of *mutual* information, i.e.

$$H^{ID}(1,2) = H(1) + H(2) - I(1;2),$$

where

$$I(1;2) = \log 2 > 0.$$

In the more general case of N *indistinguishable* particles on M cells, such that $N \ll M$, we have

$$\begin{aligned}
H^D(1,2,\ldots,N) &= \sum_{i=1}^{N} H(i) = \log M^N, \\
H^{ID}(1,2,\ldots,N) &= H^D(1,2,\ldots,N) - \log N! \\
&= \log M^N/N!, \\
I(1;2;\ldots;N) &= \log N!.
\end{aligned}$$

We conclude that the indistinguishability of the particles introduces a correlation between the particles, which causes a *reduction* of the SMI. We have calculated the mutual information of indistinguishable particles by calculating the change in the number of configurations of the system caused by "erasing the labels" on the particles. One can also give a probabilistic interpretation of the mutual information, in terms of correlation between events. We do not need this interpretation here. The reader should be convinced by checking a few examples where, whenever we "unlabel" the particles, the number of configurations or arrangements is always reduced, as a result of which the value of the SMI of the system is reduced too.

(iv) The mutual information associated with the uncertainty principle

In step (i), we saw that the continuous SMI for the location of a single particle within the "box" of length L is

$$H_{\max}(x) = \log L.$$

We also noted that if we are interested only in the location of the particle in one of the cells of length $h = L/n$, we need not use this result, but its discrete analog, which is obtained from $H_{\max}(x)$ by subtracting $\log h$, i.e. the discrete analog of $H_{\max}(x)$ is

$$H_{\max}(x) = \log L - \log h = \log n.$$

[For more details, the reader is referred to Ben-Naim (2008).]

We now do the same trick when combining the locational and the momentum SMI. In step (i), we calculated the locational SMI. In step (ii), we calculated the momentum SMI. We now want to find out the SMI associated with *both* the location and the momentum.

Classical thinking would have led us to conclude that the SMI associated with *both* the location and the momentum of a particle should be the sum of the two SMIs. However, quantum mechanics dictates that the two experiments of determining the location and the momentum of a particle are *not independent*. This is the well-known Heisenberg *uncertainty* principle. For our case the uncertainty principle states that we cannot determine *both* the location and the momentum with an accuracy of at most the order of h; here h is the Planck constant $h = 6.626 \times 10^{-34} Js$. It follows that the SMI of a particle is not the sum of the SMI associated with its location and the SMI associated with its momentum, but a corrected sum, after taking into account the uncertainty principle.

This is the same situation we encountered when passing from the continuous segment $(0, L)$ to the discrete number of cells. Here, we also divide the entire space of locations and momenta into cells of size h, where now h is the Planck constant.

In terms of SMI we can say that there is a *correlation* between the location and the velocity (or momentum) of a particle. This correlation can be cast in the form of *mutual* information, i.e.

I (uncertainty principle) $= \log h > 0$ or, equivalently,

$$H_{\max}(x, p_x) = H_{\max}(x) + H_{\max}(p_x) - \log h.$$

This is the SMI for one particle in one dimension. For the three-dimensional case we have

$$\begin{aligned} H_{\max}(x, y, z, p_x, p_y, p_z) = {} & H_{\max}(x, y, z) \\ & + H_{\max}(p_x, p_y, p_z) - 3 \log h, \end{aligned}$$

i.e. we subtract $\log h$ for each degree of freedom.

Finally, for N indistinguishable and noninteracting particles, we have

$$\begin{aligned} H^{ID}(1, 2, \ldots, N) = {} & H^{D}_{\max}(\boldsymbol{R}^N) \\ & + H^{D}_{\max}(\boldsymbol{p}^N) - \log N! - 3N \log h. \end{aligned}$$

This is an important result. To obtain the SMI of N noninteracting particles, described by their locations and momenta, we first treat the particles as being *distinguishable* and *classical*. In this case, we can sum the SMI associated with all the locations ($\boldsymbol{R}^N$) and all the momenta ($\boldsymbol{p}^N$) of the particles. Then, we correct by deducting two mutual information terms due to the fact that the particles are not classical. One is due to the indistinguishability of the particles, and the other to the Heisenberg uncertainty principle. These two corrections *reduce* the total SMI by the amount $\log N! + 3N \log h$.

(v) The entropy of a classical ideal gas

In the previous subsection, we have calculated the maximal value of the SMI of a system of N noninteracting and indistinguishable particles.

Recall that SMI may be defined for *any* distribution. It can be defined for any distribution of locations and any distribution of momenta, not necessarily at equilibrium. It can be defined for any number of particles and for distinguishable or indistinguishable particles. All these have nothing to do with the entropy. Up to this point you can rightly regard the SMI as a quantity that measures the size of the 20Q game (where are all the particles, and what are their momenta?).

In this section, we will be interested in *very special distributions*. These are the distributions of locations and momenta that *maximize* the corresponding SMI. We denoted these special distributions with asterisks, and the corresponding SMI by H_{max}. However, we also know that, starting with any arbitrary distribution of locations and momenta, the system will tend to a limiting *equilibrium distribution*: the uniform distribution for locations and the normal distribution for momenta. Therefore, we will refer to the distribution that maximizes the SMI as the equilibrium distribution.

In this section we make a huge conceptual leap, from the SMI of 20Q games to a fundamental concept of thermodynamics. As we will see soon, this leap is rendered possible by recognizing that the SMI associated with the *equilibrium distribution* of locations and momenta of a large number of indistinguishable particles is *identical* (up to a multiplicative constant) with the statistical-mechanical entropy of an ideal gas. Since the statistical-mechanical entropy of an ideal gas has the same properties as the thermodynamic entropy as defined by Clausius, we can declare that this special SMI is *identical* with the entropy of an ideal gas. This is a very remarkable achievement. Recall that von Neumann suggested calling SMI "entropy." This was a mistake. In general, SMI has nothing to do with entropy. Only when you apply SMI to *special* distributions does it become identical with entropy. Thus, we conclude that SMI is a very general concept. It applies to any distribution. Entropy is a special case of an SMI when evaluated for very special distributions.

We recall that the SMI of a system of N particles at equilibrium has two contributions due to the location and momentum, and two corrections due to the indistinguishability of the particles and the uncertainty principle. Thus, we have the following expression for the SMI of N noninteracting particles at equilibrium[18]:

$$\begin{aligned} H^{ID}(1, 2, \ldots, N) &= H_{\max}(\text{locations}) + H_{\max}(\text{momenta}) \\ &\quad - I(\text{uncertainty principle}) - I(\text{ind}) \\ &= N \log \left[\frac{V}{N} \left(\frac{2\pi m k_B T}{h^2} \right)^{\frac{3}{2}} \right] + \frac{5N}{2}. \quad \text{(I)} \end{aligned}$$

In order to obtain the expression for the entropy of an *ideal* gas, all we have to do is to use the natural logarithm and multiply H^{ID} by Boltzmann's constant, k_B:

$$S = (k_B \ln 2)H. \tag{II}$$

The multiplication by a constant k_B, as well as changing to the natural logarithm, determines the units in which we measure entropy. It does *not* affect the *meaning of entropy*, as an SMI associated with the location and momenta of a system of N noninteracting particles at equilibrium. We normally apply this identity between entropy and SMI for a thermodynamic system, i.e. when N and V are very large but the density N/V is constant.[19]

The expression (I) after the multiplication by $k_B \ln 2$ in (II) is identical with the equation obtained by Sackur and Tetrode in 1912 based on Boltzmann's definition of entropy. Table 2.1 shows the values of the entropies of some gases calculated using

Table 2.1 Calculated and measured entropies of gases at their boiling points (BP). Entropy in units of cal/mol deg.

Substance	Temperature	Entropy Theoretical	Experimental
A	BP	30.87	30.85
O_2	BP	40.68	40.70
N_2	BP	36.42	36.53
Cl_2	BP	51.55	51.56
HCl	BP	41.45	41.3
HD	298.1°	34.45	34.45
CH_4	BP	36.61	36.53
C_2H_4	BP	47.35	47.36

BP stands for boiling point.

experimental data, and the theoretical values calculated from the various degrees of freedom of the molecules. In the case of argon the main contribution to the entropy is the translational entropy which is the entropy that can be calculated either from the Boltzmann equation, the Sackur–Tetrode equation, or from the SMI. Note the excellent agreement between the theoretical values and the experimental values. It should also be noted that in many cases where there is disagreement between the theoretical and the experimental values, one can explain the difference using the so-called *residual entropy*, which is essentially the number of configurations that have nearly equal energies at very low temperature.[20]

A classic example is the case of carbon monoxide (CO), which has residual entropy of about 1.1 cal/mol deg. This is well explained by the two possibilities of arranging the CO molecule in a lattice (CO and OC), i.e. $k_B N \ln 2$. Another example is that methane with one hydrogen replaced by deuterium (CH_3D) has residual entropy of 2.8 cal/mol deg. Again, this is well explained by the number of (nearly equivalent) orientations of the molecule in the lattice, which in this case is 4.

A more challenging example of residual entropy is the case of H_2O. Here, it is more difficult to calculate the total number of arrangements of the hydrogen atoms along the O–O bonds. For a more detailed discussion of this case, see Ben-Naim (2009).

So far, we have assumed throughout that the particles do not interact with each other. Normally, the absence of interactions is considered to be equivalent to independence. However, as we have seen in this chapter, *dependence* can occur even between noninteracting particles. When the particles interact with each other, a new kind of dependence is introduced. This dependence introduces an additional correlation between the particles, which causes another reduction in the SMI. [For details, see Ben-Naim (2008).]

From now on we will *define* the entropy of an ideal gas by equation (I) (after the multiplication by $k_B \ln 2$). The *meaning* of the entropy is derived from the four terms listed in the equation above; we have the SMI associated with the locations of the particles, the momenta of the particles and the two corrections due to the uncertainty principle and the indistinguishability of the particles.

Recall that Boltzmann defined entropy as

$$S_B = k_B \ln W,$$

where k_B is the Boltzmann constant and W is the total number of states accessible to the system. In Chapter 1, we have seen that the quantity

$$\mathrm{SMI} = \log_2 W$$

may be interpreted as the amount of missing information, or the amount of uncertainty associated with an experiment having W equal probable outcomes. Thus, the relationship between S_B and SMI for this case is simply

$$S_B = (k_B \ln 2)\mathrm{SMI}.$$

In the more general case of a nonuniform distribution, the SMI is defined

$$\mathrm{SMI} = -\sum p_i \log_2 p_i,$$

whereas the statistical-mechanical entropy is

$$S = -k_B \sum p_i \ln\, p_i.$$

Therefore, the same relationship holds between the entropy S and SMI in the general case, i.e.

$$S = (k_B \ln 2)\text{SMI}.$$

2.3.2 *Properties of the fundamental entropy function $S(E, V, N)$*

In the last section, we derived the function $H^{ID}(1, 2, \ldots, N)$, which when applied to a system of N noninteracting particles in a volume V and at a temperature T was identified with the *entropy* of the system. We write this function as

$$S(T, V, N) = Nk_B \ln\left[\frac{V}{N}\left(\frac{2\pi m k_B T}{h^2}\right)^{\frac{3}{2}}\right] + \frac{5Nk_B}{2}.$$

However, for reasons discussed below, the *fundamental function* for the entropy is not $S(T, V, N)$ but the function $S(E, V, N)$, where E is the total energy of the particles. For an ideal monoatomic gas, the energy of the system is simply the total kinetic energy of the particles given by

$$E = N\frac{m\langle v^2\rangle}{2} = \frac{3}{2}Nk_B T.$$

Eliminating T from this equation and substituting in the equation for $S(T, V, N)$, we obtain the *fundamental* function

$$S(E, V, N) = Nk_B \ln\left[\left(\frac{V}{N}\right)\left(\frac{E}{N}\right)^{\frac{3}{2}}\right] + \frac{3k_B N}{2}\left[\frac{5}{3} + \ln\left(\frac{4\pi m}{3h^2}\right)\right]. \quad \text{(III)}$$

This function is *fundamental* for two reasons. First, one can derive all the thermodynamic quantities of an ideal gas from it. Second, the Second Law as formulated in terms of the entropy is valid only for this function, i.e. for the entropy function being expressed as a function of the variables E, V, N, and not in terms of any other set of variables, e.g. T, V, N.

Furthermore, those two reasons apply for *any thermodynamic system*, and not only for ideal gases. Because of their fundamental importance we repeat these two statements:

(i) For any thermodynamic system of N particles in a volume V and having a total energy E, the entropy function $S(E, V, N)$ provides all the thermodynamic quantities of that system.
(ii) For any thermodynamic system of N particles in a volume V and having a total energy E, the entropy function $S(E, V, N)$ has a maximal value at equilibrium.

Here are some clarifying comments:

(1) The N particles in the system can be atoms or molecules, and they can have translational, vibrational, rotational, etc. energies. If there are c components, we interpret N as the vector $N_1, \ldots, N_c$, where N_i is the number of molecules (or moles) of species i.
(2) The volume of the system is defined by the boundaries of the specific thermodynamic system under study. We always assume that the system is macroscopic, i.e. its dimensions are very large compared to molecular diameters of the particles. We also assume that surface effects are negligible. In addition, we assume that there are no external fields that operate on the system.

(3) The total energy of the system E includes all the internal energies of the atoms and molecules, as well as the potential energies of interactions. Actually, the internal energy of the system is always defined with respect to an arbitrarily chosen zero. The First Law is essentially a statement on the changes in the internal energy of the system caused by the exchange of either heat or work with its surroundings.

(4) Maximum with respect to what?
In calculus, when we say that a function $y = f(x)$ has a single maximum, we mean that there is a value of x such that the value of y is maximal compared to all the values of y obtained for any other value in the neighborhood of x.[21]

In some formulations of the Second Law of Thermodynamics, it is stated that the entropy of a system tends to a maximum at equilibrium. Such a statement is faulty, in two respects: first, it does not specify which *function* has a maximum, and second, it does not specify with respect to which *variable* the entropy has a maximum. Without such specification the general statement "entropy always reaches a maximum" is not valid. We shall see many examples of such statements in Chapters 3 and 4.

In thermodynamics, we assume that we have the liberty of choosing, at will, the independent variables. For instance, for a one-component system we may choose the independent variables E, V, N or T, V, N or T, P, N, etc. For each of these independent variables, we have different entropy functions: $S(E, V, N)$, $S(T, V, N)$, $S(T, P, N)$, etc.

Clearly, there are many possible choices of sets of independent variables, and the corresponding entropy functions. The Second Law of Thermodynamics, when applied to the entropy, applies *only* to the specific *entropy function* $S(E, V, N)$. That is the reason why this particular function is referred to as the *fundamental entropy function.*

Now for the question "maximum with respect to which variable?"

In the mathematical examples given above, when we write a function $f(x)$, and claim that $f(x)$ has a maximum, it is understood that the maximum is with respect to the variable x. In the case of $f(x, y, z)$ the maximum is with respect to variation of the independent variables x, y and z. In general, it is understood implicitly that a function has a maximum with respect to the *arguments* of the function, i.e. x in $f(x)$ or x, y, z in $f(x, y, z)$. In thermodynamics we must first choose the independent variables to describe the system — say, E, V, N or T, V, N. Second, unlike in the mathematical practice, the entropy function *does not* have a maximum with respect to the independent variables E, V, N but, on the contrary, these variables must be kept *constant.* What we vary to obtain the maximum of $S(E, V, N)$ is some *internal distribution* keeping E, V, N *constant.* In Figure 2.9 we saw the dependence of the entropy function $S(E, V, N)$ on the variables E, V, N, for an ideal gas.

In Chapter 1, we have seen that the SMI is maximal with respect to all possible probability distributions. The same is true of the function $S(E, V, N;\text{distribution})$. Here, we add the new argument "distribution" to the function $S(E, V, N)$. At this stage we can formulate the Second Law of Thermodynamics as follows[22]:

> For a system having fixed values of E, V and N the entropy function $S(E, V, N)$ attains its maximal value over all possible internal distributions.

Note that this law is valid for any system — ideal gas as well as nonideal gas. It is valid whether or not we have the explicit function $S(E, V, N)$. The important thing is that the variables E, V, N are kept *constant*. Such a system is referred to as an *isolated system*, i.e. a system that does not interact with its surroundings; it does not exchange energy, volume or matter with its surroundings.

We stress again that the law of maximal entropy is valid only when the entropy is viewed as a function of E, V, N, and it has a maximum with respect to the variable "distributions." It is not true for the entropy functions $S(T, V, N;\text{distribution})$, $S(T, P, N;\text{distribution})$, etc.

There are other equivalent formulations of the Second Law in terms of the Gibbs energy or the Helmholtz energy. For instance, for a system having fixed values of T, P and N, the Gibbs energy function $G(T, P, N)$ attains its minimal value over all possible internal distributions.

2.3.3 *MaxEnt or MaxSMI?*

Jaynes advanced the idea of calculating the *most unbiased* distribution by using the so-called maximum entropy principle (MaxEnt).[23] In my view, it is better to refer to this principle as MaxSMI than as MaxEnt.

The MaxEnt *principle* is based, of course, on the Second Law, where a maximum of the entropy is the *principle*. However, there is no obvious justification to apply MaxEnt to any distribution of outcomes such as dice or coins. Here, it is much more appropriate to use the SMI rather than the entropy in formulating the MaxSMI principle.

There is still the question as to why one should formulate the principle of MaxSMI. The only justification I know is that the SMI is related to the *probability* (referred to the superprobability in Chapter 1) of the probability distribution. MaxSMI corresponds to the maximum probability. Specifically, for distributions which are relevant to a thermodynamic system, the principle of MaxSMI becomes identical to that of MaxEnt. Both are based on the fact that the most probable probability distribution is the equilibrium probability distribution. As we have noted in Chapter 1, the distinction between the probability distribution and the superprobability is very important. Failing to understand this distinction is a source of much confusion regarding the Second Law.

To avoid any confusion, one should also make it clear that the "distributions" over which the entropy is maximum must be distributions of *states at equilibrium*, otherwise the entropy function is not defined. We present here a few examples.

Suppose that we start with a system as in Figure 2.16. We have four compartments, each containing an ideal gas, at equilibrium. The distribution in this case is $x_1 = N_1/N$, $x_2 = N_2/N$, $x_3 = N_3/N$ and $x_4 = N_4/N (N = \sum N_i)$. We can view the distribution as a locational distribution. The probability of finding any specific particle in a specific compartment is equal to the mole fraction x_i. Next, we remove the partitions separating

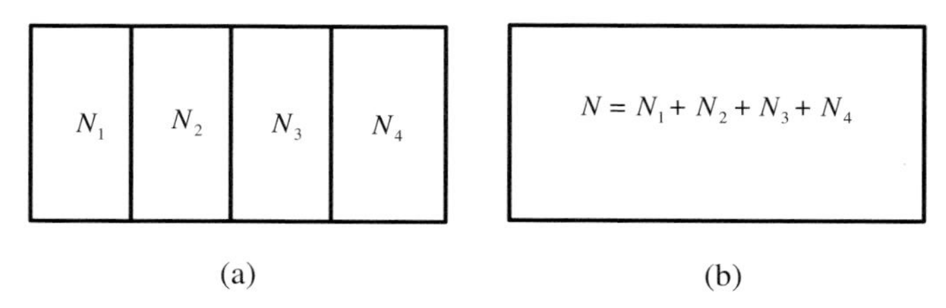

Fig. 2.16 (a) An initial distribution of particles in four compartments; (b) after removal of the partitions between the compartments, a new equilibrium state is obtained.

the four compartments. A new equilibrium will be established. The entropy of the new state will be larger than that of the initial state. In the new equilibrium state, the density in the entire system will be the same. Clearly, the locational distribution in the final state is the uniform distribution. Thus, for any initial distribution (x_1, x_2, x_3, x_4) at equilibrium (i.e. for which the entropy is defined), the entropy of the system will be *smaller* than the entropy of the system when the distribution is such that the density is constant through the entire system, i.e. when $\rho^* = N/V$, which means that $x_i^* = N_i^*/N = \rho^* V_i/N = V_i/V$, i.e. when the fraction of molecules x_i^* is equal to the volume fraction.

The second example is shown in Figure 2.17. Initially, we have four compartments, each at a different temperature. We remove the constraint on the thermal insulation between the compartments, and the entropy will increase. Initially, the momentum distribution in each compartment is different. After we replace the athermal partitions by heat-conducting partitions, a new momentum distribution is obtained which maximizes the entropy of the system. The final equilibrium momentum distribution will be the Maxwell–Boltzmann distribution with

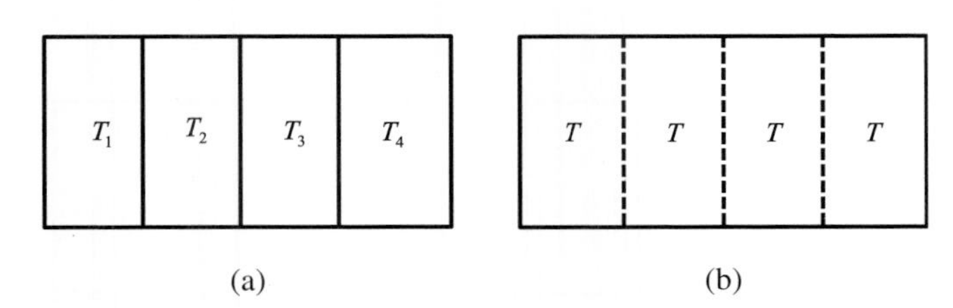

Fig. 2.17 (a) An initial state of four compartments, each at a different temperature; (b) the final equilibrium state obtained after replacement of the athermal partitions by a heat-conducting (diathermal) partitions. The velocity distribution is changed in this process.

a unified temperature T. (If all the four compartments have the same number of particles, and the same volume, then T is simply the arithmetic average of the temperature T_1, T_2, T_3, T_4. [See also Ben-Naim (2008).]

In each of the above examples we started with a given distribution of macroscopic states at equilibrium. When we remove the partitions between the compartments, a new equilibrium state will be reached for which the entropy of the entire system has increased. Having reached the final equilibrium state, we can ask: What is the probability of having the original distribution (or any arbitrary distribution)? The answer is

$$\Pr(\text{distribution}) = \exp\left[\frac{S(\text{distribution})}{k_B}\right].$$

This is essentially the same relationship we had in Chapter 1, between the superprobability and the SMI. The only difference between the two cases is that the previous one is valid for any distribution, whereas the present one is valid for a distribution of a thermodynamic system.

The third example is shown in Figure 2.18. Here, we have N_w solvent and N_s solute molecules, which can have different "states" corresponding to different angles of rotation ϕ about the C–C bond.

Suppose that we can "freeze in" the rotation about the C–C bond, and for concreteness suppose that we can make a mixture of, say, four isomers, characterized by the angles $\phi_1, \phi_2, \phi_3, \phi_4$. The system as a whole is characterized by E, V, N_1, N_2, N_3, N_4, where N_i is the number of solute molecules in the "state" ϕ_i. If there is no conversion between the different "species" of the solute, we have a mixture of N_w solvent molecules and four different solute molecules *at equilibrium*. Now suppose that we

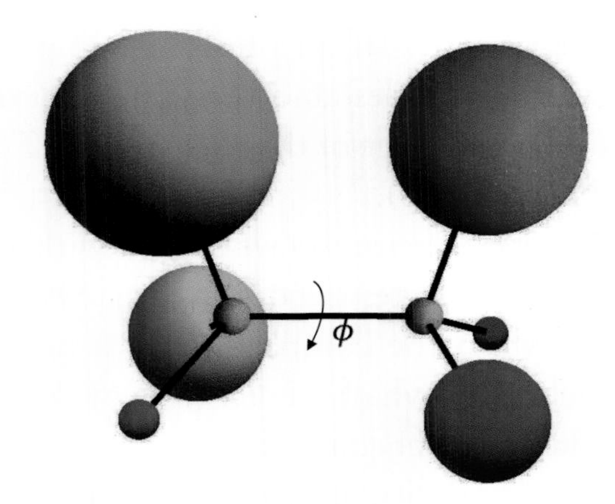

Fig. 2.18 A molecule with one internal rotation (such as a derivative of ethane).

can add some catalyst that causes a transition between all the states of the molecule. Addition of such a catalyst is equivalent to a release of the constraint on the *fixed* angles. Once a catalyst is added, a *new equilibrium* will be reached where all the angles ϕ are accessible.

In the new equilibrium state there will be a unique distribution of angles $\Pr(\phi)$, which maximizes the entropy function over all possible distributions of "frozen-in" states. It should be emphasized that such states must also be at equilibrium states, otherwise the entropy function is not defined. At equilibrium the distribution of angles is related to the entropy function by the relationship similar to the one in the previous examples, i.e.

$$\Pr(\phi) = \exp\left[\frac{S(\phi)}{k_B}\right].$$

Here, S is a *function* of the independent variables E, V, N, which were omitted from the notation above. The MaxEnt principle states that there exists a probability distribution $\Pr^*(\phi)$ which maximizes the entropy *functional* $S\left[E, V, N; \Pr(\phi)\right]$.

The extension of this principle to proteins, where instead of one angle ϕ we have many rotational angles, is straightforward. This is discussed in Ben-Naim (2013).

In concluding, we emphasize again that the maximum entropy applies when the variables E, V, N are constants (presuming no external fields and that N represents the molecular composition of the system). The maximum is with respect to all possible constrained equilibrium distributions. If a system is in a state having a distribution which is not an equilibrium distribution, then the entropy of such a system is not defined.

For any system (small or large), away from equilibrium, one can define the SMI for any given distribution. The principle of MaxEnt should be more appropriately referred to as the principle of MaxSMI. The transition from MaxSMI to MaxEnt should be made only in the state of equilibrium and for macroscopic thermodynamic systems. The multiplication by the Boltzmann constant, and the change in the base of the logarithm, merely changes the units in which the SMI is measured.

When the macroscopic state of the thermodynamic system is defined, the entropy of the system is also defined. When the state of the system changes, we can say that the SMI is changing. This is meaningful because the SMI is defined for *any* distribution. However, the entropy is *defined* only for the distribution which *maximizes* the SMI. In this sense, the value of the entropy is unique for each system.

2.3.4 *Is entropy a subjective quantity?*

Ever since entropy was interpreted as "information," people have asked the following question: Is entropy a subjective quantity? Entropy is also conceived of as something that can "flow" from one body to another, and something that can be "created." If this is the case, then entropy must be *physical.* But if entropy is information, then information must also be physical, and if information is physical, one might ask, Where does the information reside?

In my opinion, all these questions and doubts arise because of the confusion about the concepts of information and SMI. Information is an abstract concept which could be subjective or objective, important or relevant, exciting, meaningful, etc. On the other hand, SMI is a *measure of information.* It is not information (although it can be interpreted as a measure of a restricted kind of information). SMI is an attribute assigned to an objective probability distribution of a system. It is not a substantial quantity. It does not have a mass, a charge or a location. It does not reside in a definite location; nor does entropy, which is viewed as a particular example of SMI. It does not reside in any place, it is not a substantial quantity, and yet it is an attribute of a physical system.

The interpretation of entropy as a particular case of SMI rules out the subjectivity of entropy. Unfortunately, the confusion of SMI with information had led many to question the subjectivity of entropy. Perhaps the first person to relate entropy to information in an explicit way was Lewis[24]:

> The increase in entropy comes when a known distribution goes over into an unknown distribution . . . . Gain in entropy always means loss of information, and nothing more.

Note that this statement was made 18 years before the inception of information theory. Lewis is using here the notion of information in its colloquial sense, and this has led Denbigh to ask the following question:[25] How subjective is entropy?

Denbigh starts his article by first distinguishing between two different kinds of "objectivity" (or its converse, "subjectivity"):

> A weak objectivity, meaning a statement which can be publicly agreed to, and a strong objectivity, referring to statements about things which can be said to exist, or about events which can be said to occur, quite independently of man's thoughts and perceptions or of his presence in the world.

He then goes on to criticize Jaynes (1965), who claimed:

> Entropy is an anthropomorphic concept, not only in the well-known statistical sense that it measures the extent of human ignorance as to the microstate. Even at the purely phenomenological level, entropy is an anthropomorphic concept. For it is a property not of a physical system, but of the particular experiment you or I choose to perform.

Referring to the story told by Tribus and McIrvine (1971) that von Neumann suggested to Shannon to name his quantity (SMI) entropy, Denbigh comments:

> In my view von Neumann did science a disservice!

I fully agree with Denbigh's criticism of using the term "entropy" for SMI. This has led to a great confusion in thermodynamics.

As far as I could understand from Denbigh's article, his objection to calling SMI "entropy" was because he conceived of SMI as *information*, and information can be subjective. Hence, if entropy is identified with SMI, it can also be a subjective quantity.

My objection to von Neumann's suggestion is very different. As I explained in Chapter 1, SMI is not information, but a *measure* of information in a restricted sense. Therefore, unlike Denbigh (and many others who rejected the informational interpretation of entropy), I do not hold the view that SMI is a subjective quantity (presuming that we agree that the probability distributions on which SMI is defined are objective quantities). Therefore, my objection to naming SMI "entropy" is not because this makes entropy a subjective quantity, but simply because SMI is a much more general concept than entropy, and entropy is only a special case of SMI, and as such it is not a subjective quantity.

Denbigh concludes his article by saying:

> My general conclusion is that the subjectivity thesis, relating to entropy, remains unproved, and in my view can be largely dispelled by careful attention to the logic of the matter and to its linguistic expression.

In my view the relationship between entropy and SMI has already been proven! There is no relationship between entropy and subjectivity, and hence there is nothing to be proven!

The interpretation of entropy in terms of SMI was rejected by many scientists *because* SMI smacks of *information*, and information could be subjective. If that is true, then entropy also becomes subjective, which is untenable. This is the essence of Denbigh's argument.

This line of reasoning is, however, fallacious. SMI is *not* information. It is an objective quantity, as is the length of a circle, or the sum of the angles in a triangle.

It is ironic that very recently some scientists who *accept* the informational interpretation of entropy resurrected the subjectivity into entropy.

For instance, von Baeyer (2003) writes:

> It is not just the number of ways a system can be rearranged, but more specifically the number of rearrangements consistent with the known properties of the system. Known by whom? Measured by what observer?
>
> By this reckoning entropy is not an absolute property of a system, but relational. It has a subjective component — it depends on the information you happen to have available.
>
> Boltzmann understood this connection and made it more specific. He pointed out that since the value of entropy rises from zero when we know all about a system, to its maximum value when we know least, it measures our ignorance about the details of the motions of the molecules of a system. Entropy is not about speeds or positions of particles, the way temperature and pressure and volume are, but about our lack of information.

It is ironic that the same confusion of SMI with information — which caused the *rejection* of the informational interpretation of entropy — brought back the *subjectivity* as an attribute of entropy.

Besides, this quotation by von Baeyer alludes to Boltzmann's understanding of the "subjectivity" of entropy. As far as I know, Boltzmann never discussed the subjectivity of the entropy. It is not true either that "the value of entropy rises from zero when we know all about a system . . . ."

Having found numerical agreement between the calculated and the experimental values of the entropy provides a solid interpretation of the entropy. For the question as to what entropy is, we can answer that entropy is a measure of the uncertainty or of the amount of information associated with the *equilibrium* distribution of locations and momenta of all the particles.[27] It has no traces of subjectivity in it.

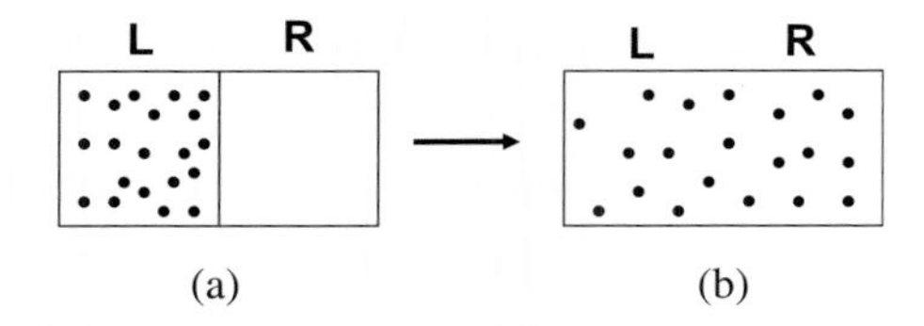

Fig. 2.19 A spontaneous expansion of an ideal gas from V to $2V$.

2.4 The Second Law of Thermodynamics

Solving the mystery associated with entropy still begs the following question: Why does entropy always increase in a spontaneous process in an isolated system?

Before we examine this question with the help of the simplest example, it should be noted that in many textbooks on thermodynamics, one can find statements referring to entropy as the *driving force* for the spontaneous processes.

Consider again the expansion of an ideal gas. (Figure 2.19).

An ideal gas is initially confined to a volume V. We remove a partition and observe that the gas will expand and fill the entire new volume $2V$.

Ask any student who has learned thermodynamics. Why did the gas expand from V to $2V$, and why will it never go back to the original volume V? The answer you are most likely to hear is that the cause of that expansion is the tendency of the entropy to increase.

If you ask: Why does the entropy increase? The immediate answer will be: That is exactly what the Second Law states!

Similarly, you might find people answering this question as follows: systems tend to evolve from a more ordered state to a more disordered state.

Here one uses the order–disorder metaphor for entropy.[28] Other statements blame entropy as the cause of the one-way direction of spontaneous processes, from the expansion of a gas, to the mixing of gases, to the splattering of a fallen egg, and even for the transition from life to death.

In one of the BBC programs on the universe, the viewer is shown a desolate village that once was bustling and prosperous. The camera pans from ruin to ruin, showing the decline, the decay, and the utter desolation and destruction. The voice-over audio explains what is unfolding before the viewer's eyes — the action of the *ravages of entropy*. This phrase also appears in many popular-science books where entropy is discussed.

It is not uncommon to read that the ravages of entropy will ultimately decimate and eventually destroy what our civilization has created, all life on earth will be gone, and the whole universe will be doomed to an eventual *thermal* death. We shall further discuss this topic in Chapter 4. We now pose the following question:

> Does the tendency of the entropy to increase *drive* the spontaneous process, or does the spontaneous process drive the entropy upward?

Before we answer this question let us examine the expansion process with different numbers of particles.

In all of the following examples we shall assume that the particles do not interact with each other (or that interaction is negligible), and that in each case only the locational distribution of the particles changes. Each particle is initially located within the boundaries of the volume V, and in the final state it is located in the larger volume $2V$ (here "located" means that it can be found anywhere within the volume V or $2V$, respectively).

There are essentially two questions associated with the Second Law that are oftentimes confused. One is: Why does a system evolve spontaneously from one state to the other? The other is: Why does the entropy always increase? The answers to these two questions are in principle different, yet they are related to each other.

The Second Law of Thermodynamics states that in any spontaneous process in an isolated system the entropy increases. It does not state anything as to *why* the entropy increases, or address the question as to why a spontaneous process occurs at all.[29]

The only answer to the second question is probabilistic. Accepting the relative frequency interpretation of probability,[30] we can conclude that a system will be found more frequently in states which have higher probability. Specifically, for thermodynamic systems, we will see that the probability of the state of equilibrium is almost one. Thus, we can say that a system will always evolve toward a state of higher probability, and eventually reach a state we call the *equilibrium state*, the probability of which is almost 1. [For details, see Ben-Naim (2008, 2012).] Thus, "probability" answers the following question: Why does the system change spontaneously?

We emphasize again that we talk about *spontaneous* processes in an *isolated* system, having a fixed energy, volume and number of particles. We assume that our systems reach an equilibrium state. We shall not consider here systems which do not reach equilibrium, or that the changes are so insignificant that we do not notice them.

Answering the question of *why* the system evolves toward the state of equilibrium leaves the question of why the entropy increases unanswered. Nevertheless, because of the intimate relationship between the SMI of the system and the probability of the state of the system, the answers to the two questions are also related to each other. This is discussed in the next subsection.

2.4.1 *What drives the system to an equilibrium state?*

In Chapter 1 we studied the process of spreading of marbles in different compartments. Here, we discuss a similar process where a system of particles spreads into a larger volume.

Consider a system of N noninteracting particles (ideal gas) in a volume $2V$ at a constant energy E. We divide the system into two parts, L and R, each of volume V, as in Figure 2.19(b). We define the *microscopic* state of the system when we are given $E, 2V, N$, and in addition we know which *specific* particles are in compartment R, and which specific particles are in compartment L. The *macroscopic* description of the same system is $(E, 2V, N; n)$, where n is the *number* of particles in the compartment L. Thus, in the microscopic description, we are given a specific configuration of the system as if the particles were labeled $1, 2, \ldots, N$. In the macroscopic description we are given the information on the number of particles that are in each compartment.

Clearly, if we know only that n particles are in L and $N - n$ particles are in R, we have

$$W(n) = \frac{N!}{n!(N-n)!}$$

specific configurations that are consistent with the requirement that there be n particles in L.

The first postulate of statistical mechanics states that *all specific* configurations of the system are equally probable. Clearly, the *total* number of *specific* configurations is

$$W_T = \sum_{n=0}^{N} W(n) = \sum_{n=0}^{N} \frac{N!}{n!(N-n)!} = 2^N.$$

Using the *classical* definition of probability, we can write the probability of finding n particles in L and $N - n$ particles in R as

$$P_N(n) = \frac{W(n)}{W_T} = \left(\frac{1}{2}\right)^N \frac{N!}{n!(N-n)!}.$$

It is easy to show[31] that $W(n)$ [hence also $P_N(n)$] has a maximum as a function of n at the point $n^* = N/2$. The maximum value of $W(n)$ (at $n^* = N/2$) is $W(n^*) = N!/[(N/2)!]^2$. The corresponding probability is

$$P_N(n^*) = \frac{W(n^*)}{2^N}.$$

Note that for any given N, there exists an n such that the number of configurations $W(n)$ [or of the probability $P_N(n)$] is maximal. Therefore, if we prepare a system with any initial distribution of particles (n and $N - n$) in the two compartments, and let the system evolve, the system's state will change from a state of lower probability to a state of higher probability. As N increases, the value of the maximum number of configurations $W(n^*)$ *increases* with N. However, the value of the maximal probability $P_N(n^*)$ *decreases* with N.

To appreciate the significance of this fact, consider the following cases:

The case of two particles: $N = 2$

Suppose that we have a total of $N = 2$ particles. In this case, we have the following possible configurations and the

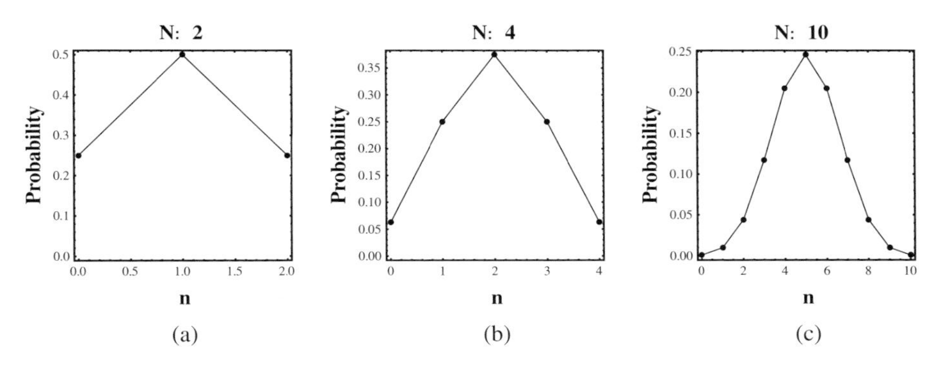

Fig. 2.20 The probability of finding n particles in the left compartment, for $N = 2, 4, 10$.

corresponding probabilities:

$$n = 0, \quad n = 1, \quad n = 2,$$
$$P_N(0) = \frac{1}{4}, \quad P_N(1) = \frac{1}{2}, \quad P_N(2) = \frac{1}{4}.$$

This means that, on the average, we can expect to find the configuration $n = 1$ (i.e. one particle in each compartment) about half of the time, but the configurations $n = 0$ and $n = 2$ only a quarter of the time [Figure 2.20(a)].

The case of four particles: $N = 4$

For the case $N = 4$, we have the distribution shown in Figure 2.20(b). The maximal probability is $P_N(2) = 6/16 = 0.375$, which is *smaller* than $1/2$. In this case, the system will spend only $3/8$ of the time in the maximal state $n^* = 2$.

The case of ten particles: $N = 10$

For $N = 10$ the distribution is shown in Figure 2.21(c). We calculate the maximum at $n^* = 5$, which is $P_{10}(n^* = 5) = 0.246$.

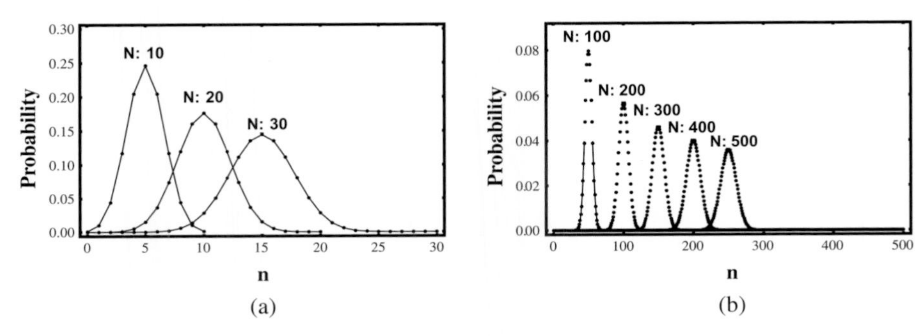

Fig. 2.21 The probability of finding n particles in the left compartments, for larger values of N.

Figure 2.21 shows the probability function $P_N(n)$, as a function of n for different values of N. As N increases, $W(n^*)$ *increases*, but $P_N(n^*)$ *decreases*. For instance, for the case $N = 1000$, the maximal probability is $P_{1000}(n^*) = 0.0252$. As N increases, the maximal probability decreases as $N^{-\frac{1}{2}}$. In practice, we know that when the system reaches the state of equilibrium, it stays there *forever*. The reason is that the *macroscopic state of equilibrium* is not the exact state for which $n^* = N/2$, but it is this state along with a small neighborhood of n^*, e.g. $n^* - \delta N \leq n \leq n^* + \delta N$, where δ is small enough, so that no experimental measurement can detect it. The probability of finding the system in the *neighborhood* of the maximum for $N = 100$ and $\delta = 0.01$ is about 0.235. For $N = 10^{10}$ particles we can allow deviations of 0.001% of N and the probability of finding the system in the neighborhood is nearly one.[32]

Figure 2.22 shows the probability of finding the system with n between $n^* - \delta N \leq n \leq n^* + \delta N$, having a deviation of $\delta = 0.0001\,N$ about the value of n^*. Plotting the probability $P_N(n^* - \delta N \leq n \leq n^* + \delta N)$, as a function of N shows that this probability tends to 1 as N increases. When N is on the

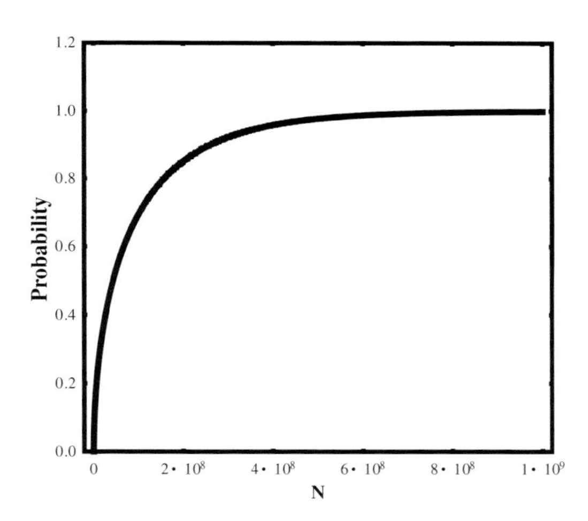

Fig. 2.22 The probability of finding about $n^* \approx N/2$ particles in the left compartment as a function of N.

order of 10^{23}, we can allow deviations of $\pm 0.00001\%$ of N or even smaller. Yet, the probability of finding the system at or near n^* will be almost one. It is for this reason that when the system reaches n^* or near n^*, it will stay in the vicinity of n^* for most of the time. For N on the order of 10^{23}, "most of the time" means *always*.

The abovementioned specific example provides an explanation for the fact that the system will "always" change in one direction, and will "always" stay in the equilibrium state once that state is reached. The tendency toward a state of larger probability is equivalent to the statement that events that are supposed to occur more frequently will occur more frequently. This is plain common sense. The fact that we do not observe deviations from either the monotonic climbing of n towards n^* or staying at n^* is a result of our inability to detect small changes in n (or, equivalently, small changes in the SMI; see below). Note that in this section we have not said anything about the entropy changes. Before turning to the calculation of the entropy changes, we repeat the main conclusion of this section. For each N the probability of finding a distribution $(n, N - n)$ in the two compartments L and R has a maximum at $n^* = N/2$. However, for a very large number of particles the probability of obtaining the exact value of $n^* = N/2$ is not very large. On the other hand, the probability of finding the system in a small vicinity of $n^* = N/2$ is almost one!

When we say that the system has reached an equilibrium state, we mean that we do not *see* any changes that occur in the system. In this example, we mean changes in the *density* of the particles in the entire system. In other experiments, when there is heat exchange between two bodies we characterize the equilibrium state as the one for which the temperature is uniform throughout the system and does not change with time.

At equilibrium the macroscopic density we measure at each point in the system is constant. In the particular system we discussed above, the measurable density of the particles in the two compartments is $\rho^* \cong N/2V$. Note that fluctuations always occur. Small fluctuations occur very frequently, but they are so small that we cannot measure them. On the other hand, fluctuations that could have been measured are extremely infrequent, and practically we can say that they never occur. This conclusion is valid for very large N.

2.4.2 *Why does the entropy always increase?*

Next, we will discuss the relationship between the probabilities of the macrostates and the formulation of the Second Law in terms of the entropy. We rewrite the essential quantities of the example discussed above in a slightly different way. Instead of n

and $N - n$, we define the fractions

$$p = \frac{n}{N}, \quad q = (1 - p) = \frac{N - n}{N},$$

where p is the fraction of particles in the L compartment and $q = 1 - p$ is the fraction in the R compartment. We now proceed to calculate the probability of this *probability distribution*. The number of specific configurations for which there are n particles in L is rewritten as

$$W(p, q) = \binom{N}{pN},$$

and the corresponding probability is

$$\Pr(p, q) = \left(\frac{1}{2}\right)^N \binom{N}{pN}.$$

This expression can be converted into[33]

$$\Pr(p, q) = \left(\frac{1}{2}\right)^N \frac{2^{N \times \mathrm{SMI}(p,q)}}{\sqrt{2\pi N p q}}.$$

This is the relationship between the SMI defined on the *state* distribution $\{p, q\}$ and the probability $\Pr(p, q)$ defined on the same distribution. Clearly, these two functions are very different. $\mathrm{SMI}(p, q)$ has a maximum at the point $p^* = q^* = \frac{1}{2}$. This maximum is independent of N. On the other hand, $\Pr(p, q)$ has a sharp maximum at the point $p^* = q^* = \frac{1}{2}$. As we have seen, the value of the maximum, $\Pr(\frac{1}{2}, \frac{1}{2})$, *decreases* with N.

From the monotonic relationship between Pr and SMI, it follows that whenever SMI increases, Pr also increases, and at equilibrium both SMI and Pr attain a maximal value. We have

Fig. 2.23 The symbolic relationship between the questions of "what" and "why."

seen that the maximal value of SMI is related to the entropy of the system. Therefore, the answer to the question of *why* the entropy increases (in a spontaneous process in an isolated system) is the same as the answer to the question of why the state of the system evolves toward equilibrium; namely, it is because the probability Pr of the equilibrium state is maximum. The connection between the questions of "what" entropy is and "why" it changes in one direction is shown symbolically in Figure 2.23.

Note carefully the two "levels" of probabilities. One is the probability distribution of a state described by (p, q). The other, Pr, is the probability of finding a state described by (p, q). To distinguish between the two probabilities I refer to Pr as a *superprobability*. Note also that the answer to the question "Why does the system evolve toward equilibrium?" is provided by the probability Pr. Because of the monotonic relationship between Pr and SMI, the answer to the question "Why does the entropy increase?" is also probabilistic. It is easy to generalize the conclusion to the case of any number of compartments. [See Ben-Naim (2008, 2009).]

Recall that the relationship between the probability Pr and SMI is the same as what we have obtained in Chapter 1 for the evolving game. Here, we have applied the same arguments to obtain the relationship between the superprobability of the distribution and the *entropy* of the system at equilibrium.

2.4.3 *Small systems and "violations" of the Second Law*

We have seen that SMI is defined for any distribution and for any number of particles. Using the Boltzmann entropy or the entropy function as derived from SMI, one could in principle also apply the concept of entropy to a system with any number of particles. For instance, one could calculate the total number of states of one particle in a box of volume V, call it W_1 and calculate the corresponding "entropy" by $S_1 = k_B \ln W_1$.

However, if we want the system to obey the Second Law, then we have to reconsider the definition of the entropy of a system having a small number of particles. In such a system we can easily observe "violations" of the Second Law.

We can demonstrate such a "violation" with the following experiment. Starting from an initial state — with, say, ten particles in the left compartment, and by removing the partition — we should initially see a net flow of particles from the left to the right compartment. However, once in a while we might find fluctuations such that the number of particles in the right (or the left) compartment might be zero, one or two, in apparent "violation" of the Second Law. As we have discussed in Subsection 2.4.1, it is true that the distribution for which there will be five particles in each compartment has the maximal probability. However, for such a system we expect large fluctuations from this maximal probability distribution. For this reason, it is advisable to use the concept of entropy and the Second Law for thermodynamic systems, i.e. systems with a very large number of particles. In such systems, deviations from the equilibrium state (i.e. configurations which are near the one that maximizes the SMI) are negligible.

It should be noted that the Clausius and Kelvin formulations of the Second Law are equivalent provided that they are applied to thermodynamic systems. If the system contains a small number of particles, then we can expect "violations" of the Second Law in both formulations. On the other hand, the Boltzmann entropy is defined in terms of the number of states. Given E, V, N we can calculate the number of states W_N for *any number* of particles N, and the corresponding Boltzmann entropy $S_N = k_B \ln W_N$.

To conclude, it is advisable to define the entropy function for a system of a large number of particles at equilibrium (i.e. in states which are near the state having the largest probability). For systems of any number of particles in any state far from equilibrium, it is wise to apply SMI rather than entropy. Using this nomenclature the Second Law is *never* violated. On the other hand, for a small number of particles, the Second Law does not apply, and hence there are no "violations." This is the reason for enclosing the word "violation" in quotation marks.

2.4.4 *How did the system evolve?*

Answering the questions "What is entropy?" and "Why does it change in one direction?" leaves open the question "How does the system move from one state to another?"

Fortunately, neither thermodynamics nor statistical mechanics deals with the question of *how* the system evolves from an initial to the final state. We always *assume* that both the initial and the final state are equilibrium states. Thermodynamics deals with the *difference* in the entropy between two equilibrium states. This is the essence of the meaning of a "state function." It is a function whose value is determined by the *state* of the system, not how the system got to that state. The answer to the "how" question is not relevant to thermodynamics.

2.5 The Association of the Second Law with the Arrow of Time

Every day, we see numerous processes apparently occurring in one direction, from the mixing of two gases to the decaying of a dead plant or animal. We never observe the reverse of these phenomena. It is almost natural to feel that this direction of occurrence of the events is the "right" direction, consistent with the direction of time. Here is what Greene writes on this matter:[34]

> We take for granted that there is a direction in the way things unfold in time. Eggs break, but do not unbreak; candles melt, but they don't unmelt; memories are of the past, never of the future; people age, they don't unage.

However, he adds:

> The accepted laws of Physics show no such asymmetry, each direction in time, forward and backward, is treated by the laws without distinction, and that's the origin of a huge puzzle.

Indeed, it is! For almost a century, physicists were puzzled by the apparent conflict between the Second Law of Thermodynamics and the laws of dynamics. As Greene puts it:

> Not only do known laws (of physics) fail to tell us why we see events unfold in only one order, they also tell us that, in theory, events can fold in the reverse order. The crucial question is, Why don't we ever see such things? No one has actually witnessed a splattered egg un-splattering, and if those laws treat splattering and un-splattering equally, why does one event happen while its reverse never does?

Ever since Eddington (1927) associated the Second Law of Thermodynamics with the "arrow of time,"[35] scientists have endeavored to resolve this paradox. On one hand, the equations of motion are symmetric with respect to going forward or backward in time. Nothing in the equations of motion suggests the possibility of a change in one direction and forbids a change in the opposite direction. On the other hand, many processes we see every day do proceed in one direction and are never observed to occur in the opposite direction. But is the Second Law really associated with the arrow of time?

We shall soon see why we *never* witness the reverse of these apparently one-way processes. Indeed, we never witness such reversed processes — but this does not mean that they *never* occur! The understanding of this conundrum hinges upon the meaning we assign to the word "never."

Our sense of the "right" direction of unfolding events is usually demonstrated by the following simple experiment. If we are shown a movie played backward, we will easily recognize that "it does not make sense." We might even find it amusing watching people walking backward or a broken egg spontaneously assembling itself and flying upward and landing intact on a table.

Why?

Because we know that this kind of process never proceeds in this direction in time.

This is what Hawking called the *psychological arrow of time*,[36] i.e. the direction in which we *feel* time passes, "the direction in which we remember the past but not the future." Hawking distinguishes between this "psychological arrow of time" and the "thermodynamic arrow of time; the direction of time in which disorder or entropy increases."

Recently, Carroll devoted a large part of his book to discussing the relationship between the arrow of time and entropy[37]:

> The mystery of the arrow of time comes down to this: Why were the conditions in the early universe set up in a very particular way, in a configuration of low entropy?

He continues:

> ... the consistent increase of entropy throughout the universe, which defined the arrow of time.

In this section, I will challenge the connection between *entropy* change and the arrow of time. In Chapter 4, I will also discuss the (meaningless) concept of the *entropy of the universe*. It will be shown that whatever the "mystery of the arrow of time" is, it has *nothing* to do with the (also meaningless) "low entropy" conditions of the early universe. Furthermore, the "increase of entropy throughout the universe," whatever it may mean, does not and cannot define the "arrow of time."

In my opinion, the association of the spontaneously occurring events with the arrow of time is merely an illusion. An illusion created by the fact that in our lifetime we have *never* seen, and we shall *never* see, a process that unfolds in the "opposite" direction. Because of the importance of the question regarding the direction of change in entropy, consider again the case of the expanding gas, but with only a few particles.

Suppose that we start with *two* particles in the left compartment of Figure 2.19. Remove the partition and the particles will expand to occupy the entire volume, $2V$. Now suppose that you take many snapshots of the particles. How often will you see the system return to its original state? If each particle has an equal probability of 1/2 to be in the left or in the right compartment, the probability of finding the two particles in one compartment — say, the left one — is

$$\Pr(N = 2) = \left(\frac{1}{2}\right)^2 = \frac{1}{4},$$

which means that on average one of every *four* snapshots will show the two particles in the left compartment, about one in four snapshots will show the two particles in the right compartment, and for about one in two snapshots one particle will be on the right and one on the left. Clearly, there is no conflict with the reversibility of the laws of motion, nor any sense of the "arrow of time." If you are shown the movie of this system, you will not be able to distinguish between the forward- and the backward-played movie.

For four particles, $N = 4$, the probability of finding all the four particles in the left compartment is

$$\Pr(N = 4) = \left(\frac{1}{2}\right)^4 = \frac{1}{16},$$

which means that, on average, for one in 16 snapshots you will find the particles in the left compartment. Again, there is no conflict with Newtonian mechanics! When the number of particles becomes very large, these probabilities become very small, *but never zero*.

Let us increase the number of particles N. Below we have the probability of finding all the N particles in one compartment for different N:

$$N = 10: \quad \Pr(N = 10) = \left(\frac{1}{2}\right)^{10} = \frac{1}{1024};$$

$$N = 20: \quad \Pr(N = 20) = \left(\frac{1}{2}\right)^{20} = \frac{1}{1048576};$$

$$N = 30: \quad \Pr(N = 30) = \left(\frac{1}{2}\right)^{30} = \frac{1}{1073741824}.$$

As you can see, the probabilities become increasingly small, but there is no conflict with Newtonian mechanics. If you wait long enough you will find all the 30 particles in one compartment, for one out of about a billion snapshots. If you are shown the movie of this system played backward, you will observe that sometimes the *rare event* of finding all the particles in one compartment will occur. But this will not be in conflict with what you would expect.

When $N = 1000$ or more, we get very huge numbers that are difficult even to write. Suppose that you take one million snapshots every second. In one year you will collect $10^6 \times 60 \times 60 \times 24 \times 365 \approx 30{,}000{,}000{,}000{,}000$ snapshots. This is a very large number of snapshots. The probability of finding all the 1000 particles in one compartment is

$$\Pr(N = 1000) = \left(\frac{1}{2}\right)^{1000} \approx \frac{1}{10^{301}}.$$

This means that on average only one in 10^{301} snapshots will show all the 1000 particles in the left compartment. Equivalently, you will have to wait $\frac{10^{301}}{3\times 10^{13}} \approx 3 \times 10^{287}$ years to see one such snapshot. The age of the universe is estimated to be about 15×10^9 years. This means that in order to observe 1000 particles in one compartment you have to wait

$$\frac{3 \times 10^{287}}{15 \times 10^9} \approx 3 \times 10^{277} \quad \text{ages of the universe.}$$

If you have enough "patience" you will see all the 1000 particles in one compartment once in billions and billions of ages of the universe! Once more, there is no conflict with Newtonian mechanics. Again, if you are shown a backward-played movie of this system, you will "never" see all the N particles in one compartment. But even if you are lucky and you *do observe* all the N particles in one compartment, this phenomenon will not last too long. Very soon the particles will spread to occupy the two compartments, and this state will remain for a very, very long time.

The calculation made above is only for a *mere one thousand* particles ($N = 1000$). For a thermodynamic system with $N = 10^{23}$, the probability of observing all the particles in one compartment is

$$\Pr(N = 10^{23}) = \left(\frac{1}{2}\right)^{10^{23}}.$$

This is an unimaginably small number. It means that if you wait billions and billions ... and billions of ages of the universe, *you will find* all the particles in one compartment! Again, there is nothing in principle that is in conflict with the laws of mechanics. The conflict is only an illusion.

Boltzmann recognized the fact that the Second Law is not an absolute law. But even if it were absolute, i.e. for any spontaneous process in an isolated system the entropy increases, it would not follow that the Second Law "defines" the "arrow of time."

The Second Law is admittedly not absolute, but no other law of physics can claim to be absolute. As I have concluded earlier (Ben-Naim, 2007):

The admitted non-absoluteness of the Second Law is in fact more absolute than the proclaimed absoluteness of any other law of physics.

Imagine a world where people live for a very long time, many times the age of the universe — say, $10^{10^{30}}$ years. In such a world, if we perform the experiment with gas expansion, and if we start with all particles in one box, we will first observe expansion and the particles will fill the entire volume of the system. But "once in a while" we will observe visits to the original state. How often? If we live for an extremely long time — say, $10^{10^{30}}$ years — and the gas consists of some 10^{23} particles, then we should observe visits to the original state many times in our lifetime. If you watch a film of the expanding gas, running forward or backward, you will feel that this is a rare event, but not an absurd one. You will have no sense of some phenomena being more "natural" than others, and there should not be a sense of the "arrow of time," associated with such processes.

It should be added here that many authors mention the direction from life to death in connection with the arrow of time and the Second Law. In my opinion the phenomena of life and death cannot be discussed in connection with the Second Law of Thermodynamics. We shall discuss further this aspect of life in Chapter 3. Similarly, some authors "predict" the eventual "death" of the universe based on the Second Law. I will contest such claims in Chapter 4.

When Boltzmann published his probabilistic theory of the Second Law, the theory was vehemently criticized by his contemporaries. First, because of the apparent contradiction between the reversibility of Newtonian mechanics and the irreversibility of the Second law. This is referred to as the *reversibility paradox*. The second paradox was based on Poincaré's theorem, which when formulated for our examples of the expanding gas states that the system will always return to its initial state (i.e. to the left compartment). This seems to conflict with the ever-increasing entropy. This is called the *recurrence paradox*.

It is most instructive to read Boltzmann's reaction to this criticism:[38]

> The reverse transition has a definite calculable (though inconceivably small) probability, which approaches zero only in the limiting case when the number of molecules is infinite. The fact that a closed system of a finite number of molecules, when it is initially in an ordered state and then goes over to a disordered state, finally after an inconceivably long time must again return to the ordered state, it is therefore not a refutation but rather indeed a confirmation of our theory. One should not however imagine that two gases in a $\frac{1}{10}$ liter container, initially unmixed, will mix, then again after a few days separate, then mix again, and so forth. On the contrary, one finds by the same principles which I used for a similar calculation that not until after a time enormously long compared to $10^{10^{10}}$ years will there be any noticeable unmixing of the gases. One may recognize that this is practically equivalent to never, if one recalls that in this length of time, according to the laws of probability, there will have been many years in which every inhabitant of a large country committed suicide, purely by accident, on the same day, or every building burned down at the same time, yet the insurance companies get along quite well by ignoring the possibility of such events. If a much smaller probability than this is not practically equivalent to impossibility, then no one can be sure that today will be followed by a night and then a day.
>
> In any case, we would rather consider the unique directionality of time given to us by experience as a mere illusion arising from our specially restricted viewpoint.

Clearly, Boltzmann did not see any conflict between the Second Law of Thermodynamics and the equations of motion or the laws of dynamics. There is no such conflict. If we live "long enough" we shall be able to observe all these reverse processes! The connection between the arrow of time and the Second Law is not absolute, but only "temporary," for a mere billion, billion . . . billion years.

It should be added that in the context of the association of the Second Law with the arrow of time, some authors invoke human experience, which distinguishes the past from the future. It is true that we remember events from the past, *never* from the future. We also feel that we can affect or influence events in the future, but *never* events in the past. I fully share these experiences. The only question I have is: What have these experiences got to do with the Second Law of Thermodynamics? No one has measured the entropy change when we *remember* the past, or do not "remember" the future. Also, no one has estimated the entropy change associated with affecting events either in the future or in the past. Therefore, such statements, though very impressive, are totally irrelevant to the Second law.

In all the examples discussed above, we focused on the possibility of *reversing* the expansion of a gas from V to $2V$. We saw that for a small N the system can be found in one compartment "once in a while." The larger N is, the rarer the occurrence of this event will be. What about the *change in time of the entropy*? In the examples of the expanding gas, if we reverse the velocities of all the particles in the system, the *state* of the system will be reversed. In particular, if we start with the gas filling the two compartments, or the two mixed gases in two compartments, and reverse the direction of the velocities of all the particles, we shall observe the gas condensing in one

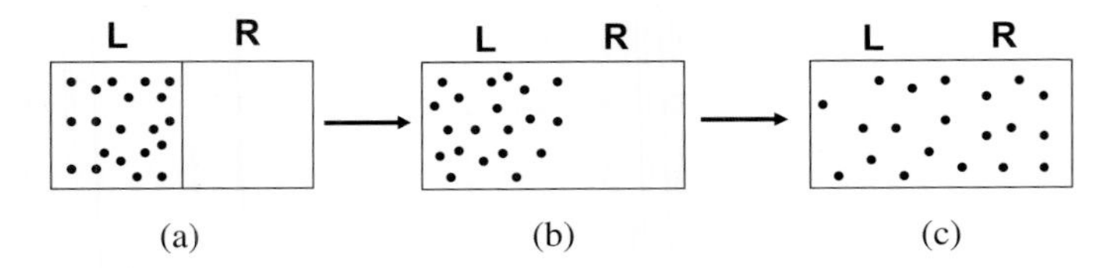

Fig. 2.24 The three stages in the expansion of a gas.

compartment, and the mixture of the two gases separating into two compartments.

Does this mean that the entropy will also decrease after the reversal of the velocities? Most of the discussions about the association of the *arrow of time* with the Second Law overlook the difference between the two questions:

(1) Will the reversal of all velocities bring the system to its initial microscopic state?
(2) Will the reversal of all velocities cause a decrease in the entropy?

Figure 2.24 shows three stages of the system: (a) the initial state, (b) a short time after removal of the partition, and (c) after a long time when the system reaches the equilibrium state.

Regarding the first question, the answer has already been provided above. The system will return to its initial state. One can ask how long it will take to get back to the initial state, but in principle we can be sure that it will visit the initial state [state (a), but of course without the partition].

Thus, while we have no difficulty in answering the first question in the affirmative, the answer to the second depends on how we define the *macroscopic equilibrium* state.

Strictly, the macroscopic equilibrium is defined as a state for which *all* the microstates (Ω) consistent with the macroscopic description of the system — say, (E, V, N) — are accessible. Within this definition, the microstate (a) (but without the partition) in Figure 2.24 belongs to Ω. This and similar microstates are rare events, but they are *included in the totality* of the microstates which we count in Ω. Therefore, when we reverse the velocities of all particles the system will continue to explore all the microscopic states, and while doing so it will also visit states in which all particles are in one compartment [e.g. (a) in the figure, but without the partition]. In this definition of equilibrium, the macrostate of the system is at equilibrium even after the reversal of the velocities. Hence, the entropy of the system is well-defined and is equal to the entropy of the system before the reversal of the velocities. In other words, the entropy does not change.

On the other hand, we might choose to *define* the macroscopic equilibrium states in terms of all the macroscopic states in the neighborhood of the microstate having the maximum probability. (See Subsection 2.4.1.) This group of states, denoted as Ω', has a probability of almost 1. Most of the time, the system at equilibrium will be visiting microstates in Ω'. Any microstate out of Ω' [e.g. (a) in Figure 2.24, but without the partition] will be considered a *nonequilibrium state*, in which case the entropy is not well-defined, and we cannot claim that it goes either upward or downward.

Thus, if the entropy is defined only for equilibrium states, then it is not a function of time. On the other hand, for nonequilibrium states the entropy is not definable, and hence we cannot say anything about the change in entropy with time.

It is here that the distinction between entropy and SMI is most crucial. SMI, unlike entropy, is definable for *any distribution*. Hence, it is also meaningful to talk about the SMI of a system in nonequilibrium states. When we reverse the velocities of all the particles the SMI associated with the *locations* of the particles might change, provided that there is a well-defined distribution of the locations of the particles. On the other hand, if we agree that the equilibrium states include all states Ω, then the entropy before the reversal of the velocities was $S(E, 2V, N)$, and after reversal it remains as such, i.e. $S(E, 2V, N)$. However, if we consider only the states Ω' as constituting the equilibrium state, then we can say that the entropy of the system started at $S(E, 2V, N)$ at the moment of the velocity reversal. Then there is a period of time when we do not know what the entropy is. After a longer period of time the system will reach again the equilibrium having the entropy $S(E, 2V, N)$.

As we have pointed out earlier, any definition of equilibrium is circular. Equilibrium is defined as a state for which the thermodynamic relationships apply. On the other hand, the thermodynamic relationship, in particular the entropy function, applies to equilibrium states. Since the equilibrium state by definition does not change with time, the entropy of the equilibrium state does not change with time either.

We have discussed the process of reversal from (c) to (a) in Figure 2.24. But the independence of the entropy on time is also valid for the process of expansion, i.e. from (a) to (c). Recall the evolution of the 20Q games in Section 1.14. There we saw that the speed at which the SMI changed depended on how vigorously we shook the system of marbles. Here, the speed of approaching the equilibrium state depends on the temperature (or, equivalently, the average speed of the molecules).

In the expansion of the gas from volume V to $2V$, we start with an initial equilibrium state defined by (E, V, N) for which the entropy is $S(E, V, N)$. After the expansion has occurred we have a new equilibrium state and a new value of the entropy $S(E, 2V, N)$, which is higher than the value $S(E, V, N)$. In this process the value of the entropy did increase from the initial value of $S(E, V, N)$ to the final value $S(E, 2V, N)$, in some period of time — say, Δt. However, there exists no *function* $S(t)$ that describes the dependence of entropy on time. In the particular example of the expansion process, the time interval Δt, or the "speed" of the expansion process, could depend on the size of the opening between the two compartments. The smaller the opening is, the longer it will take for the system to get from the initial equilibrium state to the final equilibrium state, but the change in the value of the entropy ΔS will be the same: $R \ln 2$. The speed could also depend on the temperature, but again ΔS will be the same. Similarly, in the heat transfer from a body of high temperature to a body of low temperature, the entropy change of the entire system is fixed by the initial and final equilibrium states. On the other hand, the time it takes to proceed from one equilibrium state to the other would depend on the thermal conductivity of the partition separating the two bodies.

Next, we might ask: At what time does the entropy change from $S(E, V, N)$ to $S(E, 2V, N)$ in the expansion experiment?

At the very moment when we remove the partition in the expansion process, the system is not an equilibrium state, and therefore it is meaningless to talk about the entropy of the system at that moment. The variation of the entropy with time will look like the one in Figure 2.25(a), where t_0 is the time we removed the constraint (here the partition), and t_1 is the time the system reached a new equilibrium state.

We can do the expansion processes in very small steps, each time opening a small window on the partition for a short period

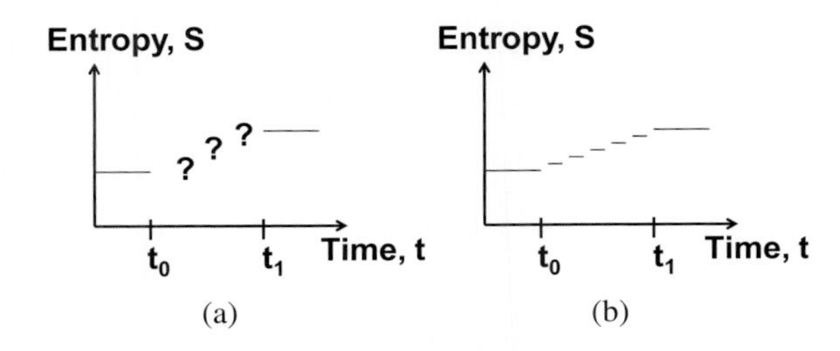

Fig. 2.25 Stepwise changes in the entropy in the expansion gas.

of time, then closing it and waiting until equilibrium is reached, then opening it again, and so on. In this case the entropy will change with time as in Figure 2.25(b). Note, however, that the stepwise functional dependence of S on t is not due to the existence of the function $S(t)$, but because we chose to open and close the window in an arbitrary sequence of times. Thus, it is reasonable to conclude that the entropy changes at the very moment we remove the partition, or at each moment we open the small window.

Whatever one concludes regarding the change in entropy in this and similar processes, one still cannot conclude that *the entropy of the universe always increases with time*.

Saying that the entropy of the universe always increases and reaches a maximum is a logical extrapolation from what we observe in well-defined thermodynamic systems. But this extrapolation is a wild one, and as long as one cannot provide a way of measuring the entropy of the entire universe or calculating it based on the physical parameters of the universe, it is meaningless to talk about the entropy of the universe, or the fate of universe. (We shall further discuss the "entropy of the universe" in Chapter 4.)

As we have seen in this section, neither entropy nor the Second Law is associated with the arrow of time. Some authors even go further and "identify" *time* with *entropy*. Such an identification raises the following question: *What is time?*

This question in my opinion, is not a scientific question but a philosophical one. Carroll[37] devotes most of his book to the "theory of time."

Here are some quotations from Carroll:

> The most mysterious thing about time is that it has direction: the past is different from the future. (*Page* 2)

In my opinion this is not a mystery at all. Indeed, time has direction. The past is different from the future. That is how time is! Is it a mystery that matter has positive mass or positive volume?

> The question of our low-entropy beginning takes on a different cast: not "why did the universe start out with such a low entropy?" but rather "why did our part of the universe pass through a period of such low entropy?" (*Page* 4)

Both of these questions are meaningless as long as there exists no way to *define* the "entropy of the universe" — neither in the beginning nor in the present universe. No one knows what the entropy of the universe was, is or will be. Asking "why did our part of the universe pass through a period of low entropy?" is as meaningful as asking why the universe started with such a *low love*, *low beauty*, or whatever you want.

> Entropy increases throughout, as the system progresses from order to disorder. Whenever we disturb the universe, we tend to increase its entropy. (*Page* 30)

Again, this is a doubly meaningless statement about the "entropy" of the universe, and its association with disorder. I have no idea what "disturbing" the universe means. When a child is born, is the universe disturbed? Does the entropy of the universe increase or decrease at the moment of birth?

> The beginning of our observable universe, the hot dense state known as the Big Bang, had a very low entropy. (*Page* 32)

This is a very common, yet meaningless, statement. We cannot *define* the entropy of the universe at the Big Bang, let alone claim that the universe "had a very low entropy."

Here is another fancy, impressive but unfounded, statement:

> The arrow of time manifests itself in many ways — our bodies change as we get older, we remember the past but not the future, effects always follow causes. It turns out that all of these phenomena can be traced back to the Second Law. Entropy quite literally makes life possible. (*Page* 38)

In my view these phenomena have nothing to do with the Second Law. The claim that *all* these phenomena *can* be traced back to the Second Law is unwarranted at best. No one has ever showed any connection between what we remember and what we forget with the Second Law. I doubt that there is any connection. The claim that "entropy makes life possible" is a grandiose — yet vacuous — statement. It is unfortunate that such statements have impressed so many readers of Carroll's book. We shall further discuss entropy and life in Chapter 3.

Pause and think

Here are a few more statements from Carroll's book (I urge the reader to read carefully each of these statements, and examine their meanings in light of what we know about information and entropy):

> Ultimately, the reason why we can form a reliable memory of the past is because the entropy was lower then. (*Page* 40)
>
> That one extra bit of information, known simply as the "Past Hypothesis," gives us an enormous leverage when it comes to reconstructing the past from the present. (*Page* 43)
>
> The ultimate explanation of the arrow of time as it manifests itself in our kitchen and laboratories and memories relies crucially on the very low entropy of the early universe. (*Page* 45)

This is a small sample of statements taken from Carroll's book. The book is extremely repetitious and replete with statements about the connection between entropy and the universe, black holes, life phenomena and life itself. Although the author occasionally admits the "*hypothetical*" nature of some of the statements, it does not seem to me that he questions the *meaning* of his statements. We shall discuss further some of the issues associating entropy with life and the universe in Chapters 3 and 4.

2.6 Does the Maxwell Demon Defeat Entropy or Is It Defeated by Entropy?

Let us start with a quotation from Maxwell himself [James Clerk Maxwell (1867), "The Theory of Heat"]:

> One of the best established facts in thermodynamics is that it is impossible in a system enclosed in an envelope which permits neither change of volume nor passage of heat, and in which both

the temperature and the pressure are everywhere are the same, to produce any inequality of temperature or of pressure without the expenditure of work. This is the Second Law of thermodynamics, and it is undoubtedly true as long as we can deal with bodies only in mass, and have no power of perceiving or handling the separate molecules of which they are made up. But if we can conceive **a being whose facilities are so sharpened** that he can follow every molecule in its course, such a being, whose attributes are still as essentially finite as our own, would be able to do what is at present impossible to us. For we have seen that the molecules in a vessel full of air at uniform temperature are moving with velocities by no means uniform, though the mean velocity of any great number of them, arbitrarily selected, is almost exactly uniform.

Now let us suppose that such a vessel is divided into two portions, A and B, by a division in which there is a small hole, and that a being, who can see the individual molecules, opens and closes this hole, so as to allow only the swifter molecules to pass from A to B, and only the slower ones to pass from B to A. He will thus, without expenditure of work, raise the temperature of B and lower that of A, in contradiction to the Second Law of thermodynamics.

This is only one of the instances in which conclusions which we have drawn from our experience of bodies consisting of an immense number of molecules may be found not to be applicable to the more delicate observations and experiments which we may suppose made by one who can perceive and handle the individual molecules which we deal with only in large masses.

In dealing with masses of matter, while we do not perceive the individual molecules, we are compelled to adopt what I have described as the statistical method of calculation, and to abandon the strict dynamical method, in which we follow every motion by the calculus.

As is evident from the above quotation, Maxwell did not "invent" the "Maxwell demon." He referred to a "being" who

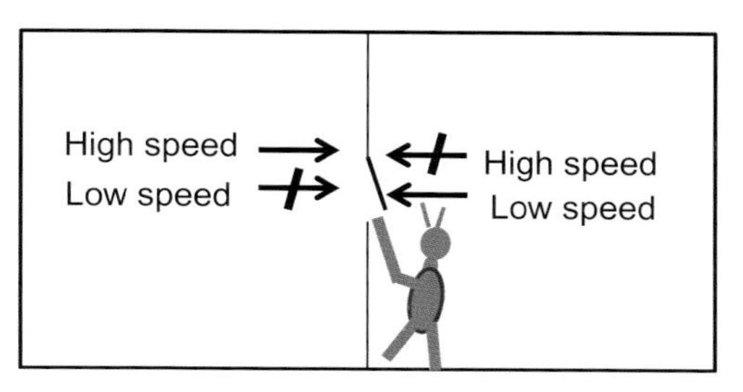

Fig. 2.26 Maxwell's demon allowing high speed molecules to enter the right compartment, and low speed molecules the left. The net effect is heating the right compartment and cooling the left compartment.

can "see individual molecules." It was only later that this "being" was referred to as an "intelligent being," and then as a demon.

The original idea of Maxwell was to envisage a "being" who can see the individual molecules, measure their speeds, and discriminate between high-speed and low-speed molecules.

Suppose that one starts with a system at a uniform temperature (Figure 2.26), and suppose that there exists a being (or a demon, or whatever you want to call it) which can distinguish between fast-moving molecules and slow-moving molecules. Such a demon could control a tiny door between the two compartments in Figure 2.26. For instance, a fast-moving molecule approaching the door from the left side will be allowed to pass through the door. A slow-moving molecule approaching the door from the left side will be denied passage through the door. Similarly, the demon could allow slow-moving molecules approaching from the right to pass, but deny the passage to fast-moving molecules approaching from the right side.

Clearly, if we start the experiment with a uniform temperature in the two compartments, after some time the average speed of molecules in the right compartment will increase, and at the same time the average speed of the molecules in the left

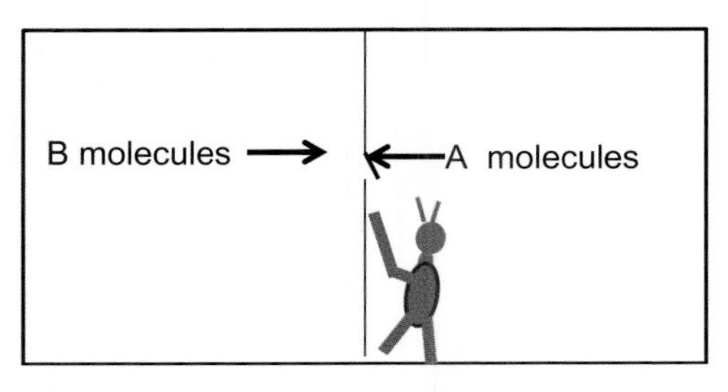

Fig. 2.27 Maxwell's demixing demon, allowing A molecules to pass from right to left and B molecules to pass from left to right.

compartment will decrease. This effect is equivalent to raising the temperature of the right compartment and lowering the temperature of the left compartment.

Maxwell, as well as many others, concluded that if such a being existed it would have contradicted the Second Law of Thermodynamics.

Before we examine the question of whether or not the demon violates the Second Law, let us describe the two "simpler" demons.

One can think of a demon which can distinguish between two molecules A and B (Figure 2.27). This demon can allow the passage of A molecules to the left compartment and B molecules to the right compartment. Let us call it the *demixing demon.* Starting with equal numbers of A and B molecules in each compartment, after some time one would get pure A in the left compartment and pure B in the right compartment, in effect achieving separation of the two gases in apparent violation of the Second Law.

One can think of a more primitive (perhaps less intelligent?) demon operating between the two compartments separating two quantities of the same molecules A (Figure 2.28). Its task is not to measure the speed of the molecules, or even to distinguish

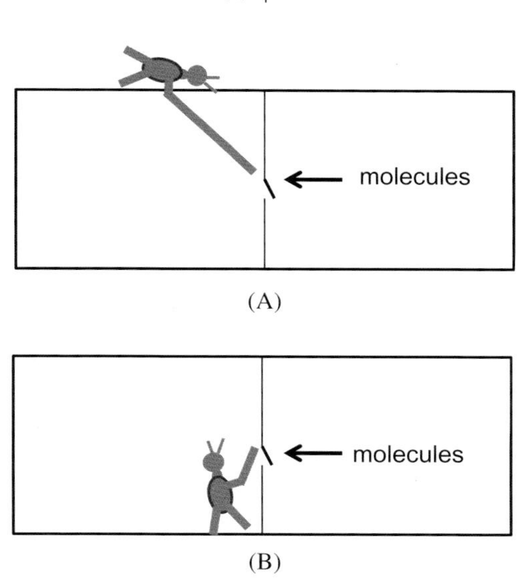

Fig. 2.28 Maxwell's condensing demon. It allows gas molecules to cross in one direction only — from right to left. (A) The demon is not part of the system — it operates on the gas molecules outside the system; (B) the demon is part of the system.

between A and B, but much simpler. We start with $2N$ molecules at equilibrium in the system of volume $2V$, i.e. about N molecules in each compartment. We put a partition with a little door at the center of the system (Figure 2.28). The task of the demon is the following:

The demon will allow every particle knocking on the door from the right side to pass. On the other hand, particles knocking from the left side will not be allowed to pass. Clearly, after some time all the particles which were initially in the right compartment will be transferred to the left compartment, and the entropy of the gas will change by the amount $-Nk_B \ln 2$. This is simply a reversal of the expansion of a gas from V to $2V$. Let us call this demon the *condensing demon*. As far as the entropy of the gas is concerned, this demon has achieved an apparent violation of the Second Law.

Let us discuss in more detail this more "primitive" demon. Our conclusion will be valid for any other demons.

Does the demon violate the Second Law? Of course not! The Second Law states that for a spontaneous process in an *isolated* system, the entropy increases. Before we analyze whether the demon violates the Second Law, we must first decide whether the demon is part of the system or not.

Focusing on the simplest process in Figure 2.28, we consider the following two cases:

A. The demon is not part of the system [Figure 2.28(A)]

If the demon is *not* part of the system, and since we have started with an equilibrium state (all the $2N$ particles in $2V$), and ended with a new equilibrium state (all the N particles from the right compartment were transferred to the left compartment), the entropy of the *gas* obviously decreased in this case by the amount $-k_B N \ln 2$. However, this was *not a spontaneous process in an isolated system*. Thus, there is no violation of the Second Law. Any agent from outside the system can cause a reduction in the entropy of the gas. You do not need any unspecified and ill-defined demon to do this job. I can do this, and you can do it as well. Thus, in this case there is no violation of the Second Law.

B. The demon is part of the system [Figure 2.28(B)]

This case is more complicated. Suppose that the gas molecules plus the demon are considered to be a single isolated system. In this case the system is *isolated*, but it is not in an equilibrium state — neither in the initial nor in the final state. Therefore, the entropy of such a system is not defined. We can still talk about the SMI of the system. Again, we have to distinguish between two cases:

(i) If the gas and the demon are *independent* in the sense discussed in Chapter 1, then the SMI of the entire system is the *sum* of the SMI of the gas and of the demon. In this case, the SMI of the gas had *decreased* by the amount $N \log_2 2$ (due to the change in the locational SMI). On the other hand, we cannot calculate the change in the SMI of the demon simply because its states are not defined. In fact, if the gas and the demon are independent, the demon would not "know" what to do. Therefore, we must consider the more "realistic" case — that what goes on in the demon and in the gas are dependent events.
(ii) If the gas and the demon interact, i.e. they are *not independent*, then the SMI of the combined system is not the sum of the SMI of the gas and the SMI of the demon. Here, we must take into account the change in the locational SMI of the gas, the change in the SMI of the state of the demon, and the *mutual information* due to the interaction between the demon and the gas molecules. Clearly, we can calculate only the change in the SMI of the gas, but we have no idea how to calculate the change in the SMI of the demon, or the mutual information due to the interaction between the demon and the gas.

Thus, we see that neither the SMI nor the entropy change of the system can be calculated in the case where the demon is part of the system.

Many scientists discussing the Maxwell demon have argued that the demon must use *information* to lower the entropy of the gas. The reasoning is based on the accepted idea that the entropy of the entire universe must increase in the process. Since the entropy of the gas decreases, it follows that the entropy of the demon must have increased. But how can one measure the change in the entropy of the demon? One way out of this dilemma is to say that the demon has *used information* to reduce the entropy of the gas. This kind of argument sounds very appealing, particularly in the recent literature where the identification of entropy with *information* is taken for granted.

Let us repeat the argument in more detail:

We can calculate the entropy of the gas before and after the process, and we find that the difference is negative (say, for 1 mole of gas: $-R \ln 2$). Next, we make the following statements:

(i) The Second Law states that the entropy of the entire universe cannot decrease. Therefore, one concludes that the entropy of the universe must have increased by more than the amount $R \ln 2$.

(ii) If only the gas and the demon were involved in this process, then the entropy of the demon must have increased by more than the amount $R \ln 2$.

(iii) Finally, if one identifies (erroneously) the change in entropy as a change in the demon's information, then one concludes that the demon must have used *information* (or neg-entropy) to separate the gases or to condense the gas into one compartment.

All of these statements are fallacious. The entropy of the gas did change by the amount $-R \ln 2$. However, we do not know anything about the entropy of the universe. The Second Law, as formulated in terms of "the entropy of the universe cannot decrease," simply does not apply. The entropy of the universe is not even defined, and the same is true of the entropy of the demon.

We shall further discuss the question of the entropy of the universe in Chapter 4.

There is a huge literature on the Maxwell demon. Two books have been devoted to the demon [Leff and Rex (2007, 2010)]. We shall present here only two quotations from this vast literature.

Brillouin, who pioneered the application of information theory to thermodynamics, writes[39]:

> The origin of our modern ideas about entropy and information can be found in an old paper by Szilard (1929), who did the pioneer work but was not well understood at that time. The connection between entropy and information was rediscovered by Shannon, but he defined entropy with a sign just opposite to that of the standard thermodynamic definition. Hence, what Shannon calls entropy of information actually represents neg-entropy.

In my view, this is a gross exaggeration. As we have seen earlier in this chapter, entropy is related to SMI, *not to information*. Furthermore, Szilard did not *discover* information theory. He only *hypothesized* that since the entropy of the gas is lowered, the entropy of the demon must have increased, and that increase is referred to (by Brillouin) as *neg-entropy*, or information. We have already commented on the unfortunate usage of the term "entropy" for SMI in Chapter 1. It is even more unfortunate to refer to *information* as neg-entropy. We shall further discuss neg-entropy in connection with life in Chapter 3.

Landsberg wrote the demon's obituary, saying, "Maxwell's demon died at the age 62 (when a paper by Leo Szilard appeared), but it continues to haunt the castles of physics as a restless and lovable poltergeist." Personally, I was never haunted by Maxwell's demon. A real demon violating the Second Law never existed — was never born and never died. Figuratively, of course, the demon was born in Maxwell's mind, but whether or not it violates the Second Law is not an answerable question in physics.

Here is another quotation associating Maxwell's demon with information[40]:

> Sadly, Boltzmann didn't live to help overcome Maxwell's demon; he succumbed to the struggle with his own. Boltzmann was often prickly and antisocial, and his novel ideas made him powerful enemies. On top of that, he was prone to bouts of depression and exhaustion. He hanged himself never knowing the secret that would lead physics to victory over Maxwell's demon. Ironically, the formula at the center of that victory was inscribed on Boltzmann's grave: $S = k \log W$, the formula for the entropy of a container full of gas. But it wasn't entropy that defeated Maxwell's demon. It was information.

It is strange that Seife, who in most of the book identifies entropy with information, and the Second Law with the laws of information, claims that it is *information* rather than *entropy* which defeated the demon.

In my view such statements, impressive as they may sound, are *meaningless.* Neither entropy nor information defeated the demon, nor any other agent that can separate two gases, or heat one part of a system and cool the second part.

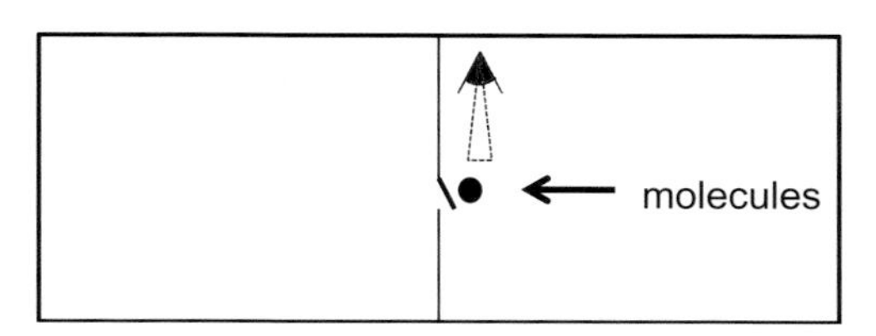

Fig. 2.29 An electronic eye which, when detecting a molecule approaching the door from the right, opens the door and lets the molecule pass through.

2.6.1 *An unintelligent "demon"*

Consider the following simplified versions of the Maxwell demon. Suppose that the system is as in Figure 2.28. We start with a system of ideal gas in two compartments with equal numbers of particles — say, N.

We now replace the "intelligent" demon by a simple device, as shown in Figure 2.29. Instead of an intelligent being, we have an electronic eye, which, when detecting a molecule approaching the door from the right, opens the door to let the molecule pass through. Let us call this device the Robot, and place it on one side of the partition — say, the right side.

Starting with the initial condition as in the case of Figure 2.28, i.e. N particles on each side of the system, we will find after some time that all, or nearly all, the molecules will pass from the right to the left compartment.

Note that initially the gas is in an equilibrium state, and the Robot is "activated" at some time, and we observe what happens to the gas. The entropy of the *gas* will be reduced by $k_B N \ln 2$.

Does this process contradict the Second Law?

The answer is NO! If we consider the gas alone as our system, then it is not an *isolated* system, and the Second Law does not apply. If we consider the gas and the Robot as our *isolated* system, then the process occurs in an isolated system (the gas and the Robot). However, in this case we do not know what the initial state of the system is, and whether or not it is an equilibrium state. Therefore, the entropy of this (combined) system is not defined at the initial state, and we cannot say anything about the change in the entropy of the system in the process. It would also be meaningless to claim that the Robot is an "intelligent" being, or that it has used "information" or "neg-entropy" to reduce the entropy of the gas.

It is instructive to further explore several different initial states:

(i) Suppose that we start with all the $2N$ particles in the *left* compartment. Clearly, nothing will happen to the gas in this case, as the Robot will not "see" any molecule approaching the door from the right side. Therefore, we can conclude that the Robot did not cause any change in the entropy of the gas.

(ii) Initially, we have N particles in each compartment. In this case, the Robot will let all the N particles pass from the right to the left compartment. The entropy of the gas will *decrease* (due to the change in the locational SMI of the gas molecules). As we have noted earlier, the Robot *did* change the entropy of the gas, but we cannot tell the entropy change *in* the Robot.

(iii) Now, we start with all the $2N$ particles in the right compartment. If we wait long enough we shall see that all the $2N$ particles have moved from the right to the left compartment. Clearly, the entropy of the gas *does not* change in this process. The Robot certainly *did* something. In fact, it did twice as much as it did in case (ii) and infinitely more than in case (i). Yet, in all cases, if we view the Robot as being part of the system, we cannot tell the change in the entropy of the system in any of the above processes simply because the initial state of the combined system is not well-defined thermodynamically.

As a more challenging case, suppose that the same Robot we have employed above in Figure 2.29 is working only within a limited period of time — say, it operates on batteries which can last for only a few minutes. Once the batteries are drained, the Robot stops functioning. We start with, say, 10^{23} in the right compartment. Assume that on average about 10^{22} particles pass the eyes of the Robot in 1 min. Thus, if the batteries of the Robot last for 1 min, 10^{22}, or one tenth of all the particles, will pass from the right to the left compartment. Now we can calculate the change in the entropy of the gas for each *lifespan* of the batteries, *tb*. If the batteries of the Robot last for *tb* min; ($t = 0, 1, 2, \ldots, 10$ min), we can plot the entropy change of the gas as a function of *tb*. Clearly, the longer the lifespan of the batteries, the "harder" the Robot's works. Yet, the entropy of the gas initially *increases* and then *decreases* when *tb* changes from 0 to 10 min. (At $tb = 0, \Delta S = 0$, for $tb = 1, 2, 3, 4, \Delta S > 0$, it reaches a maximum at $tb = 5$, then for $tb > 0, \Delta S < 0$.) In any of these cases, we cannot tell the entropy change in the Robot, and certainly we cannot claim that it uses *information* to achieve these changes in the entropy of the gas.

Pause and think

Consider the three cases in Figure 2.30.

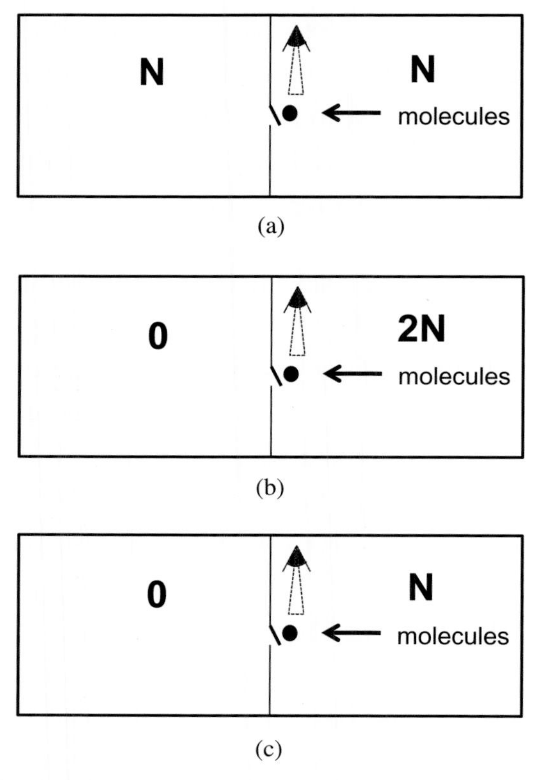

Fig. 2.30 Three experiments with the electronic eye: (a) initially N molecules in each compartment; (b) initially $2N$ molecules in the right compartment and none in the left; (c) initially N molecules in the right compartment and none in the left. The electronic eye operates until N molecules are transferred from right to left.

We have three initial states of the gas:

(a) N molecules in each compartment;
(b) $2N$ molecules in the right compartment;
(c) N molecules in the right compartment;

In each of these cases the Robot is allowed to "work" for t seconds, enough time to let N particles move from the right to the left compartment.

In case (a) the entropy of the gas *decreases* by the amount $k_B N \ln 2$. In case (b) it *increases* by the amount $k_B N \ln 2$. In case (c) the entropy of the gas does not change.

What is the change in the entropy of the Robot?

What is the change in the entropy of the universe?

Did the Robot use *information* to do the *same* job, and how much information did it use in each case?

My answers are given in Note 41.

Finally, let us discuss an even more primitive demon which can also achieve a reduction in the entropy of the gas, and for which we *can* calculate the entropy change of the demon. In this case the demon cannot be said to be an intelligent being, or to use *information* to reduce the entropy of the gas.

Our demon in this example is a piece of ice at a very low temperature — say, 1 K. The system is described in Figure 2.31. The gas is initially as in Figure 2.31(a), with N water molecules in each compartment. In addition, the very cold ice cube is placed next to the left wall of the left compartment, and it is initially insulated. Thus, the initial states of the gas and the ice

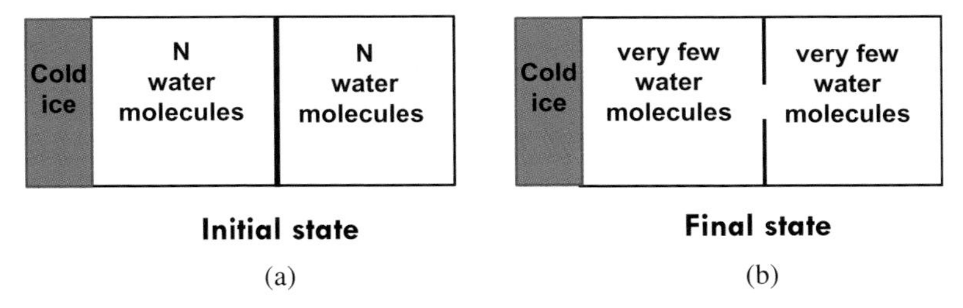

Fig. 2.31 Two compartments, each initially having N water molecules. On the left hand side we have a very cold cube of ice. Initially, the two compartments are separated by a partition, and the cold ice cube is insulated from the gas.

are in equilibrium states, and each has a well-defined entropy. The entire system is isolated.

Now, we remove the partition between the two compartments, as well as between the gas and the ice [Figure 2.31(b)]. What will happen in this isolated system? We assume that the ice is very cold and that it has many more molecules than the gas in the two compartments.

After some time *almost* all the water molecules in the two compartments will be condensed and stick to the surface of the ice. As in the case of Figure 2.28, nearly all the water molecules in the right compartment will move to the left compartment, reducing the entropy of the *gas.* Furthermore, there will be an additional decrease in the entropy of the gas molecules which will be condensed on the surface of the ice. In the final equilibrium state, only a very small number of water molecules will be left wandering in the two compartments (the vapor pressure of the ice at such a low temperature is negligibly small).

Clearly, our primitive demon (the ice) has achieved an impressive *lowering* of the entropy of the water molecules in the gaseous phase which were initially in the two compartments. Did this demon use information to achieve that? Certainly not. The ice *did not have* any information on the gas molecules and did not *use* any information to achieve that task. Fortunately, and unlike the previous examples, here we have clear initial and final equilibrium states of the entire system (the gas and the ice). We found that the entropy of the gas decreased, both because of the flow from the right to the left compartment and because of the cooling of the gas molecules. We can also calculate the change in the entropy of the ice, and we will find that its entropy has increased (when gas molecules condense on the surface of the ice, they release their kinetic energy, which causes an increase in the temperature of the ice, and hence also the entropy of the ice).

To conclude, in this example we have a *spontaneous* process in an *isolated* system (including the gas and the ice) where the entropy of the gas *decreases* but the entropy of the entire system has increased. Unlike in the process involving the Maxwell demon, here we can calculate the change in the entropy of both the gas and the solid ice. We will find that the entropy of the combined system has increased in accordance with the Second Law. Note that in this case we applied the Second Law to the isolated system, and not to the entire universe. Also, our "unintelligent" Robot did not use any "information" to achieve the change in the entropy of the gas.

The conclusion of this section is very important. For over a hundred years scientists were daunted by the Maxwell demon. Much ink was wasted on trying to solve the enigma — did the demon violate the Second Law? In my opinion, much of this effort was in vain. As long as we cannot define the entropy of the demon, or the entropy of the universe, there is no point in speculating over the question of whether or not the demon violated the Second Law. It is also meaningless to claim that the demon was defeated by either entropy or information. We will further discuss the relevance of entropy and information to life and the universe in Chapters 3 and 4.

2.7 The Change in Entropy for Some Simple Processes

In the previous sections we discussed the concept of entropy, its derivation from purely thermodynamics, and then from molecular considerations, and finally from SMI. In this section we present a few very simple examples of *processes* for which we can calculate the changes in the entropy. We then compare the results with those we calculated in Section 1.11 for processes involving marbles. Some of the processes are spontaneous, and some are not. In all of the examples discussed below we calculate the change in entropy from the *entropy function* of the ideal gas. This procedure demonstrates that, unlike in thermodynamics, we do not need to devise any "reversible" path in order to calculate the entropy difference. Instead, we simply calculate the difference in the entropy function between the two states. The comparison between the changes in the entropy and the changes in the SMI for the similar processes is very instructive.

2.7.1 *Expansion of an ideal gas*

This is the simplest conceivable process. We have calculated the change in the SMI for a system of N marbles in Subsection 1.11.1. Here, we have an ideal gas (we consider only a classical system of N noninteracting particles) in a volume V and having a fixed energy E (Figure 2.32). Such a system as a whole is said to be isolated.[42]

We remove the partition between the two compartments and observe that the system will always expand to occupy the entire new volume, $2V$. Here, we ask: What is the change in the entropy

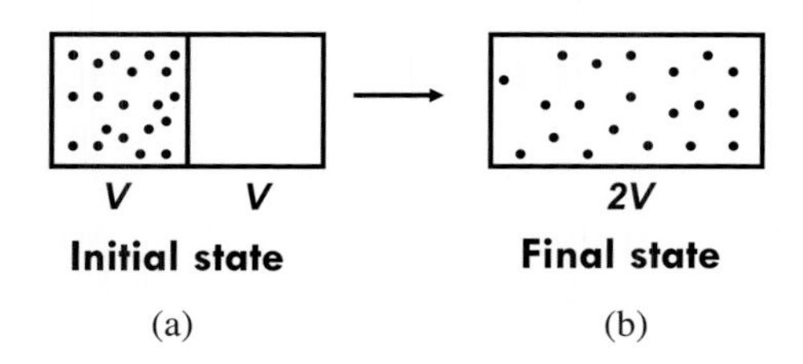

Fig. 2.32 Expansion of gas from volume V to $2V$.

of the system after we remove the partition between the two compartments?

As we have noted earlier, thermodynamics has developed an ingenious method for calculating the changes in entropy for such a process. Recognizing that the entropy is a state function, i.e. for each *state* characterized by E, V, N the entropy is well-defined, we can write the total differential of the entropy, and then by integrating from V to $2V$ we obtain the result[43] $k_B N \ln 2$.

However, if we already have the equation of the entropy as a function of the relevant variables, we do not need to resort to a *quasistatic* process. We recognize that the entropy is a state function, and therefore it has a value for any *state* of the system. In particular, for the difference in entropy between the two states in Figure 2.32 we can use the equation of the entropy function to calculate the difference in entropy,

$$\Delta S(\text{expansion from } V \to 2V) = k_B N \ln 2, \qquad (1)$$

which is the same as what one obtains in thermodynamics (see Note 43).

Exercise

Calculate the change in entropy for the following three processes:

(i) The change in volume from V to $5V$;
(ii) The change in volume from V to $1000V$;
(iii) The change in volume when we remove the partition and let the gas escape to occupy the entire universe.

The first two exercises are trivial. The third needs some thinking; do not rush to extrapolate. We will come back to this process in Chapter 4.

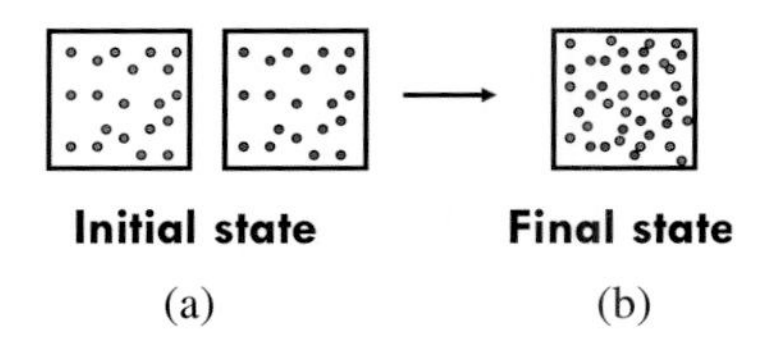

Fig. 2.33 A process of "pure" mixing.

2.7.2 *The pure mixing process of ideal gases*

In Subsection 1.11.2 we calculated the change in the SMI for the process of mixing marbles. Here, we do the same for two ideal gases — say, argon and krypton. The initial and final states are shown in Figure 2.33. Note that this process is not a spontaneous one. However, we can perform the process by means of semipermeable partitions, and find the change in entropy in it by thermodynamic means only. [See Ben-Naim (2008).]

Alternately, we can use the entropy function for a mixture of two ideal gases and calculate the change in entropy simply by knowing the initial and the final state (without resorting to devising a quasistatic process leading from the initial to the final state).

Initially we have N argon atoms in a volume V and having energy E, and N krypton atoms in another volume V and energy E. In the final state we have a mixture of N argon atoms and N krypton atoms in the same volume V, and total energy $2E$. (Note that we could use the temperature instead of the energy for this example. The temperature is simply proportional to the average kinetic energy of the particles in the system.)

The entropy change in this process is

$$\Delta S(\text{pure mixing}) = 0. \tag{2}$$

In this example, we also have a simple molecular interpretation for the zero change in entropy. In both the initial and the final state, each molecule is confined to volume V. Thus, there is no change in the *locational information* of all the particles in the system. Also, the average kinetic energy per particle is unchanged in the process. This is equivalent to the statement that the temperature does not change in the process. Therefore, the momentum (or velocity) distribution does not change either. This *explains* why there is no change in entropy in this process. Note that the result we obtain here is exactly the same as for the change in SMI in the process of mixing two different colored marbles (see Subsection 1.11.2). In Chapter 1, we were concerned only with the locational information. In the marble experiment we could use any number of marbles. Here, we use an enormous number of particles. We also assume that both the initial and the final state are equilibrium states. Note also that we see *mixing* in this process. But the mixing has no effect on the entropy change. [For more details, see Ben-Naim (2008, 2012).]

2.7.3 *The pure assimilation process of ideal gases*

This is an important process. The reader is urged to review the experiment on "pure assimilation" with marbles in Subsection 1.11.3. As we will soon see, this process gives a *different* result from the case of marbles. It also reveals the important difference between the process of assimilation in a thermodynamic system and in a marble system with a small number of marbles.

The process is described in Figure 2.34. It is essentially the same process as in Figure 2.33, but the two gases are *identical* — say, argon in both of the compartments.

The calculation of ΔS by thermodynamics means is quite simple. We can start from the left hand side (a), bring the two

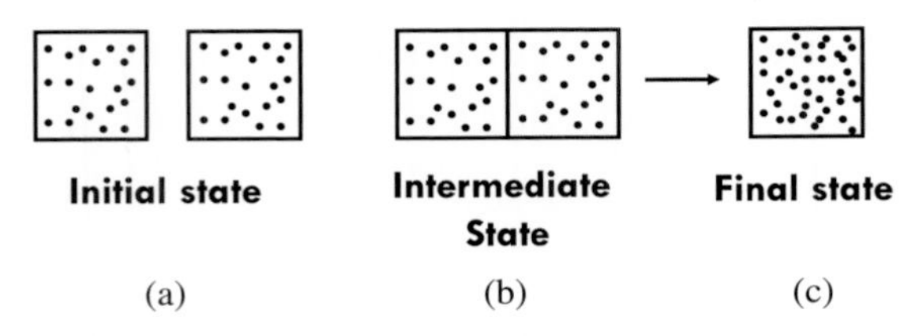

Fig. 2.34 A process of pure assimilation: (a) the initial state, (b) an intermediate state, (c) the final state.

compartments into contact (b), remove the partition and then compress the system from $2V$ to V (c). The change in entropy can be calculated from thermodynamics. The result is as in the expansion experiment, but with a negative sign, i.e.

$$\Delta S(\text{pure assimilation process}) = -k_B 2N \ln 2. \tag{3}$$

Note that within *thermodynamics* the process in Figure 2.34 is the same as in the case of *compression* of an ideal gas, i.e. reversing the expansion process described. In the figure we added an intermediate step between the initial and the final state.

The molecular interpretation of this process, as well as the interpretation of the change in entropy, is very different from the thermodynamic interpretation.

We can calculate the change in entropy in the process by subtracting the entropy function of the system on the left hand side (a) from the entropy of the system on the right hand side (c). The result is

$$S(\text{final}) - S(\text{initial}) = k_B \ln \frac{(N!)^2}{(2N)!}. \tag{4}$$

This result is very different from the thermodynamic result in the equation (3). However, quite surprisingly, it is the same result as that we obtained for the marble experiment in Subsection 1.11.3.

Is the result (4), which apparently differs from the thermodynamic result (3), correct?

To understand the difference between the results (3) and (4), we first note that the process depicted in Figure 2.34 is referred to as *pure assimilation*, for the following reason. First, each particle is confined to a volume V in both the initial and the final state. This means that there is no change in the *locational* SMI in this process. Also, the energy per particle does not change in the process. Since all the energy is due to the kinetic energy, this means that there is no change in the distribution of momenta either, and hence no change in the *momentum* SMI. The only change in going from the left (a) to the right (c) in Figure 2.34 is that on the left we have *two groups* of N *indistinguishable particles*. On the right we have one group of $2N$ indistinguishable particles. We call this process "*pure assimilation*" because here N particles are "mixed" with the *same kind* of particles (as Gibbs called such a process). We prefer not to use the term "mixed" here, because there is no conspicuous mixing. Instead, we say that N particles are *assimilated* into another group of N indistinguishable particles. Thus, the result (4) is the correct result. It is the same result as that we got for the change in SMI for the process of assimilation of marbles in Subsection 1.11.3. However, the result (4) is valid for N of the order of the Avogadro number. Thus, for very large N it is easy to see that the equation (4) tends to the result (3), i.e.

$$\Delta S(\text{pure assimilation process}) = k_B \ln\left[\frac{(N!)^2}{(2N)!}\right] \xrightarrow{N\to\infty} -k_B 2N \ln 2. \qquad (5)$$

This is the same as the result (3). Thus, the result (4) is correct for the change in the SMI for the process of assimilation for

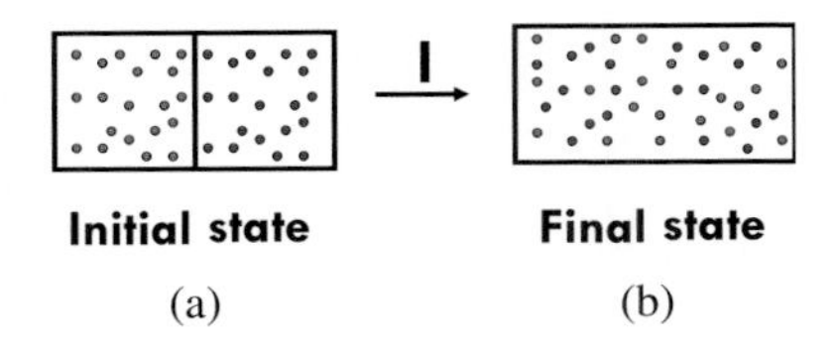

Fig. 2.35 A process of mixing and expansion.

any number of particles (or marbles). It tends to $-k_B 2N \ln 2$ only in the limit of a very large number of particles, in the thermodynamic limit.[44] It should be emphasized that this effect has its origin in quantum mechanics. It cannot be explained within classical thermodynamics.

2.7.4 *The process of mixing and expansion*

This process is a classical textbook example. It is usually referred to as a *mixing process* and the corresponding entropy change is referred to as the "entropy of mixing." As we will soon see, this process does involve mixing. However, the *mixing* in this process has no effect on the entropy change. The process is depicted in Figure 2.35. Compare it with the process discussed in Subsection 1.11.4. The initial state is the same as in Figure 2.33. We initially have N particles in a volume V and another N particles of a *different* kind in a volume V. The particles do not interact with each other so that we have two ideal gases. We remove the partition between the two compartments and observe *mixing*. Unlike the process in Figure 2.33, this process is spontaneous. It involves mixing of two kinds of particles. Again, thermodynamics teaches us how to calculate the change in entropy for this process. The result is

$$\Delta S(\text{process I in Figure 2.35}) = 2k_B N \ln 2. \qquad (6)$$

Thermodynamics does not tell us why the entropy increases in this process. Since we *see* mixing occurring in the process, it is natural to conclude that the positive change in entropy is due to the mixing. This is also consistent with the very common interpretation of entropy as a measure of *disorder*. Mixing is certainly conceived of as an increase in the disorder. Therefore, one concludes that mixing is a spontaneous irreversible process, and involves an increase in entropy. This conclusion appears in most textbooks on thermodynamics.[45]

As usual, thermodynamics does not reveal to us the molecular reasons for the change in the entropy. Gibbs himself was puzzled by the fact that the entropy of mixing as calculated in the equation (6) is *independent* of the *type* of molecules. Take argon and methane, propane and water, benzene and neon, etc. for as long as they are ideal gases; the entropy change is the same as in the equation (6). Why is it so? Gibbs did not answer this question. He only pointed out that when the two gases are "of the same kind" the entropy change is zero. We will soon see that the molecular interpretation of entropy does explain why the entropy of mixing is independent of the type of molecules, as well as why there is no entropy change when the two gases are of the same kind.

The molecular interpretation of the entropy change is very different from the thermodynamic one. Again, using the entropy function for the initial and the final state, we find that the entropy change in process I of Figure 2.35 is

$$\Delta S(\text{mixing and expansion}) = k_B N \ln \frac{2V}{V} + k_B N \ln \frac{2V}{V} = 2k_B N \ln 2. \quad (7)$$

This result is identical with the equation (6). However, it was calculated by using the entropy function. This calculation immediately reveals the *source* of the change in entropy. The gas on the right hand side was initially confined to volume V and in the final state it is confined to a new volume, $2V$. Therefore, the change in entropy due to the *expansion* of this gas is $k_B N \ln 2$. The same is true of the gas on the right hand side, which contributes the same change in entropy due to the expansion from V to $2V$. As you can see, the mixing that we *see* in this process does not contribute anything to the change in the entropy. This process is equivalent to the two processes of expansion. Viewing this process, which we have referred to as mixing and expansion, as purely an expansion immediately explains the independence of the *kind* of molecules which puzzled Gibbs. Clearly, the change in entropy for the expansion, as in the process of Figure 2.35, is due to the change in the *locational* information of each molecule in the system. This change is independent of the *kind* of molecules, or of the marbles. In the case of marbles we obtained $2N\log_2 2$, or one bit per particle. Here, we get the same result but in different units. The interpretation of ΔS in this process is simple — it is due to the change in the *locational* information of all the $2N$ particles. The mixing that we observe has no effect on the thermodynamics of this process. This removes one mystery; why ΔS in process I is independent of the kind of molecules as long as they are different. The second mystery is discussed in the next subsection.

2.7.5 *Assimilation with expansion*

Perhaps the simplest of all the processes we have discussed in this chapter is the process II shown in Figure 2.36. It is the simplest

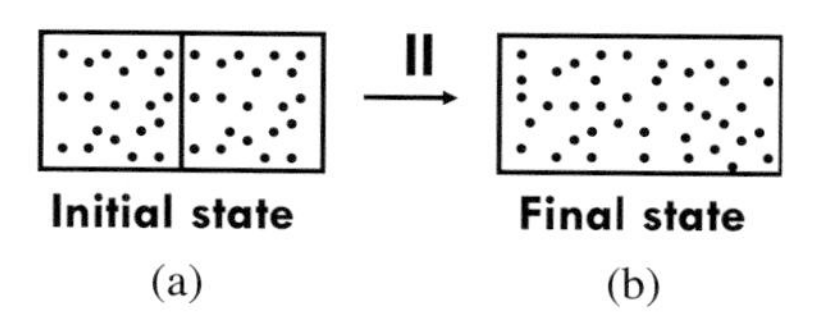

Fig. 2.36 A process of assimilation and expansion.

from the thermodynamic point of view, but far from simple from the molecular point of view. Process II is exactly the same as process I in Figure 2.35 except that the two gases are of the same kind.

From the thermodynamic point of view we *see* nothing happening. We remove the partition between the two compartments and we do not *see* any change in density, temperature, volume, etc. It is natural to conclude that if nothing happens, then the entropy should not change either. This is exactly what thermodynamics tells us — no change in the state of the system, no change in the entropy.

The molecular view is quite different. Using the entropy function for the initial and the final state, we can calculate the difference in the entropy. The result is

$$\Delta S(\text{assimilation and expansion}) = 2k_B N \ln 2 + k_B \ln \frac{(N!)^2}{(2N)!} > 0. \quad (8)$$

It is easy to prove that this quantity is always positive.[46]

Note also that this result is, apart from a multiplicative constant, the same as that we obtained in Subsection 1.11.5 for the experiment with marbles.

The two terms in the equation (8) correspond to *two processes* that occur in the seeming "nonprocess" II in Figure 2.36. First, the volume available for each particle increases from

V to $2V$. This contributes the first term on the right hand side of the equation. Second, the two groups, each of N indistinguishable particles, are now assimilated into one group of $2N$ indistinguishable particles. The contribution of this process is the second term on the right-hand side of the equation. The sum of the two terms is always positive, for any number of particles (or marbles). However, when N is very large so that we can use the Stirling approximation in the form[47] $\ln(N!) \approx N \ln N - N$, we get a different result.

In this approximation, the two terms on the right hand side of the equation (8) cancel each other out. We can thus conclude that, first, there are two processes going on in II — both expansion and assimilation. Second, the entropy change is always positive in this process. And, finally, when N is very large, the difference between the two terms on the right hand side of the equation becomes insignificantly small compared with each term, and therefore can be neglected. It is only in this limit that we get the same result as in thermodynamics, i.e. no change in the entropy. It should be noted that Gibbs was the first to discuss the two processes I and II. At that time, it was not clear why the entropy change in I is $2Nk_B \ln 2$, and zero for process II. Gibbs realized that the so-called "entropy of mixing" in process I is independent of the kind of molecules in the two compartments. He failed to see that this "entropy of mixing" is nothing but the "entropy of expansion." In classical thinking, one could imagine a continuous change in the *kind* of molecules — say, from A to B. If such a change were possible, then we could mix A and B, and get an entropy of mixing, $2Nk_B \ln 2$. Now, let A be changed gradually to B. The entropy of mixing will remain the same, but will change abruptly when A becomes identical to B. This discontinuous change in the entropy of mixing was

Table 2.2 Comparison of the changes in SMI and the changes in entropy for some processes.

Process	Change in SMI in the experiment with marbles	Change in entropy in thermodynamic systems
Expansion	$N \log 2$	$k_B N \ln 2$
Pure mixing	0	0
Pure assimilation	$-2N \log 2$	$-k_B 2N \ln 2$
Mixing and expansion	$2N \log 2$	$k_B 2N \ln 2$
Assimilation and expansion	0	0 for $N \to \infty$

considered a paradox and referred to as the Gibbs paradox. In fact, there is no paradox at all. The presumed *continuous* change from A to B is not feasible. This nonparadox, and other erroneous conclusions reached by Gibbs, are discussed in Ben-Naim (2008).

Table 2.2 compares the results we obtain for the four processes discussed above with those obtained for the experiments with marbles (Section 1.11).

We see that except for the factor k_B and the change in the base of the logarithm, the results for the changes in SMI are identical with the results for the changes in entropy. However, remember that the results for the SMI changes apply for *any* system with *any number* of particles, whereas the results for the entropy changes apply to a thermodynamic system having a very large number of particles, and for the *equilibrium distributions.*

The results in Table 2.2 cannot be used to "prove" that the entropy is the same as the SMI. As we have seen in Section 2.3, the entropy is a particular case of an SMI applied to a thermodynamic system at equilibrium (i.e. having the distribution of locations and momenta that maximize the SMI). The differences between the general SMI and the entropy should be kept in mind when we discuss "information" or "theory of information" in connection with living systems, black holes or the entire universe; see Chapters 3 and 4.

2.7.6 *A spontaneous process of demixing*

In process I of Figure 2.35 we saw that the mixing of two different gases was a spontaneous process. Here, we consider a spontaneous process of *demixing*.

We start with a mixture of N_A particles of type A and N_B molecules of type B in a small volume, v. We replace one wall of the system with a partition permeable only to particles of type A, and the opposite wall is replaced by a partition permeable only to particles of type B. The whole system is isolated, and the volume adjacent to each of the two permeable walls is V (See Figure 2.37.)

Upon replacing the impermeable walls by the two permeable walls, a spontaneous process will ensue; A particles (blue) will flow to the right hand side and B particles (red) to the left hand

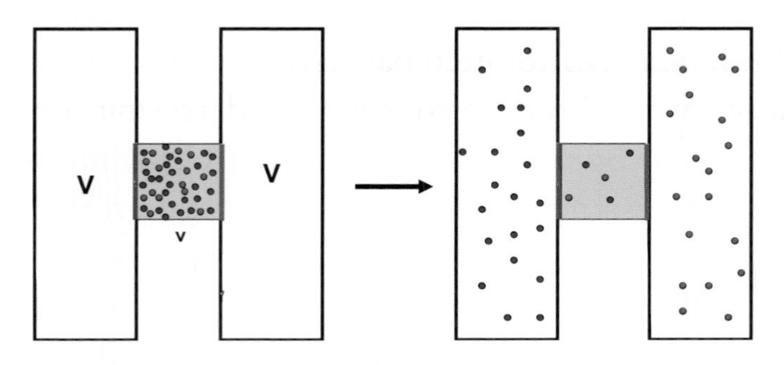

Fig. 2.37 A process of demixing and expansion.

side. Clearly, the larger the volume V, the larger the number of particles that will flow from the smaller volume v to the larger volume V. If we choose $V \gg v$, then most of the A particles (say, N'_A) will flow to the right-hand-side volume V, and most of the B particles (say, N'_B) to the left-hand-side volume V. Clearly, the larger the volume V, the more complete the demixing of the initial mixture of A and B.

We now close the impermeable walls and calculate the change in entropy for this process either from thermodynamics or from the entropy function. The result is

$$\Delta S = N'_A k_B \ln \frac{V}{v} + N'_B k_B \ln \frac{V}{v}. \tag{9}$$

Thus, we can say that we have achieved an almost complete *demixing* of the mixture in a spontaneous process in an isolated system. Note that when $V \gg v$, we can neglect the difference between N_A and N'_A, as well as the difference between N_B and N'_B, and rewrite instead of (9) the approximate result

$$\Delta S = N_A k_B \ln \frac{V}{v} + N_B k_B \ln \frac{V}{v} > 0. \tag{10}$$

The interpretation of this entropy change is the same as in the case discussed in Subsection 2.7.4, i.e. it is due to the change in the *locational* SMI for each particle.

Had we focused on the conspicuous demixing process that occurred in this process, we might have reached the erroneous conclusion that the demixing is a spontaneous process and therefore causes an increase in the entropy. The point to be emphasized here is that in both process I of Figure 2.35 and the present process (Figure 2.37), the *cause* of the change in the entropy is the increase in the *locational* SMI. The fact that we *see* mixing in process I and *see* demixing in Figure 2.37 is

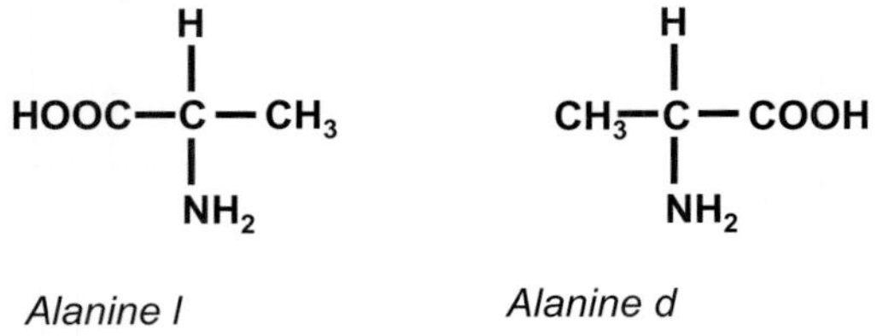

H
HOOC—C—CH_3
NH_2

H
CH_3—C—COOH
NH_2

Alanine l *Alanine d*

Fig. 2.38 Two enantiomers of alanine, *l* and *d*.

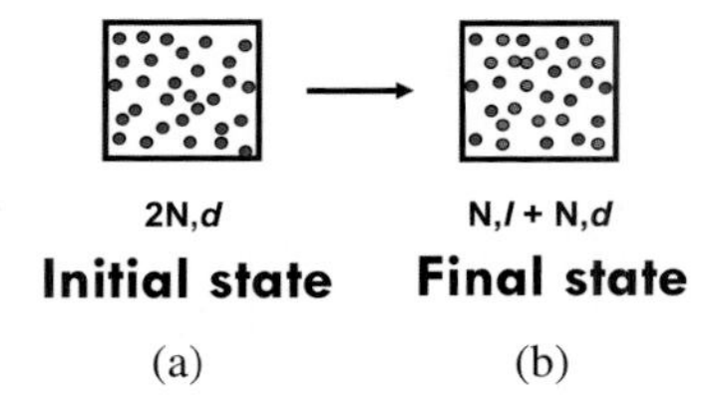

Fig. 2.39 A process of racemization.

inconsequential to the change in the entropy in these processes. [For more details, see Ben-Naim (2008).]

2.7.7 *A spontaneous process involving pure deassimilation*

Consider a molecule having a chiralic center — say, the amino acid alanine (Figure 2.38), with two different enantiomers, *d* and *l*. Then consider a process where initially we have $2N$ molecules of pure *d*. Upon the addition of a catalyst, about half of the molecules will be converted from *d* to *l* form. This process is called "racemization" (Figure 2.39).

The SMI interpretation is straightforward. We start with $2N$ indistinguishable particles and we end up with N indistinguishable particles of one kind (*d*), and another N indistinguishable particles of a different kind (*l*). Since the *volume accessible to each particle* does not change in this process, there is no change in the *locational* SMI. In this process the only change that occurs is in the *number of indistinguishable particles*. Therefore, this process has been referred to as a process of *pure deassimilation*. The informational interpretation of this process has been discussed in detail in Ben-Naim (2008).

2.7.8 *Spontaneous heat transfer from a hot to a cold body*

One of the classical examples of a spontaneous process used in the formulation of the Second Law of thermodynamics is the process of heat transfer from a hot to a cold body.

Within macroscopic thermodynamics it is easy to calculate the change in entropy in the process of heat transfer between two bodies A and B at two temperatures T_1 and T_2, with $T_2 > T_1$. Assuming that a small amount of heat dQ is transferred from B to A, we find, using Clausius' definition, that the entropy change for the combined system is

$$dS = dS_1 + dS_2 = \frac{dQ}{T_1} + \frac{-dQ}{T_2} = \frac{dQ(T_2 - T_1)}{T_1 T_2} > 0.$$

This calculation only confirms what we already *know*: that heat is transferred from a hot to a cold body, i.e. $dQ > 0$ and $T_2 - T_1 > 0$. Hence, the entropy change is positive. Thermodynamics does not offer any *interpretation* for the increase in entropy.[48] The informational argument is quite different but lengthy. We provide some details in Note 49.

Thus, we have seen that in the spontaneous process of heat transfer (Figure 2.40) the SMI has increased. It should be noted, however, that the positive change in the SMI as calculated above is not an *explanation* of why the process goes spontaneously from the initial state to the final state. In both the initial and the final

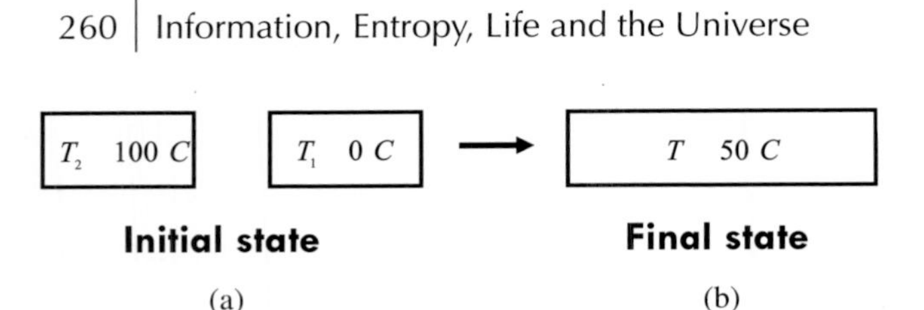

Fig. 2.40 A spontaneous process of heat flow from a hot body to a cold body.

state, the momentum distribution is such that it maximizes the SMI of the system. Since both the initial and the final state are equilibrium states, the change in the SMI is the same as the change in the entropy of the system (after multiplying by k_B and changing the base of the logarithm). The *reason* for the spontaneous process is probabilistic, i.e. the final momentum distribution is far more probable than the initial distribution. The positive change in the entropy is the *result* of the spontaneous process, not the *cause* of it. [More details can be found in Ben-Naim (2008).]

2.7.9 *Some "pathological" processes of "expansion"*

Consider the three thought experiments depicted in Figure 2.41. In (a) we start with a system in which all the particles are moving in concert along the z axis. We assume that the walls of the container are perfectly smooth and perpendicular to the z axis, so that whenever a particle collides with a wall it will be perfectly reflected and will continue to move along the z axis, but in an opposite direction. In Figure 2.40(b) the particles are moving in concert back and forth along the x axis. In Figure 2.40(c) we start by vigorously stirring the gas in the left compartment.

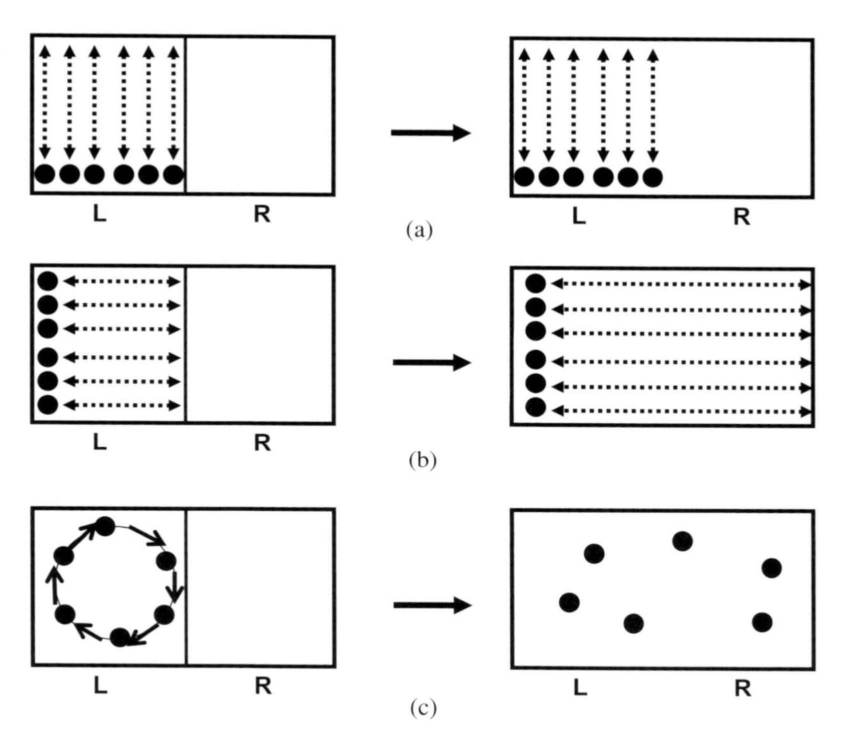

Fig. 2.41 Three processes of "expansion." Initially, all the N particles are in the left compartment, and then we remove the partition between the compartments.

Next, we remove the partition as we did in the expansion experiment in Figure 2.32. What is the change in entropy in these processes?

Before we attempt to answer this question, we should make a clear distinction between the following two questions: What will happen in the system? What is the entropy change in the process? Very often, these two questions are given the same answer.

The answer to the first question is simple for the three processes (a), (b) and (c). In process (a), assuming that the particles are moving up and down in concert indefinitely, we do not expect to see any change in the state of the system. Though the volume accessible to each particle has increased from V to $2V$, none of the particles will actually access the extra volume available. In case (b), the particles will expand into the newly accessible volume. They will be moving back and forth along the x axis in the entire volume $2V$. In the third process, we start with the gas under vigorous turbulent motion and remove the partition. After some time the turbulent motion will subside, and the particles will expand to occupy the entire volume $2V$ and reach an equilibrium state, as in the expansion process in Figure 2.32.

The next question is: What is the change in entropy in these processes? We might tend to assume that if nothing happens in process (a), then the entropy of the system will not change. One can also argue (erroneously) that the particles are moving in perfect "order" (low entropy?), and since the "order" did not change, the entropy should not change. In the second example (b), the particles do "expand" to occupy the new volume. Should the entropy increase? Arguing from the order point of view, there is still perfect order of motion. Perhaps we can use the "information" argument. We know that all particles were in volume V. Now they are *spread* in a larger volume, $2V$. Perhaps the entropy should increase. What about process (c)? Here, we definitely see the expansion process and in the final state the system will reach an equilibrium state with a uniform locational distribution and Maxwell–Boltzmann velocity distribution. Should we conclude that in this case the entropy change will be as in the expansion process, namely $k_B \ln 2$ per particle?

My answer to all three questions about the entropy change is simple: I do not know. Note that the answers given about the *change* of the *state* of the system in the three experiments are correct. On the other hand, the change in the entropy of the system in these processes is not even *defined*, let alone estimated. Recall that the entropy of the system is defined for a macroscopic system at equilibrium. Neither the initial nor the final states in processes (a) and (b) are equilibrium states. Hence, the entropy function is not defined for these states. Therefore, I decline to answer the question about the entropy change. Note, however, that it is meaningful to define the locational and the momentum SMI for each state in experiments (a) and (b), but whatever the values of the SMI are, these will be irrelevant to the *entropy* of the system.

In the third example, the system in the final state is an equilibrium state. However, the initial state is not an equilibrium state, and therefore I cannot give you an answer to the question of how much the entropy has changed in process (c). It should be noted, however, that we could adopt the Boltzmann definition of entropy, where W is the *total* number of accessible states in the system. In this case, the initial state of the system can be viewed as a subset of *all* possible states, and hence the change in entropy in this view is $k_B N \ln 2$. However, from the thermodynamic point of view the initial state of the system is not an equilibrium state. Therefore, we cannot assign an entropy value to the initial state.

These processes have been deemed to be "pathological" in this section. In all these processes the initial and final states are well-defined. Yet, the entropy changes for these processes are not defined. We used the word "pathological" as a prelude to the life processes discussed in the next chapter. It is unfortunate that entropy changes are discussed in the context of life processes, which are far more "pathological" than the processes discussed in this section. Here, of course, we use "pathological" not in the ordinary, medical sense of the word.

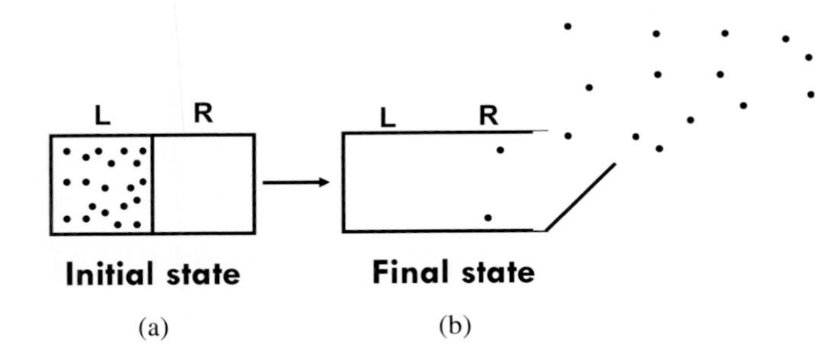

Fig. 2.42 Expansion from volume V to infinite volume.

Finally, let us discuss the analog of the process discussed in Subsection 1.11.6 (see Figure 2.42).

Instead of expansion from M cells to kM cells, as in Figure 1.34, we have an expansion process from an initial volume V to $k \times V$. The change in entropy is $k_B N \ln k$. In the "expansion" process in Figure 2.42 we open the system so that the gas can access all of the universe. In the process described in Subsection 1.11.6 we found that when $k \to \infty$ the change in the SMI becomes infinity.

For the "expansion" process in Figure 2.41 we tend to conclude that since $k \to \infty$ (presuming that the universe is infinitely large) the change in entropy will also be infinity. However, I am personally reluctant to reach this conclusion. Once the particles are allowed to leave the container, each particle leaving the container must have a component of its velocity toward the right direction (Figure 2.42). After some time all the particles will leave the container. They will *never* reach an equilibrium state. Therefore, the final state of the system is imaginable, but the entropy of the system in this state is not calculable. This example is a prelude to the discussion of entropy in the context of the entire universe, which will be further discussed in Chapter 4.

2.8 Can Entropy Flow Be Created or Destroyed?

In most textbooks on thermodynamics one may find the following equation regarding the change in entropy of a system:

$$dS = d_eS + d_iS,$$

where d_eS is due to the flow of a small amount of heat into the system at a given temperature T. Thus, $d_eS = dQ/T$. When heat flows *into* the system ($dQ > 0$), the entropy increases. When heat flows out of the system ($dQ < 0$), the entropy decreases. d_iS is referred to as the *internal* change in entropy due to an irreversible process occurring in the system.

According to Clausius' definition, $dS = dQ/T$. When heat *flows* into the system at constant T, the *change* in entropy is dS. However, accepting the informational interpretation of entropy, it would be awkward to say that entropy *flows* into the system. This would be as awkward as saying that the *number* of particles flows into the system when we add a quantity of matter into the system. Likewise, it would be awkward to say that entropy is *created* or *destroyed.* Indeed, in the case of heat flows into the system, thermal energy flows into the system. As a result of this thermal energy flow, the entropy of the system *increases.* However, the entropy as an SMI does not flow. Similarly, the association of the words "created" and "destroyed" with entropy is inappropriate. Entropy is not *created,* as much as the length of a rubber band is not created but simply increases or decreases. Similarly, when heat flows *into* the system the distribution of kinetic energies of the molecules broadens.[50] As a result of this change the entropy of the system increases.[51]

2.9 Entropy as a Measure of Disorder

There are many qualitative interpretations of entropy. Perhaps the oldest and the most popular interpretation is in terms of order and disorder. In my previous books,[52] I have discussed at length this interpretation of entropy. I have shown that this interpretation is inadequate even in qualitative terms. It is true that there are many spontaneous processes which look like proceeding from a more ordered to a less ordered state. However, this correlation is not always valid; besides, a descriptor of the *state* is not necessarily the same as a descriptor of entropy.

In many textbooks you might find the process of the melting of ice as an example of a spontaneous process involving increase in entropy, and interpreted in terms of increase in disorder. Clearly, this increase in entropy is due to the heat flow into the system, and not to disordering of the system. One can take a glass of water and bring it into contact with an environment at a temperature of, say, $-10°$. The water will freeze, and the system will spontaneously become more *ordered*, and its entropy will decrease. In both cases, one finds the claim that the overall change in the entropy of the system and the rest of the universe must be positive. I do not agree with such assertions. I will discuss the reasons for this in Chapter 4.

Here, I would like to present two processes of expansion. In one the temperature decreases, and in the other the temperature increases. These two processes cannot be interpreted by any one of the descriptors of entropy (disorder, information, spreading, information, freedom, chaos, etc.).[52] On the other hand, SMI can provide a quantitative explanation of these processes.

The first process is a spontaneous expansion of gas in a column of air in a gravitational field.

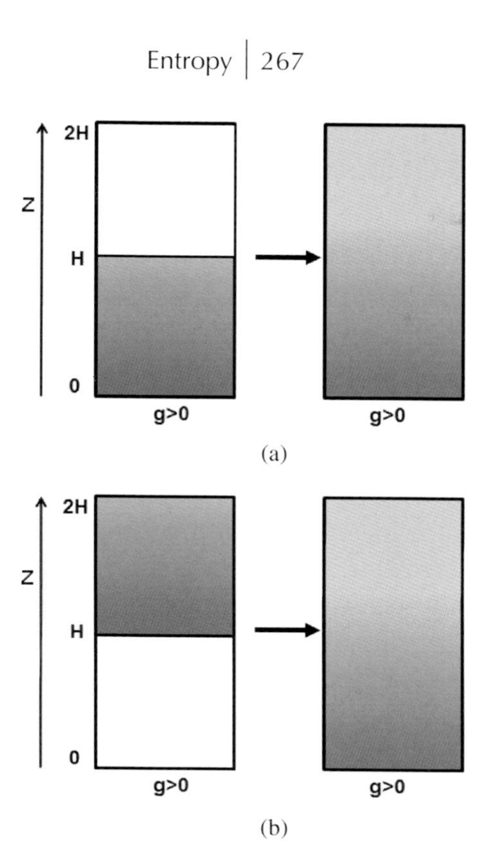

Fig. 2.43 Expansion of an ideal gas in a gravitational field from V to $2V$: (a) upward expansion, (b) downward expansion.

Figure 2.43 shows two processes of expansion. In Figure 2.43(a), the gas expands from V to $2V$ *upwards*. In Figure 2.43(b), the expansion is *downward*. In both cases the entropy change due to the expansion may be interpreted in terms of the change in the *locational* SMI. However, in addition to the increase in the accessible volume for each particle from V to $2V$, there is a change in the temperature of the system. In the expansion in Figure 2.43(a), the gas will be cooled, i.e. the temperature will *decrease*. The reason is that some of the particles will move higher in the gravitational field. Since the system is isolated, the energy for this elevation of the particles must come

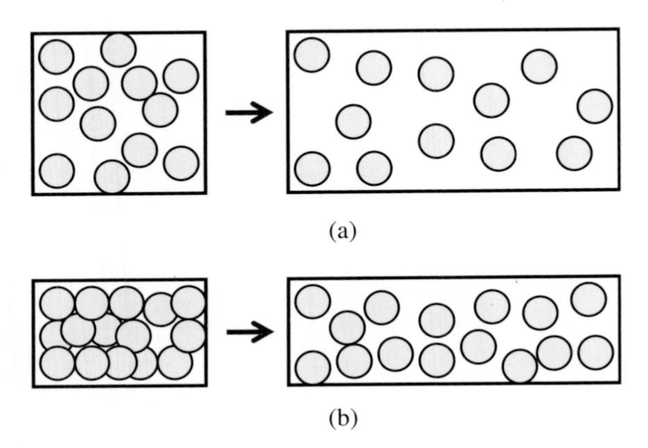

Fig. 2.44 Two processes of expansion of a nonideal gas from V to $2V$: (a) expansion involving a decrease in the average interaction, hence lowering the temperature; (b) expansion involving an increase in the average interaction, hence elevating the temperature.

from the kinetic energy of the particles. In the process shown in Figure 2.43(b), we will observe an *increase* in the temperature, simply because some of the particles will flow downwards in the gravitational field, and hence their kinetic energy will increase. Clearly, none of these two processes can be explained in terms of increase in disorder.

The second example is shown in Figure 2.44. Here, we have interacting particles expanding from V to $2V$ in an isolated system. As we know, this expansion for noninteracting particles (ideal gas) will involve a change in entropy of $k_B N \ln 2$. Also, the temperature of the system will not change.

In the case of Figure 2.44(a), the expansion will cause an increase in the average distance between the particles, and hence their average interaction energy will decrease. One must *invest* energy to separate the particles. Since the system is isolated, the kinetic energy of the particles will decrease, and hence the temperature will *decrease*.

Figure 2.44(b) shows the same process, but the initial state is at very high pressure, so that particles are penetrating into the repulsive region of the intermolecular distances. Upon expansion, energy is released in this process, and hence the temperature will *increase*. Again, one cannot explain these processes by using the order-to-disorder metaphor. [For a quantitative discussion, see Ben-Naim (2008, 2012).]

Consider again the process of heat flow from a hot body to a cold body. The cold body's temperature will increase. Hence, its kinetic energy will be more widely spread. On the other hand, the hot body will be cooled — its kinetic energy will be less widely spread. Neither the disorder nor any other metaphor can account for the net change in entropy in this process. On the other hand, the SMI can be shown to be higher in the final state in this process.[52]

2.10 Summary of Chapter 2

We started this chapter with two apparently very different definitions of entropy: the macroscopic and the microscopic. The first is based on experimental facts and measurements; the second was suggested heuristically by Boltzmann. These two definitions seem to be completely unrelated. However, all the calculations based on Boltzmann's definition of entropy proved to agree with those obtained by means of the experimental entropy.

We have also shown how entropy can be obtained from SMI. This derivation of entropy provides a simple and appealing interpretation of entropy. It should be emphasized, however, that SMI is a far more general concept than entropy. SMI may be defined on a distribution of the outcomes of a dice, or of the location and momentum of a single particle. Entropy, on the other hand, is defined for very special distributions of macroscopic systems at equilibrium. It is identical to SMI (up to a multiplicative constant) only when the distribution is an equilibrium distribution. Equivalently, entropy is the value of SMI attained for that distribution which maximizes the SMI. Both SMI and entropy can be interpreted as a measure of information. However, this is a very special *measure* of a very special kind of information.

In Chapter 1, we discussed the confusion between colloquial information and SMI. In the present chapter, we have introduced entropy as a special case of SMI. It is clear that SMI is neither *information* nor *entropy*. Yet, many people use the concepts of entropy and information interchangeably — and some even confuse information with bits.

The Second Law is concerned with the question of why entropy always changes in one direction. The answer to this question is probabilistic. In this sense the Second Law is not really a law of physics, but a law of probability — or, if you like, a law of common sense.[53] It states that a system will spend more time in states having a larger probability. Specifically, for macroscopic thermodynamic systems, "larger probability" becomes "certainty," and "more time" becomes "*almost always.*" The relationship between the probability of the state of the system and the SMI of the system was discussed in Section 1.14.

Confusing entropy with SMI, and SMI with information, leads to confusing entropy with information. Entropy is a physically measurable quantity. Information is in general not a physical quantity, nor is it measurable. This confusion leads to claims that thermodynamics is part of information theory, and that information is subject to physical laws, etc.

The most important message of this chapter may be summarized as follows:

(1) Information theory is not part of thermodynamics. The laws of thermodynamics do not govern *information*.
(2) SMI is not *information*, but a specific measure of a specific kind of information.
(3) SMI is a very general concept; it is certainly not entropy.
(4) Entropy is a particular case of SMI. It is defined on a particular set of probability distributions of a thermodynamic system at equilibrium (i.e. the distributions that maximize the relevant SMI). In this sense entropy is a very special kind of measure of information defined for a very special set of distribution functions.

It should be emphasized that although entropy is a special case of SMI, it does not follow that "thermodynamics is part of information theory" [as claimed by Seife (2010)]. In other words, the fact that thermodynamics and information theory share the same concept of SMI does not make the two theories related to each other. This is trivial — yet, surprisingly, an entire article was devoted to this claim.[54] In Note 55 we provide a few more quotations where information is confused with SMI, and SMI is confused with entropy.

Figure 2.45 summarizes, in a very qualitative manner, the relationships between the general concept of information (GI), Shannon measures of information (SMI) and entropy. The three regions in the figure should not be understood in the strict set-theoretical sense. The region denoted as "general information" (GI) includes every item of information. Although we do not have a definition of the term "information," we have an intuitive understanding of what it means. The SMI region includes

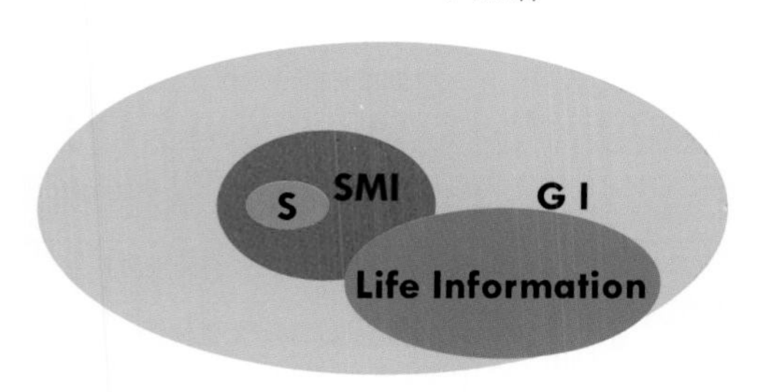

Fig. 2.45 Schematic relationships between the "regions" of general information, SMI and entropy.

points, each representing a well-defined distribution, either discrete or continuous. It includes all possible distributions, and hence it is a well-defined set of objects. Because of the interpretation of SMI as *an amount of information contained in or associated with a distribution*, we view the set denoted as SMI as a subset of "general information." It is a subset only in a conceptual sense, i.e. each piece of information contained in a distribution "belongs" to the general information, conceptually. But each point in the region GI is not necessarily associated with a distribution. The region denoted as "entropy" is a genuine subset of the set denoted as "SMI". That is so because entropy is defined on a subset of *all* distributions. Therefore, the distributions on which entropy is defined belong to the region denoted as "SMI". However, the region denoted as "entropy" is only a subset of GI in a conceptual sense, in the same sense that SMI is a subset of GI.

I have added in this figure a region called "Life" which intersects the SMI and GI regions, but not the entropy region.

We shall further discuss the application of the concepts of information, SMI and entropy to life in Chapter 3.

Finally, I owe you an answer to my question in the Prologue. If you do not know how to answer the questions, you are in good company. No one can give meaningful answers to these questions.

Chapter 3

Life

I do not know of any other topic in our lives which is more interesting, more enthralling and more mysterious than life itself. We know a great deal about many aspects of life which includes biochemistry through molecular biology, biology, medicine, psychology, sociology, and much more. Yet, we know that we still do not know many things, and there are probably many other things that we do not know that we do not know, and there are also things that perhaps we will never know.

Since time immemorial people have been puzzled by the phenomena of life — their own lives, as well as the lives of other living things. What is this thing which distinguishes a living system from inanimate objects? Any child can tell when a cat is alive or dead, yet it is almost impossible to *define exactly* this transition from being alive to being dead.

I was always fascinated by the question "What is life?" Three times in my life I had near-death experiences. In all of those cases, I remember concentrating on the very moment of death, to try to experience, perhaps also to understand, that transition point from life to death. Luckily, I missed that point. The first time I asked myself the "scientific" question "What is the exact physical process that defines the transition from being alive to being dead?" was when I watched my father as he lay on his deathbed.

In the throes of death, my father sat on his bed, his back supported by a pile of pillows. Although breathing laboriously and in great pain, he was still lucid. Speaking in Hebrew, I asked him whether he wanted to lie down, and he said "No." My sister-in-law asked him in Ladino (a Spanish dialect) whether he wanted something to eat, or drink, and with much effort he said "No" again. As the minutes ticked away, his breathing became increasingly heavy and noisy. I watched him closely and I knew that the "angel of death" was hovering above him, and that he was on the edge of the cliff of life, about to fall into the abyss of death. With bated breath, I fully concentrated on what was going on. Would I catch that very specific transition?

His breathing became much more labored and he had a fixed gaze. And then . . . stillness, and a deafening silence. He had ceased to breathe and had stopped moving. His eyes stayed open, still bearing the same expression of excruciating pain.

A few seconds passed, and when I realized that his candle of life had been snuffed out, I called my sister-in-law and told her that my father was probably dead. She looked at him, asked him a few questions, and when no answer came, knew that he was dead. With reverence, she removed all the pillows which had propped him up, and laid him in bed. I heard my older brother say, "*Shema-Yisrael.*" This is a Jewish prayer as one goes to sleep, as one wakes up, and when one faces imminent death.

It was clear to me that my father had passed away. But what is that transition from life to death that we call "passing away?" Is there any physical description of the *state* of the body which transforms from the state of being "alive" to the state of being "dead," or is it something unphysical which we refer to as the *soul* leaving the body at the moment of death?

Every culture or religion has its own version of characterizing and distinguishing between the states of being "alive" and being "dead." In the Book of Genesis, we find the following two versions of the creation of man.

In the first chapter (Genesis 1:26), we find:[1]

> Then God said, Let us make man in Our image, according to Our likeness; let them have dominion over the fish of the sea . . . and over every creeping thing that creeps on earth.

In this version of the story, man is created differently from other living things ("Our image"), but there is no mention of the soul. However, in the next chapter (Genesis 2:7), we find:[2]

> And the Lord God formed man of the dust of the ground, and breathed into his nostrils the breath of life; and man became a living being.

Here, it is clearly said that the *breath* transforms from the inanimate *ground* to become a *living being*.

Similarly, in the Book of Ezekiel (37:5–37), we find:[3]

> Thus says the Lord God to these bones: "Surely, I will cause breath to enter into you, and you shall live.
>
> I will put sinews on you and bring flesh upon you, cover you with skin and put breath in you; and you shall live. Then you shall know that I am the Lord.
>
> Indeed, as I looked, the sinews and the flesh came upon them, and the skin covered them over; but there was no breath in them. So, I prophesied as He commanded me, and breath came into

> them, and they lived, and stood upon their feet, an exceedingly great army.

Is there such an entity (which we call the "soul") which is "breathed" into the body to make it alive, and which leaves the body when a person dies?[4]

A few day after my father's death, we were sitting at the *shiva* (the Jewish weeklong mourning period) when I heard my brother say that he had actually *seen* the angels take my father's *neshama* (soul) to heaven. I was not fortunate enough to witness that. I was more fascinated by the question "What exactly is the event which marks the transition from being alive to being dead?" Perhaps I was so immersed in trying to *understand* the transition from life to death that I missed *seeing* it.

There are several criteria for determining the state of being dead — stoppage of breathing, cessation of brain function, etc. All these are operational definitions and they work well most of the time. However, there are some exceptions where a person is pronounced dead but then shows signs of life (a famous case is the former Egyptian president Hosni Mubarak). There have even been cases where some people were pronounced dead and then came back with stories of seeing a white, blinding light which led them to a brief visit to *heaven*.[5]

To the best of my knowledge, no one who was pronounced dead ever came back with a story of visiting *hell*. Perhaps a visit to hell is really an irreversible process. The reader is referred to a very interesting discussion about "Belief in Afterlife" by Shermer (2011).[6]

In the Hebrew language, addressing someone as "my *neshama*" is a term of endearment. It is similar to saying "my soul," "my darling," "honey" or "sweetie" in the English language. The root of the word "*neshama*" is "*nasham*" (נשם), which means "breath or breathing." The air we breathe is the closest thing for describing the entity we call the "soul," which distinguishes the living from the dead.

The following story demonstrates the double meaning of "*neshama*", as a nickname and the soul.

A couple who had loved each other for many years were accustomed to calling each other "my *neshama*," or simply "*Neshama*." One evening, the husband woke up in the middle of the night and saw an apparition entering their bedroom. He opened his eyes, and rubbed them in disbelief as he tried to focus on what had quickly transformed into an ominous figure. Although terrified, he managed to ask, with a trembling voice, "Who are you and what are you doing in our bedroom?"

"I am the angel of death, and I have come to take your *neshama*," the ominous figure with a cold, hollow voice and fiery-red eyes answered the husband.

Without any hesitation, the husband woke up his wife and told her, "*Neshama*, wake up, someone has come to take *you*!"

This is the most I can say about the *neshama*, or the soul, that entity which is supposed to define the difference between *life* and *death*; yet it has so far eluded its own definition. At the moment it is not even clear whether this entity, which we call the soul, is *physical*, or has any mass.[4]

In the following sections of this chapter we will discuss attempts to characterize life and define life. In doing so we will see that the concepts of *information* and *entropy*, and the Second Law, play prominent roles. We will examine critically the applicability of these concepts to life phenomena.

3.1 Can Life Be Reduced to Physics and Chemistry?

As we will see throughout this chapter, there are many theories dealing with specific aspects of life. Thermodynamics is successfully applied to many specific processes — chemical reactions, conversion of chemical energy to mechanical or electrical energy, and so on. However, it is not clear whether thermodynamics or any other theory of physics or chemistry is applicable to an entire living organism. In other words, it is not clear what a theory of life entails.

There are essentially three groups of opinions regarding the reduction of life phenomena to physics and chemistry:

(i) Physics and chemistry as we know them today are, in principle, sufficient for explaining life phenomena. There is no need to add any metaphysical concept, such as "life principle," vitalism or life force.

(ii) Physics and chemistry as we know them today are unable to explain life. However, it is hoped that some future extension (or revolution) in physics and chemistry will be able to explain life.

(iii) Life cannot be explained and understood without accepting some concept which is beyond physics and chemistry. Whether we call it "vitalism," "life force," "life principle," or any other concept, it will not be under the control of the laws of physics and chemistry.

The first two groups of opinions are sometimes referred to as *reductionism* or *materialism*, i.e. life phenomena can, or can in principle, be *reduced* to physics and chemistry, and all the characteristics of life, such as consciousness, thoughts or feelings, can be explained and understood in terms of atoms and molecules (matter) and there is no need to introduce vague and undefined nonmaterialistic concepts. The third group of views is referred to as *dualism*, *antimaterialism*, *antireductionism*, *vitalism*, etc.

In fact, there is a whole spectrum of views between these three views. For instance, some people believe that life phenomena cannot be reduced to the laws of *classical physics*[7] but there are some characteristics of *quantum physics* that can handle life phenomena.

Finally, there is also the possibility that we, living systems, will *never* be able to understand the living system which we are ourselves. It is also not clear whether we will ever *know* if we can or cannot understand living systems, and life phenomena will forever remain a mystery.

Personally, I believe that life phenomena will eventually be understood by the laws of physics and chemistry. Perhaps we will need a few extensions of the laws of physics, but we will not need any entity which is not beyond physics and chemistry.

Recently some authors have claimed that information theory is already providing answers to the question of what life is. We will discuss some of these views in the following sections of this chapter.

At present it is safe to say that there exists no theory, molecular (microscopic) or thermodynamic (macroscopic), which can either describe or explain life phenomena.[7] It should be stressed that there is a fundamental difference between describing some characteristics of life and describing life itself. There are many aspects of life which are now well-understood. There are many characteristics of life which are sometimes used in *defining* life.

The most complete list of the characteristics of life was recently published by Pross (2012) in his delightful book *What Is Life? How Chemistry Becomes Biology.*

In Chapter 1 of that book, Pross surveys some of the characteristics of life which make life *different* from inanimate matter:

(i) Life's organized complexity
(ii) Life's purposeful character[8]
(iii) Life's dynamic character
(iv) Life's diversity
(v) Life's far-from-equilibrium state
(vi) Life's chiral nature

In an attempt to define life, many authors are in effect listing some characteristics of living systems, such as self-organization, metabolism, reproduction or evolution. Unfortunately, none of these can be reckoned to be a definition of life. There are many living systems which do not have one or more of these characteristics. A mule is an example of a living organism which does not reproduce.

As we will discuss later in this chapter, it is not clear at all whether we will ever have a definition of life. It is not clear whether such a definition, if it exists, will be *universal*, i.e. pertaining to *all* living organisms. Perhaps different definitions will be needed for different species, and of course a very special definition for human beings.

It might be the case that in the future we will find how "life" emerges from a complex system. We might find that perhaps every atom or every molecule has *some degree of life* (of consciousness, of free will, etc.), and what we recognize as living organisms have just "a little more," and human beings have much more of this thing we call "life."

Crick (1981) wrote a short but illuminating book, *Life Itself*, where he discussed in plain language the origin of life, the nature of life and the definition of life. What I liked most about this book is the complete absence of any discussion of *entropy* and the *Second Law* in connection with life. Crick does not explain why, but I guess he realized that both entropy and the Second Law are irrelevant to the central question "What is life?" This will be further discussed in the following sections of this chapter.

Regarding a "definition" of life, Crick writes:

> It is not easy to give a compact definition of either "life" or "living."

In the following sections we will discuss only a few aspects of life that were directly or indirectly associated with *information* and *entropy*. We will discuss exactly those topics which were omitted from Crick's book, i.e. entropy and the Second Law in relation to life. We will also discuss some aspects of life which are involved in storing, processing and transmitting information. We will distinguish between information *about* the living system and information that is *contained* in some molecules, organs or tissues, and sometimes the entire living organism.

3.2 The "Book of Life"

The "atom" of all living systems is the *cell*. At the heart of a cell is the nucleus, and at the heart of the nucleus resides the so-called "book of life," a sequence of letters which contains the *information* on that particular organism. This information is written in a four-letter language: A, T, C, G. Each of these

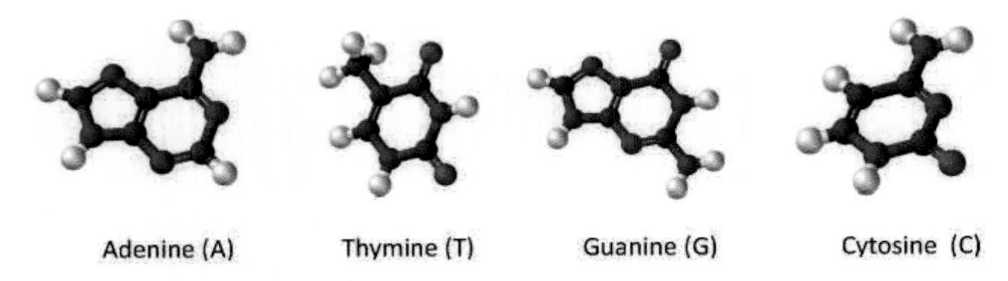

Fig. 3.1 The "letters" of the DNA: A, T, G and C.

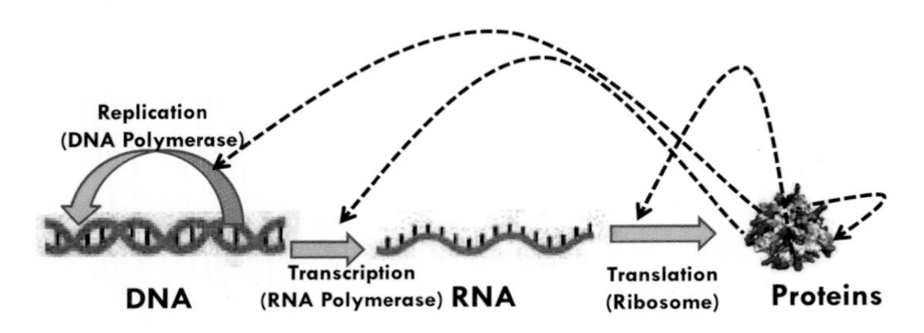

Fig. 3.2 The central dogma of molecular biology.

letters is a different molecule (base) (Figure 3.1), and the entire "book" is the DNA molecule. The whole sequence of letters, typical of each organism, is well-packed and well-protected at the chromosomes. The DNA of human beings contains billions of letters or bases.

The DNA is perhaps the molecule most associated with *information*, both in the colloquial sense and in the theoretical-informational sense. One can also assign *entropy* to a segment of the DNA or to the entire DNA. However, this entropy has *nothing* to do with the "information" content of the DNA.

Figure 3.2 shows schematically the so-called "central dogma" of molecular biology. The information stored in the DNA is read and executed in essentially two major ways. In one, the "message" written in the DNA is replicated. This serves as the main mechanism for the transmission of the information from generation to generation. This is shown by the curved heavy arrow on the left hand side of the figure. The second way is the *translation* of the information from the DNA language to the protein language. This is shown by the horizontal heavy arrows pointing from the left to the right.

It is often said that the "information" written in the "book of life" is preserved, and is transmitted from generation to generation. However, no one can tell you what the "information" written in the DNA is. People talk about the "book of life," the "blueprint of the animal," "the recipe for building up a specific organism" and so on. All these are figures of speech. There is no doubt that the DNA contains information, but it is not clear how to define or describe this type of information.

The Bible contains many stories. Some may be true, some fictitious. Before the invention of printing, people had the painstaking task of copying each letter of the text by hand. What was conserved was the *pattern* of letters. When read by a person who is familiar with the language, this pattern of letters tells a story — or conveys information.

Nothing like this exists in the sequence of bases in the DNA. If you just read the sequence, you do not get any information beyond the sequence itself. Indeed, the pattern of letters in the DNA is conserved (to some extent) by replication. It is also translated into a different pattern of amino acids in proteins. Again, *reading* this pattern of letters does not tell any story, or provide any explicit information. Yet, it is true that the DNA *contains* information on the specific animal. We can also say that this *information* "tells" the ribosomes which sequence of amino acids it has to construct!

In the following sections we will describe first the replication of the information on the DNA. Next, we will discuss how the message in the DNA is *translated* into proteins, which are the building blocks as well as the builder-robots which construct the entire organism. Finally, we will describe the "translation" of the one-dimensional "information" in the sequence of amino acids into a functional protein. The overall result of these processes is the formation of a whole living system. In this sense we can say that the *information* contained in the DNA has been translated into the structure, behavior, etc. of a living system.

However, this translation is not a simple translation from one language to another. The eventual result of the translation depends on the *environment* in which it is carried out, which includes the pressure, the temperature and many chemicals. Different environments might lead to different outcomes or no outcome at all.

3.2.1 *Replication of the information in the DNA*

Whatever the *information* contained in the DNA is, it has to be transmitted as accurately as possible, from generation to generation. The mechanism of this remarkable process was discovered only in the middle of the 20th century.

The DNA is written in a *four-letter* language. This simply means that the DNA is a sequence of four letters denoted as A, T, G, C. It is also known that A binds to T and G binds to C. (see Figure 3.3.) This pairing of letters points to the mechanism of replication of the DNA. Note that the mechanism of *replication* is independent of the *information* carried by the DNA. The process simply replicates the sequence of symbols, whether or not that sequence carries any information.

The mechanism is briefly described in Figures 3.4 and 3.5. A piece of DNA is shown on the left hand side of Figure 3.4. It consists of a pair of strands, each having the four letters A, T, G, C. The two strands are bound to each other in such a way

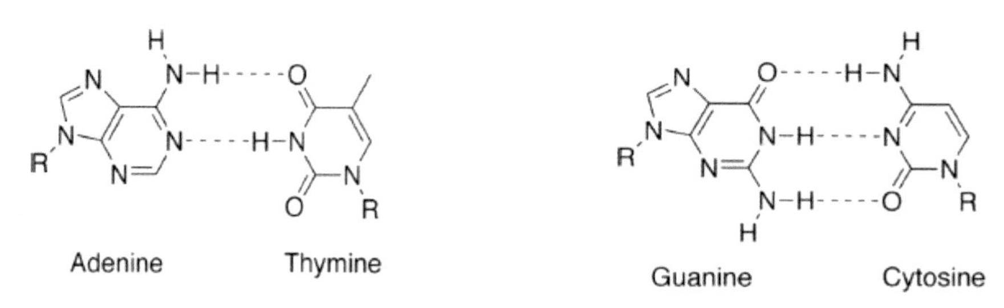

Fig. 3.3 Binding of A (adenine) to T (thymine), and G (guanine) to C (cytosine).

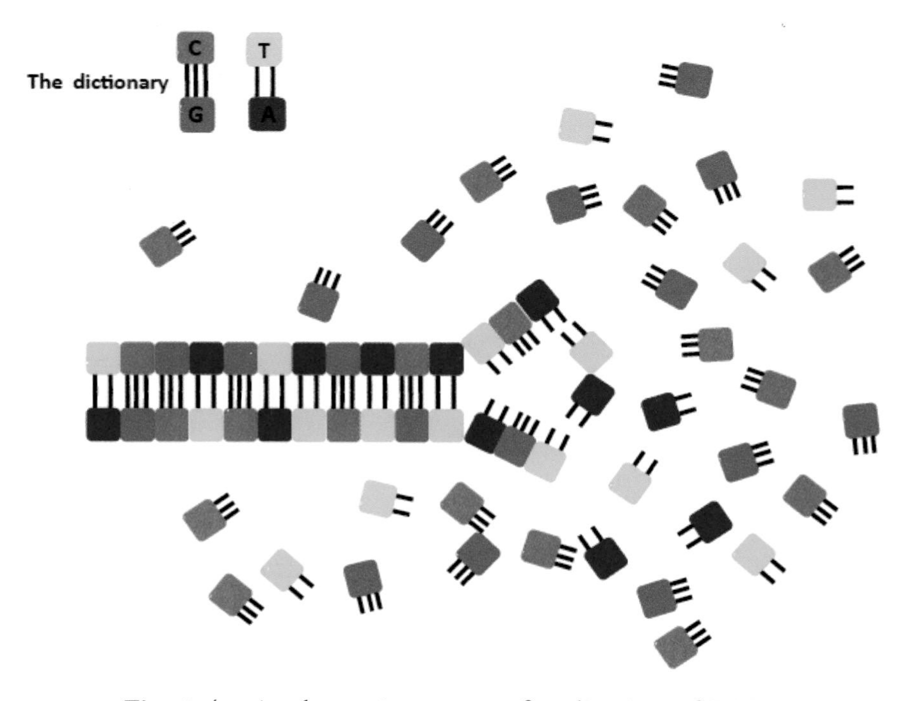

Fig. 3.4 A schematic process of replication of DNA.

that A binds to T, and G binds to C. Figure 3.6 shows a model of this double-stranded DNA. When this double strand opens up, it exposes the single letters A, T, G, C to the solvent, which contains free letters A, T, G, C floating around.

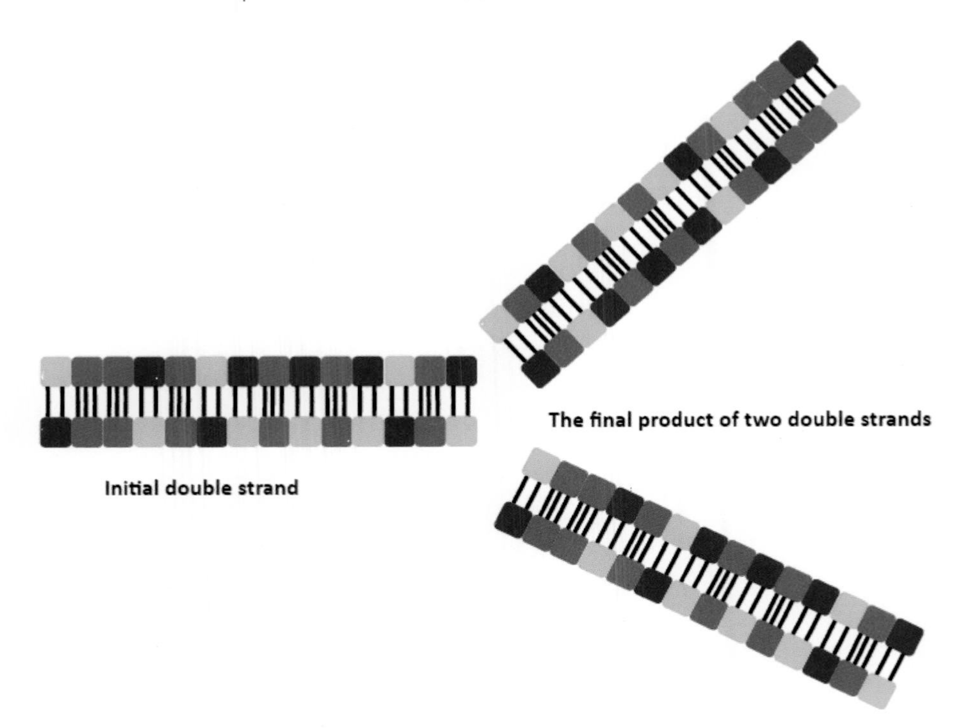

Fig. 3.5 The replication process; one double strand of DNA is transformed into two identical copies.

Each letter on one of the strands of the DNA which becomes exposed to the solvent is now able to bind with only one letter in the solvent: A to T, T to A, G to C and C to G. As the original double strand opens up, the exposed letters form new connections with letters from the solvent, and thereby gradually two pairs of a double strand sequence are formed. In effect, the original double strand sequence has been replicated into two identical double strand sequences (Figure 3.5). Of course, the actual mechanism of opening up the DNA, and the attachment of new letters to the single strand, is a highly complicated process which is controlled by many enzymes, but the net effect of this process is the replication of the original sequence.

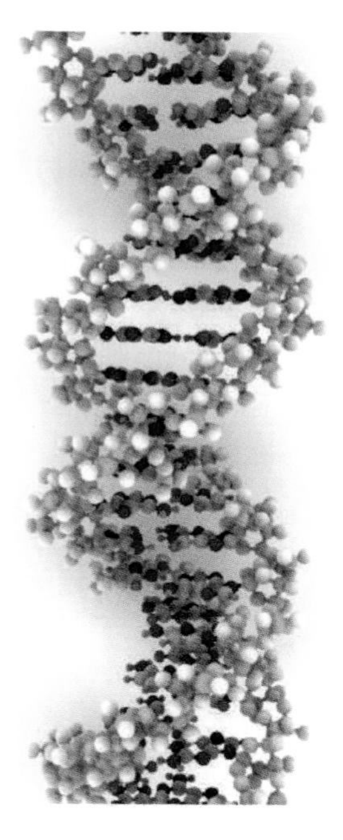

Fig. 3.6 A molecular model of DNA.

It should be emphasized again that this process of replication has nothing to do with the "information" carried by the sequence of letters. In what sense this sequence carries information will be discussed shortly. At this point it should be clear that the sequence does not contain information in the colloquial sense of the word, nor is SMI involved in this process. Each organism has its own unique sequence. You can think of the DNA as a little book which contains a long sequence of letters. Each organism carries its own unique book. Each cell in the organism contains the entire book of that organism. When a cell replicates, its DNA is also replicated. At this stage, it does not matter whether or not the book contains any meaningful message. It is like a unique identity card carried from generation to generation, or like a barcode on an identity card which contains a sequence of symbols. It reveals the identity of the person only when the sequence is translated into a meaningful message. In the case of the DNA, the translation of the sequence of letters A, T, G, C into a meaningful message is more complicated and involves several stages. At the final stage of this translation process, the sequence does not reveal its meaning, but it leads to the *production* of the organism, and thereby reveals the *identity* of the organism.

3.2.2 *Translation of the information from the DNA into proteins*

As in the previous section, we say that the *information* contained in the DNA is *translated* from the language of DNA into the language of protein. The information itself is not specified. It is certainly not the kind of information you read in a book. Even without specifying the kind of information which is written on the DNA, we use the term "flow of information" to describe the translation of whatever is written on the DNA into a new language. In fact, the book of life consists of "chapters." Each chapter, referred to as a *gene,* is translated into one protein. Here, we will discuss only that part of the DNA which is translated into proteins. There are vast regions in the DNA which are not translated into proteins. Some of these regions are involved in regulation and control processes, and some are still not fully understood.

The translation of the sequence of letters in the DNA into a new sequence of amino acids in a protein is also very complicated, and involves many intermediate steps, all executed and controlled by many enzymes. We will focus only on the formal translation from the 4-letter language A, T, G, C into the 20-letter language of amino acids, skipping the actual mechanism of this process of translation. The letters in proteins are also different molecules, which are shown in Figure 3.7.

Clearly, one cannot translate each letter in the DNA into a unique letter in the protein. Also, pairs of letters in the DNA will not be sufficient for translation into 20 letters of the protein. However, triplets of letters in the DNA are more than enough for translation into 20 letters of the protein. The "dictionary" for this translation is known as the genetic code.

This code is shown in Table 3.1. (To read the code word for an amino acid, look at the letter in the first, second and third positions, then look at the amino acid at the center. For instance, to TTT corresponds the amino acid phenyl alanine, which is abbreviated to phe; to the triplet CCC corresponds proline, abbreviated to pro.) The actual translation process occurs in the ribosomes. The mechanism is schematically shown in Figure 3.8. First, the information on the DNA is *transcribed* into mRNA. This step is essentially the same as the replication process (see Figure 3.2). The information contained in the DNA is simply transcribed into the messenger RNA (mRNA), which carries the same information. Then, each triplet on the RNA is "read" by a molecule (tRNA) which on one side binds to the triplet of the RNA, and on the other side binds to one amino acid. As the sequence of letters on the RNA is read, triplet after triplet, a corresponding sequence of the protein letters is produced on the other side. The net result is a sequence of amino acids which forms a protein.

The reader might get the impression that all these processes of transcription and translation are simple. They are anything but simple. The discovery of these processes took many years

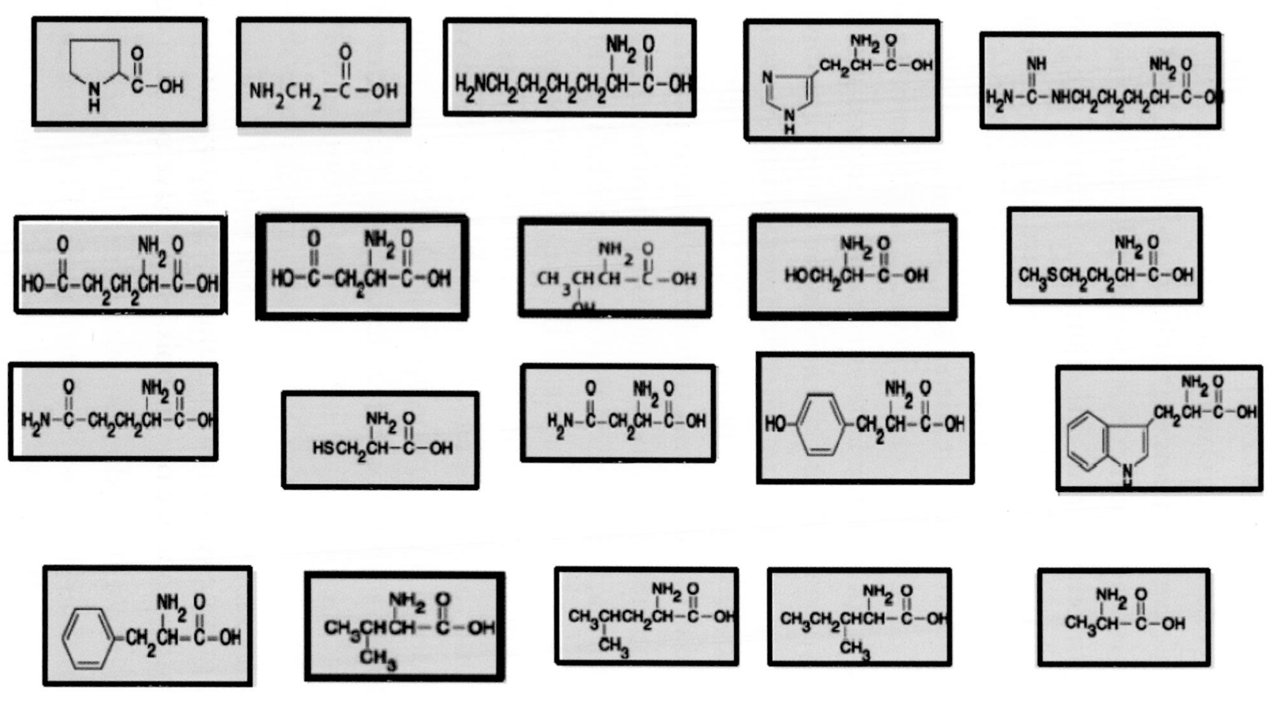

Fig. 3.7 The 20 letters (amino acids) of proteins.

Table 3.1 The genetic code.

First position	Second position				Third position
	T	C	A	G	
T	Phe	Ser	Tyr	Cys	T
	Phe	Ser	Tyr	Cys	C
	Leu	Ser	Stop	Stop	A
	Leu	Ser	Stop	Stop	G
C	Leu	Pro	His	Arg	T
	Leu	Pro	His	Arg	C
	Leu	Pro	Gln	Arg	A
	Leu	Pro	Gln	Arg	G
A	Ile	Thr	Asn	Ser	T
	Ile	Thr	Asn	Ser	C
	Ile	Thr	Lys	Arg	A
	Met	Thr	Lys	Arg	G
G	Val	Ala	Asp	Gly	T
	Val	Ala	Asp	Gly	C
	Val	Ala	Glu	Gly	A
	Val	Ala	Glu	Gly	G

of hard work by many scientists. There are still gaps in our understanding of all the molecular details of these processes.

We stress again that at this stage the translation from one language to another does not depend on the *information* which is carried by the sequence of symbols. As we have pointed out in the previous subsection, whatever information is carried by the DNA, it is not *information* in the colloquial sense, nor is SMI involved in the process.

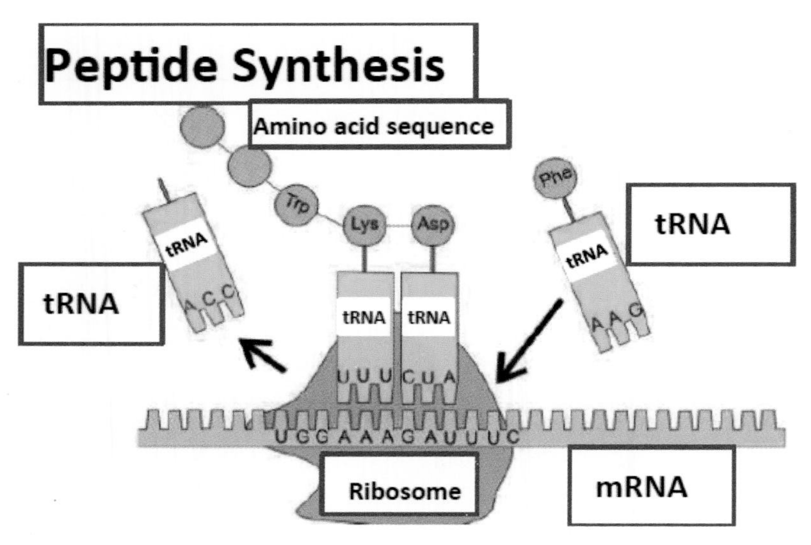

Fig. 3.8 The translation of information from the DNA language (via the mRNA) to the protein language.

3.2.3 *Execution of the information in the process of folding the protein*

Once proteins are formed, the processing of the *information* that they got from the gene continues, but now the information is translated into "executable orders": to build a tissue, to accelerate a reaction, to regulate a process, and so on. All these processes constitute the totality of activities of a living organism, which we identify as life.

The sequence of amino acids that has just been synthesized must first fold into a unique, stable and functional 3D structure.[9] This process is nowadays referred to as *protein folding*, (Figure 3.9). It is considered one of the most challenging, daunting problems in molecular biology.[10]

How does the sequence of letters of amino acids "know" how to fold into a precise 3D structure which is stable enough to survive and, long enough so that the protein can perform

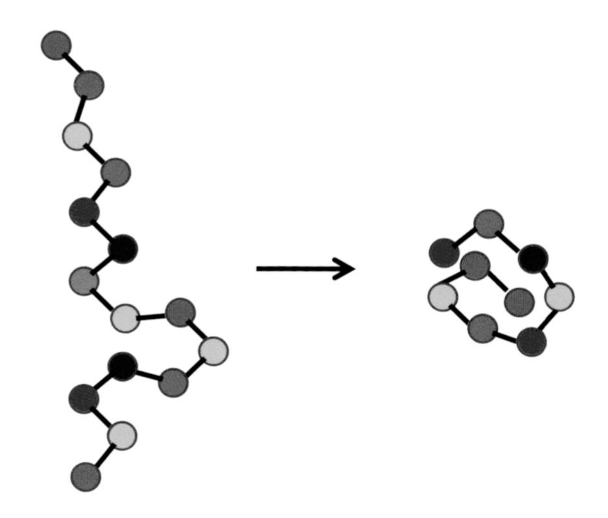

Fig. 3.9 A schematic process of protein folding.

a function in the organism? This problem was listed as one of the 125 "big unknowns" of science by the editors of *Science* in 2006.[11]

> Can we predict how protein will fold? Out of a near infinitude of possible ways to fold, a protein picks one in just tens of microseconds. The same task takes 30 years of computer time.

There are essentially three problems involved in the protein folding problem. The first is: How does the sequence "know" how to fold into the correct 3D structure? The second is: What makes the 3D structure stable so that it can function? The third is: How does it fold so rapidly into the 3D structure?[10]

In all of these questions the term "*information*" has been involved, not in its precise sense of a message consisting of the letters of the alphabet, but in a more colloquial sense. People have speculated that the "information" contained in the sequence of amino acids is sufficient for determining the 3D structure of the

Table 3.2 List of amino acids and their code letters.

Amino acids		
Ala	Alanine	(A)
Arg	Arginine	(R)
Asn	Asparagine	(N)
Asp	Aspartic acid	(D)
Cys	Cysteine	(C)
Glu	Glutamic acid	(E)
Gln	Glutamine	(Q)
Gly	Glycine	(G)
His	Histidine	(H)
Ile	Isoleucine	(I)
Leu	Leucine	(L)
Lys	Lysine	(K)
Met	Methionine	(M)
Phe	Phenylalanine	(F)
Pro	Proline	(P)
Ser	Serine	(S)
Thr	Threonine	(T)
Trp	Tryptophan	(W)
Tyr	Tyrosine	(Y)
Val	Valine	(V)

protein. This *informational language* probably originated from Anfinsen's *thermodynamic hypothesis*.

The relevant part of the thermodynamic hypothesis is[12]:

> ... the native conformation is determined by the totality of the inter-atomic interactions and hence, by the amino acid sequence, in a given environment.

Many scientists interpreted Anfinsen's hypothesis as if it implied the existence of a *folding code*. One can find statements referring to the "second half of the genetic message," "cracking the second half of the genetic code" or "the second quarter or the second fifth. . . . " of the genetic code. All of these statements allude to the possibility that there exists some "code" which, similar to the genetic code, translates from one language to another — in this case, from the sequence of amino acids to a specific 3D structure of protein.

If such a "code" could ever be found, it would open the door to a whole new area of discovery of newly designed proteins which could perform new and desirable functions, but which were not "discovered" by the process of natural evolution. Such hopes were largely frustrated, however. No such code was found, and in my opinion it is extremely unlikely that such a code will ever be found.[10]

Indeed, the sequence of amino acids along the protein *determines* the 3D *structure* of the protein, and the eventual *function* of the protein. However, as Anfinsen realized, the "reading" of the information contained in the sequence, and "executing" the folding process, depends not only on the sequence of amino acids but also on the "given environment." The main component in the "environment" which is relevant to "reading" and "executing" the information is most likely the water molecules.[13]

Figuratively, one can say that water molecules first "read" the "information" in the sequence of amino acids. This information is then translated into a specific 3D structure. Again, this is only a figure of speech. Water molecules *interact* with the various amino acids. They exert different forces on different groups along the protein. A given pattern of amino acids "dictates" a unique pattern of forces. When these forces are strong enough (stronger than the random thermal forces), the protein will fold along a (nearly) unique path to form the native 3D structure. There is no *information* involved in this process.[10]

Thus, unlike a book which tells a story, or an instructional manual which provides instructions, the protein does not contain any information of this kind. Instead, water molecules exert strong forces on various groups in the protein. These forces *force* the folding of the protein into the final structure.

One can find many statements in the literature where intermolecular interactions are interpreted as "exchange of information." For instance, two argon atoms at a distance of about 3–5 Å *attract* each other. One can say that one atom "sends a message" to the other atom, which says, "Come closer to me." When they are at very short distances, they repel each other. Again, one can say that one atom sends a "message" to the other, saying, "Get out of here."

Similarly, one says that some solutes (hydrophobic) "fear" or "hate" water. Other (hydrophilic) solutes are "loved" by water. All these personification expressions are only figures of speech. There are no messages exchanged between atoms and molecules, or feelings of love and hate between molecules.[10]

Although I have nothing against using these terms in connection with atoms and molecules, I believe that one should be careful not to get carried away and apply the laws of physics to such abstract "information" carried by molecules, or to the "love" and "hate" feelings between molecules.

This is not a joke. People do apply the *laws of thermodynamics* to "information" and talk about the laws of information. In fact, there is not a single law of information which applies to information, or to SMI. There are of course some theorems involving SMI, but these have nothing to do with thermodynamics.

Without going into any details of how water molecules do the *reading* and *executing* of the information written in the sequence of amino acids, the final result is a unique protein (in a specified environment) which can perform a unique *function.*[14] The proteins can go on to associate with other proteins (see next subsection), and can accelerate a chemical reaction, transport oxygen from one part of the organism to another, and perform a myriad of other functions. One can ask the following question: Is the information on the function of the protein "written" in its structure, which was written in the sequence of amino acids, which was written in the DNA? As before, we talk about the information *contained* in the protein, but this is only a figure of speech. The protein does not have the "knowledge" about how to fold into a precise 3D structure. The resulting 3D structure does not have the "knowledge" about its mission in the organism. Instead, the protein which has just been folded goes on to do whatever it can do, not because it *knows* what to do but because it *can* do what it does.

Regarding the stability of the 3D structure of proteins, Volkenstein (2009) writes:

> The information contained in an albumin molecule is not limited to its bare molecular structure ... an albumin molecule has a globular spatial structure determined ultimately by the entropic properties of water, that is, by hydrophobic forces.

This is a vague, unfounded and meaningless statement. Hydrophobic forces are *not* "entropic properties of water," and the "entropic properties of water" — whatever that means — do not determine the structure of proteins.

It is sometimes explained that proteins were *designed* by evolution to perform some function. For example, if in the course of evolution the organism (or Nature or a Creator) needed some agent to transport oxygen, or to catalyze a chemical reaction, then an appropriate molecule was "designed" to fulfill this function.

However, this is not the way evolution works. Evolution does not *design* what it needs — a single molecule, or a whole organ. Evolution "acts" on whatever already exists. During evolution, molecules, organs and entire organisms are produced. A molecule which happens to have the capability of doing some useful function is selected. Evolution does not design anything to do a job it needs. It makes use of whatever job an evolved molecule or an organ is able to do. Wings were not *designed* to enable birds to fly. Wings evolved in some organisms. These organisms exploited the capability of the wings to enable them to fly. Likewise, hemoglobin was not *designed* to transport oxygen (efficiently!). It was synthesized, and once the organism "recognized" its ability to transport oxygen, it created the job of an oxygen carrier and transporter.[15]

Along the entire process leading from the DNA to the protein to a specific function, and ultimately to specific living organisms, we use the phrase "information contained" in the DNA, and translated into protein, into structure, and into function. However, care must be exercised to distinguish between this "information" and the information handled by *information theory*. Failing to make this distinction is a source of great confusion leading to application of the "laws of information" or the "laws of thermodynamics" to the genetic information to which these laws do not apply.

3.2.4 *Execution of the information in the process of self-association and molecular recognition*

Once proteins fold into their native structure, they can either go ahead and carry out a specific *function*, or associate with other molecules to form a larger assembly of macromolecules which carry out a different function. The secrets of the spontaneous assembly of proteins, and the specific mode of association also known as *molecular recognition,* were also deemed to be one of the 125 big unknowns of science.[11]

The association of *information* with the recognition process is different from the way information is translated from the DNA to a sequence of proteins to the structure of protein. Here, the two proteins have to "recognize" the correct binding mode, and this binding should be unique and stable enough so that the resulting aggregate will be able to function.

Molecular recognition is very common in biochemistry, specifically in enzymatic reaction, in immunology, in the translation of the genetic information, and so on. There are two main principles for molecular recognition. The dominant paradigm is the so-called "lock and key" model. The idea here is that the solute "recognizes" a specific site on the protein. The recognition is realized by the total *direct* interaction energy between the solute (ligand) and the site. In the simplest case, maximum interaction involves geometrical fitting into the site. As can be seen in Figure 3.10(a), the solute (L) fits into the binding site on the protein (P). Better fitting is equivalent to the maximum interaction energy between the solute and the site. We may refer to the lock and key model as a geometrical fitting. In Figure 3.10(b) we also show a possibility of fitting due to a

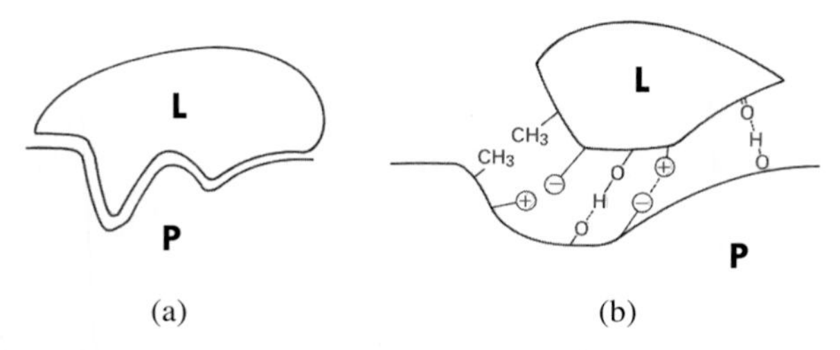

Fig. 3.10 Lock-and-key mechanism of recognition: (a) geometrical fit; (b) fitting by means of patterns of functional groups.

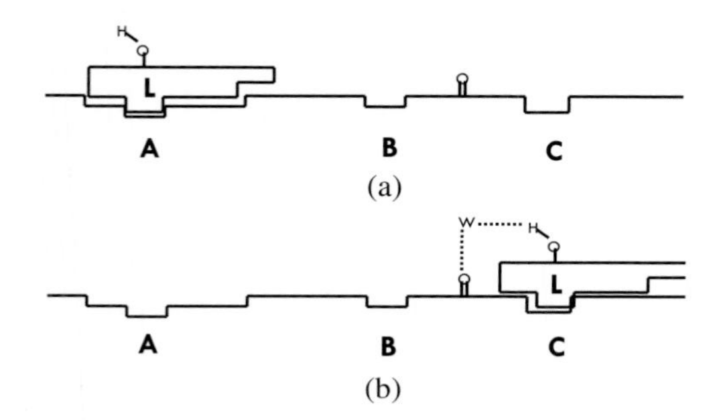

Fig. 3.11 (a) Best fitting of L to site A; (b) stronger binding to site C due to the hydrogen bond bridge.

pattern of functional groups, such as charges of opposite signs or, hydrogen bonding. In all of these cases we may say that the solute "reads" the surface of the protein and selects to bind to that site which better fits either its geometrical structure or its pattern of functional groups.[15]

A completely different mechanism of molecular recognition occurs when there are functional groups that are not *in* the binding interface, but can form hydrogen-bonded bridges. We demonstrate this method of recognition in Figure 3.11. In Figure 3.11(a) we show a ligand L and three sites: A, B and C. Clearly, according to the lock and key model, site A will be the preferred binding site simply because the geometrical fitting between the ligand L and site A is the best. We can also say that the probability of binding to site A is larger than that of site B, and the probability of binding to site B is larger than that of site C. This is true either in vacuum or in a simple solvent.

On the other hand, in water, the preference for binding to the different sites may change. We illustrate this possibility in Figure 3.11(b). Here, at site C, a water molecule can form a hydrogen-bonded *bridge* between the ligand L and the protein. Note that the two functional groups — here two carbonyl groups — do not belong to the interface between the solute and the protein. If such a hydrogen-bonded bridge is possible, then the geometrical fit might become negligible or even irrelevant. In Figure 3.11(b) we show that the possibility of a hydrogen-bonded bridging could *reverse* the preference for one site over the other. In such a case, we may say that the solute "recognizes" the site at which it can form the maximum number of hydrogen-bonded bridges. In this case, the probability of binding to site C will be larger than for either site A or site B.

In Section 3.3 we will discuss another aspect of the binding phenomena involving "communication" between sites, i.e. transmitting information from one site to another.

It is interesting to note that Pauling thought that the specificity of binding phenomena was an essential feature of life. In his biography we find the following comment[16]:

> Life at the molecular level, Pauling began to realize, was largely a matter of specificity, of molecules in the body being able to recognize and bind to specific antigens, enzymes to specific substrates, genes, in some mysterious manner, both to each other and their specific protein products. The mechanism of this exquisite biological specificity was unknown. But Pauling thought that his approach, based on precisely complementary shapes, was the key.

Pauling believed that the laws of chemistry and physics would eventually answer the question of what life is. His idea was that the specificity of the binding — say, between antibody and antigen — could explain much of life's phenomena. Personally, I agree that molecular recognition, i.e. the specificity of binding phenomena, is ubiquitous in biological systems. I cannot accept the view that these phenomena fully characterize life, and they certainly do not answer the question "What is life?"

> Life, in Pauling's view, could be reduced to the possession of specific characteristics and the ability to produce progeny to which these specific characteristics are passed on. It was a matter of molecular specificity, and it could be adequately explained by the principles of chemistry.

Indeed, the specificity of the binding *can* be explained by the principles of chemistry. In fact, the lock and key model is already well explained by the totality of the interactions between atoms on the ligand and atoms on the protein or the DNA. When the binding occurs in aqueous solutions, the strength (or the Gibbs energy) of the binding is also explained by the principles of chemistry. Here, the main contribution to the binding strength is mediated by the solvent, as was explained above, specifically by water molecules forming hydrogen-bonded bridges between groups that are referred to as hydrophilic groups (carbonyl, hydroxyl, etc.).[10,15]

3.3 Application of Information Theory to DNA

Perhaps the first application of information theory to DNA was carried out by Gatlin (1972). In her book *Information Theory and the Living System*, the opening sentence is[17]:

> Life may be defined operationally as an information processing system — a structural hierarchy of functional units — that has acquired through evolution the ability to store and process the information necessary for its own accurate reproduction. The key word in the definition is information.

It is certainly true that living organisms process information, but this is not the *only* characteristic function of the organism. In general, one cannot use a *characteristic* property of a living system as a *definition* of a living system. A living organism can also be said to be an extremely sophisticated factory of processing chemicals. This is also true, yet far from being a *definition* of a living organism.

Gatlin continues:

> It is obvious from a simple inspection of the language of the biologist that the word information is indispensable. It appears on page after page of any modern biology textbook. For example, Stent (1963) states that mutations are: "a sudden change in informational content of the hereditary substance"; "bacterial viruses are the carriers of a complex hereditary apparatus whose informational content must be very high"; "these viruses bring into the host cell sufficient genetic information for the construction of all enzymes . . . "; etc. The word is vital to the vocabulary of the biologist.

Genetics is, of course, not the only branch of biology where the term "*information*" is used very frequently. Much of what is going on in the brain may also be referred to as information processing. Furthermore, *transmission of information* between single molecules, between cells, between organs, etc. is a common way of "communication" between molecules, cells or organs. At higher levels, we also speak of exchanging information between animals, and at the highest levels, between human beings. It was the last kind of information exchange for which Shannon developed information theory.

Gatlin treats the sequence of bases in the DNA in the same manner as any sequence of symbols, or letters in any language:

> Every individual living system has its own unique sequence of bases along its DNA chain, which we may regard as a sequence of symbols that stores information in the same manner as a sequence of letters in any language. In fact, one could regard language, an ordered sequence of symbols with a definite meaning, as the basis of all life, and the base sequence of DNA as a message in the genetic language.
>
> Since there are four kinds of DNA bases, over 4^{10^9} base sequences are possible for present-day organisms. This number is greater than the estimated number of particles in the universe.

Based on calculations made by Hoyer and Roberts (1967), Gatlin concludes:

> That if we wished to compare two or more sequences, we could possibly do this manually. It was also concluded that it appears unlikely at the present time that we will ever study the properties of the total DNA of living systems, particularly of higher organisms, by simply determining their base sequences.

This was in 1972. It is now more than 40 years since the publication of Gatlin's book, and much progress has been achieved, far beyond what was anticipated in 1972.

It is interesting to note that at the time of publication of Gatlin's book (1972) it was quite clear that the *genetic information* is stored in the DNA. On the other hand, it was far from clear how information is stored and processed in the brain — whether it is by means of a sequence of letters as in DNA, or as patterns of connections between the neurons.

Gatlin devotes a considerable part of the introduction to discussing the question of whether or not modern physics could handle living systems. She quotes Elasser (1958, 1966) and Wigner (1967), who both attempted to find out whether living systems obey the fundamental "laws of chemistry and physics." Neither Elasser nor Wigner believes that quantum-mechanical laws are adequate for explaining the living system. Elasser (1958) postulated the existence of higher "biotonic" laws which govern living systems.

The most forceful spokesman for the antireductionist viewpoint was Michael Polanyi (1967). He draws directly on the *laws of information theory* in his attempt to explain why quantum mechanics is inadequate for explaining the living system. He states that "all objects conveying information are irreducible to the terms of physics and chemistry." The reason is that we cannot understand an information-processing machine, or any machine for that matter, merely from a description of its hardware.

The major part of the book is concerned with the application of SMI to the sequence of the DNA. First, Gatlin makes a rough estimate of the maximum information content of a DNA, and asks, "Would this information be enough to describe the entire living organism?"

The answer to this question is that we *do not need* all the information to describe a living system. The information written in the DNA is enough for creating proteins. From then on the information "evolves" in the sense that the proteins "fold spontaneously" (with or without the aid of chaperons), then "self-assemble" spontaneously, and these new macromolecules "know" what they have to do to construct the entire animal.

In the last part of the introduction the author introduces the concept of entropy, as used in information theory:

> The unifying thread of our story is the entropy concept. As *Homo sapiens*, we have always believed that we are higher organisms. After all, we are more complex, more differentiated, more highly ordered than lower organisms. As thermodynamicists, we recognize these words and realize that the concept of entropy must somehow enter into our explanation. We have always had the vague notion that, as higher organisms have evolved, their entropy has in some way declined because of this higher degree of organization. For example, Schrödinger made his famous comment that the living organism "feeds on negative entropy."

In my view, Gatlin confuses here thermodynamics entropy with informational entropy (SMI). Furthermore, the author expresses the common misconception about "higher organisms," "higher degree of organization" and "lower entropy." I will further discuss the idea of feeding on "negative entropy" in Section 3.5.

Notwithstanding these misconceptions about entropy and information, Gatlin was undoubtedly a pioneer in the application of SMI to DNA. We present here only the rudiments of her calculations.

First, note that since the letter A (adenine) pairs with T (thymine), and the letter C (cytosine) pairs with G (guanine), the probability of the occurrence of A must be the same as for T, and similarly for the probabilities of C and G.

However, the probabilities of the occurrence of A and C (or of T and G) will depend on a particular organism, much as the frequencies of letters in the English alphabet would depend on a specific author or a specific book.

From the base composition in the DNA of *Micrococcus lysodeikticus* (ML), one finds that

$$\Pr(A) = \Pr(T) = 0.145,$$
$$\Pr(C) = \Pr(G) = 0.355.$$

Hence, the corresponding SMI is

$$\text{SMI}^{(1)}(\text{ML}) \approx 1.87 \text{ bits}.$$

On the other hand, for *E. coli*, one finds that

$$\Pr(A) = \Pr(T) = \Pr(C) = \Pr(G) = \frac{1}{4}.$$

Hence

$$\text{SMI}^{(1)}(\text{E. coli}) = \log 4 = 2 \text{ bits}.$$

Note that the superscript (1) refers to the value of the SMI for the distribution of *single* letters of the alphabet.

The *divergence* from the uniform case is defined by

$$D_1 = \text{SMI}^{(1)}(\max) - \text{SMI}^{(1)},$$

which for *E. coli* is zero, but for ML is

$$D_1 = 2 - 1.87 = 0.13 \text{ bits}.$$

In the second step Gatlin defined the SMI of pairs (or doublets) of bases in the DNA. This is the same as analyzing the dependence between consecutive letters in any alphabet.[18]

From the SMI of single bases and the SMI of pairs of bases, Gatlin also computed the *redundancy* in the DNA of several organisms — from viruses to bacteria to invertebrates and vertebraes. The results obtained are not conclusive. The values of the redundancy seem to vary widely for phages and bacteria, whereas for vertebrates the variations are much more moderate. It is not clear what these results mean in terms of the information contained in the DNA. This conclusion is similar to that reached for the redundancy in different texts or different languages.

Gatlin also used Shannon's second theorem (which deals with the question of how to transmit a message through a noisy channel with minimal error) for the transition of the information from the DNA to proteins. In this case the *source* is the DNA and the target is the sequence of amino acids in proteins. Then she speculated about the connection between the redundancy in the DNA and the probability of its selection in the process of evolution.

Since Gatlin's pioneering work many scientists have used information theory in connection with the information in the DNA. [For a review, see Erill (2012).]

Recently, Adami (2004) has published a few interesting articles discussing the application of information theory to biology. He pointed out that the genetic information in the DNA is transmitted from generation to generation. In this sense the information is conserved. On the other hand, the information itself is the subject of evolution. Adami also estimated the information content of proteins, as well as the conditional information, or the mutual information between pairs of amino acids.

Finally, I feel it is my obligation to science to comment on a recent book titled *Genetic Entropy and the Mystery of the Genome,* by Sanford (2005). This book uses the term "entropy" just for *embellishing* its title. The book itself is a flagrant misuse, or perhaps abuse, of that term. I have added some further comments on this book in Note 19.

3.4 Transmitting Information Between Molecules — Direct and Indirect Interactions

In the previous section we discussed the SMI associated with the DNA, and how this "information" is transmitted from generation to generation in the process of evolution. We now turn to discussing "transmitting information" in a more colloquial sense, which has nothing to do with information theory.

Any two or more molecules which approach each other within a short range of distances interact with each other. They can attract or repel each other. This interaction may be viewed as a result of exchange of "messages" between the two molecules. We may say that the two molecules "communicate" with each other by *direct* intermolecular interaction. The most conspicuous kind of communication occurs when the two particles — say, two argon atoms — approach each other within a very short distance — say, about 3.4 Å (1 Å equals 10^{-8} cm). The two molecules strongly repel each other. One atom cannot penetrate the spherical region around the second molecule, (Figure 3.12). As if each atom were claiming, "Here I am, this is my territory, no other molecule is allowed to enter."

At larger distances, the molecules can attract each other and sometimes repel each other (say, if they carry electric charge of the same sign); see Figure 3.13.

In biological systems we also talk about communication between molecules which is not due to *direct* intermolecular interaction. The best-known example is the "communication" between two oxygen molecules bound to hemoglobin.

The story of hemoglobin is quite long.[15] I will discuss only one aspect of hemoglobin which is relevant to the

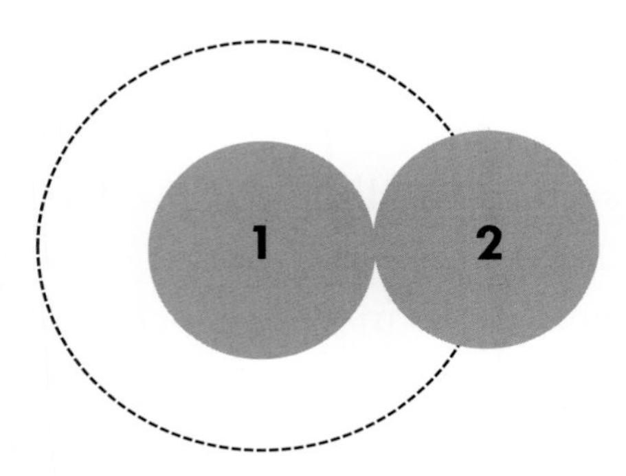

Fig. 3.12 Repulsion at very small separation. An atom (2) approaching another atom (1) cannot penetrate an excluded volume (dashed circle) produced by the repulsion between the two atoms.

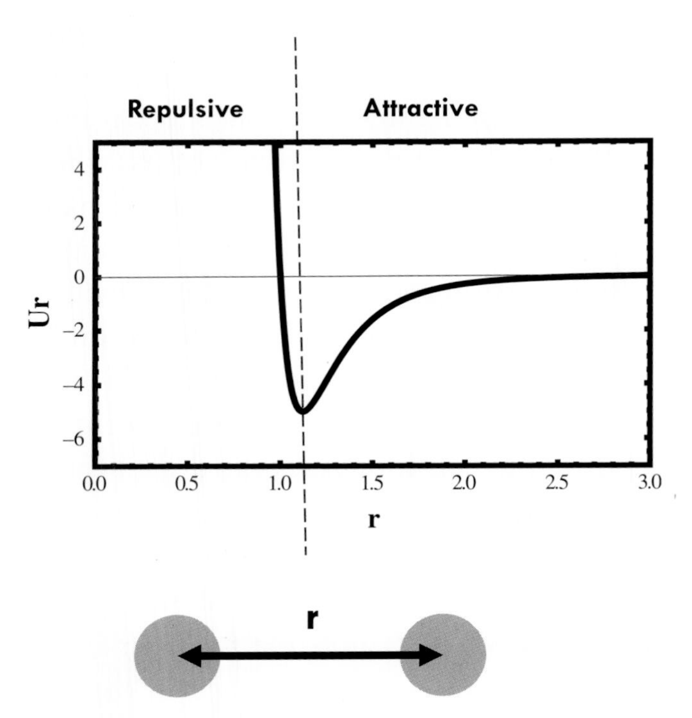

Fig. 3.13 Attractive and repulsive regions as a function of the distance between two atoms of unit diameter.

Fig. 3.14 A protein (P) with two binding sites for a ligand (L).

communication between two or more oxygen molecules bound to the four sites on hemoglobin.

Soon after the discovery of the function of hemoglobin as a carrier of oxygen molecules, it was found that the binding of the oxygen molecules is *cooperative*. For the purpose of this section, *cooperative binding* means that the probability of binding on one site depends on whether or not a second site is empty or occupied. It is here that the metaphorical description of "communication" enters. How does one oxygen molecule on one site "know" the state of occupancy on another site?

When there is a strong interaction between the binding solutes, the answer to this question is simple. If the solutes are charged, then clearly one solute "feels" the presence of the other solute through the Coulombic interaction; it can be either attractive or repulsive.

In Figure 3.14 we show two binding sites on a protein (**P**). Suppose that the two sites are identical and a solute (a ligand L) has the probability p of binding to any one of the sites 1 and 2. Once a site is occupied, the probability of binding to the second site depends on whether the two solutes attract or repel each other. When the binding curves of oxygen to hemoglobin were analyzed, it was found that the binding is highly cooperative.

We know that the oxygen molecules are not charged, and the distance between two binding sites on hemoglobin is larger than 20 Å. At this distance the oxygen–oxygen interaction is negligible. We can say that two oxygen molecules on different

Fig. 3.15 Hemoglobin molecules have four binding sites for oxygen molecules. These are shown as black rectangles in the figure.

sites do not "see" each other. Yet, there seems to be strong cooperativity between the oxygen molecules.

Pauling (1935) was puzzled by this apparent strong "interaction" between the two oxygen molecules on two different sites. At that time the structure of hemoglobin was not known. It seems that the two oxygen molecules "communicate" with each other. The first oxygen has some probability Pr(1) of binding to any of the four sites on the hemoglobin (Figure 3.15). The second oxygen, approaching any of the other unoccupied sites of the hemoglobin, has a *large* probability of binding — say, Pr(2) > Pr(1) — as if they sense that one site is already occupied. Alternatively, we can say that when an oxygen molecule binds to one site, it sends a *message* to all other sites saying, "Here I am, you can bind to the remaining sites with greater probability or with stronger binding energy."[20]

This kind of "communication" continues for the third oxygen, which binds with a different probability, Pr(3) and the fourth oxygen, with probability Pr(4).

How is this "communication" achieved? It is known that the distance between any two oxygen molecules on two sites is quite large, so that *direct* interaction between the molecules is negligible at such distances.

The secret underlying this type of communication is referred to as the allosteric effect.[15] The two (or more) oxygen molecules "cooperate" not by *direct* interaction, but through *indirect* "communication" mediated through the hemoglobin molecules. The molecular mechanism of this type of communication is discussed in great detail in Ben-Naim (2001). Very briefly, once an oxygen molecule binds to one site, the *structure* (conformation) of the protein changes. Therefore, the second oxygen approaching the hemoglobin "sees" a different site, and hence the binding energy is now different. Figure 3.16 shows schematically the two structures of hemoglobin, referred to as oxyhemoglobin and deoxyhemoglobin. Thus, by changing

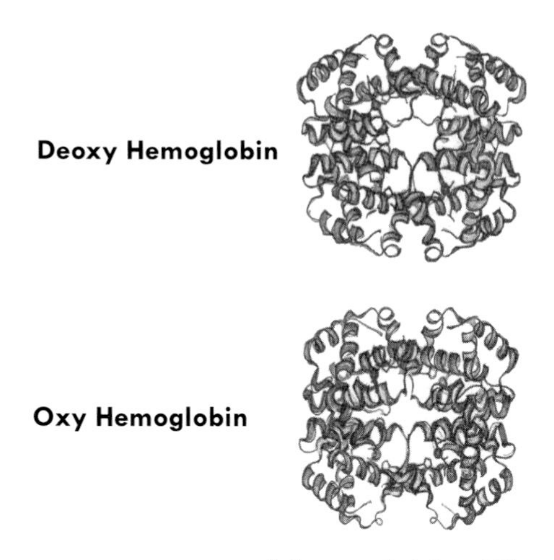

Fig. 3.16 The two structures of hemoglobin. The oxy hemoglobin is obtained when the concentration of oxygen is high. At low oxygen concentration the deoxy structure is obtained.

the structure of the hemoglobin molecule, the two oxygen molecules "communicate" between themselves. Of course, this communication is different from the type of communication we use in daily life.

It was found that this type of communication is very common in other biological processes, from enzymatic activity (or inhibition) to the translation of the information on a specific gene.[15]

Perhaps one of the most exciting cases of cooperative binding is associated with the translation of specific genes in specific cells. The entire DNA of an organism is contained in the nucleus of each of the cells of that organism. However, not all the genes in each cell are expressed and translated into protein. How does a cell "know" which "chapter" in the DNA to "read" or not to "read"? The mechanism for such communication was proposed by Jacob and Monod.[21,22]

In each cell are specific proteins called *repressors* which bind to a specific site on the DNA (referred to as the *operator*), and this binding turns on and off the "reading" of the genetic information on the following gene. This binding is highly cooperative and was referred to as a *genetic switch*.[23]

To summarize, any two molecules are said to *communicate* with each other through their interactions. In biological systems, we find many examples where this kind of communication is *indirect* and mediated by a protein or a DNA molecule.

3.5 Information Storage and Processing in the Brain

In Section 3.2 we discussed the storage and processing of information in the DNA. The brain is also said to store and process information. However, the information stored and processed in the brain is very different from that stored in the

DNA. For the present discussion we exclude the processes in the DNA of the neuron, and focus on the mechanism of storing and processing this information by the combined effects of many neurons. Recent models of brain functions suggest that the information stored in the brain is not by means of an "alphabet," but in specific patterns of *connections* between the neurons where it is stored either by creating new neuron–neuron connections or by strengthening already existing connections. The number of neurons in the brain of a human being is about 100 billion. It is estimated that each neuron is connected to some 1000 other neurons. Thus, with such a huge neural network we can have an immense number of patterns of connections. This number exceeds by far the number of neurons. It is difficult to estimate the capacity of the brain to store "information." The different estimates available are highly speculative. It is clear, however, that the storage of information is different from the way a computer stores information. It is probably similar to an analog format rather than digital.

Once a message is received in the brain, the brain interprets this pattern of connection as a color, smell or taste. In Chapter 1 we discussed the sense of *heat* and *cold* and how it is transmitted from the skin to the brain. The storage is not in terms of frequencies of electromagnetic waves, or by molecules that have created the sense of hearing or smelling; nor is it in terms of a sequence of letters like "red" or "bitter." When we retrieve that information, it is not in terms of a sequence of letters, but as *experience* of "redness" or "bitterness," etc. One of the most interesting features of memory is its "associative" character. You see or hear something and you recall something *associated* with what you have seen or heard. Sometimes a smell might remind you of a food you ate a long time ago. A story you hear might remind you of a similar or related story. How many times have you forgotten a person's name, then by recalling his first name you easily remembered his last name? All these are explained as a result of similarity between the two patterns associated with the two things that you see or hear.

The processing of information in the brain is again very different from the processing of information in the DNA. The processing of information in the brain is not in a digital language. (Note, however, that models of brain function are based on a digital language of zeros and ones for "firing" and "not firing" neurons, respectively.) We are far from understanding how this processing of information is carried out. For instance, how do we reach a conclusion to elect a specific person to be a president? Suppose that all the people in a country get the *same information* about a certain candidate. Each person processes this information differently and comes to a different conclusion. The reason is that the processing of information depends on the individual, who already has a large storage of other information, as well as a way of processing the information which is unique to each person.

There are many processes involving information whose mechanism we do not know. How does the brain prove a mathematical theorem? How does it create a fictitious story which people enjoy reading (and different people respond differently to the same story)? How does an artist create a drawing or a sculpture or a piece of music which evokes excitement and admiration in others (and again different people respond differently to seeing the same picture or hearing the same symphony, depending on their unique "state of mind")?

Whatever the mechanism of processing the information in the brain is, it is quite clear that it is different in an essential way from processing the information in the DNA. It is most unlikely to involve SMI — not even what we call information theory, or the science of information.

The processing of information by the brain is also very different from that by a computer. If you ask a person, "What is the color of the sky?", he or she *understands* the question, then gives you the *information* relevant to the question — say, "The sky is blue."

If you search Google, or any other search engine, for the same question, the program does not *understand* the question, nor does it process the question by using information theory. Instead, it examines the pattern of words in the question, compare it with the data stored on the Internet, and responds giving you links to the sites where this pattern of words or similar patterns appear.

If you ask for the weather in New York on a specific day, the program might give you the correct answer because it has built-in answers to specific answers. But if you ask Google, "What is the color of an apple?", you might get links to articles about colors, apples, food and, of course, the colors of iPhones and iPads.

Finally, if you say, "Name a president of the United States who is not buried in that country," you will get, among other information on the burial sites of various presidents, a reference to vos Savant's book (1972), where this question in answered.

The claim that "the new science of information is explaining everything in the cosmos, from our brains to black holes"[24] is in my opinion an unjustifiable exaggeration. It would be more accurate to — say that the "science of information," or information theory, *does not explain* anything in particular, *the brain's way of working*.

3.6 What Is Life?

When discussing the question "What is life?" one cannot avoid starting with the most famous book, written by Schrödinger.[25]

This book is based on lectures delivered by Schrödinger in Dublin in 1943. It was most influential for a long time and probably laid the cornerstone for the creation of the whole field of molecular biology. It has also encouraged many physicists to apply the methods of physics to biology. In this section we will present some of the highlights of Schrödinger's book.

In Chapter 1, the author correctly points out that "the physicist's most dreaded weapon, mathematical deduction, would hardly be utilized in explaining life. The reason for this was not that the subject was simple enough to be explained without mathematics, but rather that it is too much involved to be fully accessible to mathematics." Then Schrödinger makes the following statement:

> The obvious inability of present-day physics and chemistry to account for such events is no reason at all for doubting that they can be accounted for by those sciences.

It is clear from this and other statements that Schrödinger is a reductionist, in the sense discussed in Section 3.1. He attempts to explain the source of the difficulty in applying the methods of physics and chemistry to living systems. The fundamental difference between a living system and any piece of matter that physicists and chemists have ever handled is in the structure, or in the arrangement of atoms and molecules in the organism, which differs fundamentally from that of a system dealt with in physics and chemistry. It seems to me that Schrödinger, at least in this stage of the book, believes that once physicists enter into biology and apply their powerful arsenal of physical methods and theories, they will eventually answer the question posed in the title of the book.

Then the author asks a series of questions, such as:

> "Why are the atoms so small?" "Why must our bodies be so large compared with the atom?" "Must that be so? Is there an intrinsic reason for it? Can we trace back this state of affairs to some kind of first principle, in order to ascertain and to understand why nothing else is compatible with the very laws of Nature?"

In my view all these questions have no answer within physics or chemistry. Perhaps these questions belong to philosophy rather than to physics. In many places throughout the book Schrödinger uses the terms "orderliness," "organization" and the like to characterize life.

My impression is that Schrödinger uses the terms "orderly thing," "orderliness," "physical organization," "well-ordered organization" and similar terms in connection with life phenomena in anticipation of his use of the Second Law of Thermodynamics in later chapters. He discusses the statistical laws which govern the behavior of "an enormously large number of atoms." This is of course true of any system consisting of a large number of atoms and molecules, whether it is a piece of wood, a rock or a living system. He provides a few examples where there is competition between a "directing force" like a magnetic field or a "gravitational field" on one hand, and a random counteraction "heat motion" on the other hand.

Although he does not say it explicitly, it seems to me that Schrödinger attempts to use the example of a system of magnets, in a magnetic field, to explain life; as in a magnetic system, the *magnetic field* tends to *order* the molecules, whereas the random "*thermal motion*" tends to randomize or *disorder* the system.

Likewise, in a living system there is always the *random thermal motion* on one hand, and there must be some "ordering force" that competes with this random motion, to produce "order" and "organization" in a living system, on the other hand. But what is this "ordering force," or the "ordering field," which operates in a living system which competes with the random thermal motion? As far as I can tell, the author does not provide an answer to this question.

Thus, although Schrödinger explicitly describes himself as a reductionist, he does not explain how such a reduction can be achieved.

Chapter 6 is titled "Order, Disorder and Entropy." He starts with the very common interpretation of the Second Law in terms of the tendency of a system to go *from order to disorder*:

> It has been explained in Chapter 1 that the laws of physics, as we know them, are statistical laws. They have a lot to do with the natural tendency of things to go over into disorder. Life seems to be orderly and lawful behavior of matter, not based exclusively on its tendency to go over from order to disorder, but based partly on existing order that is kept up.

The idea that life somehow withstands the "natural tendency to go from order to disorder" appears very frequently in the literature; "life withstands the ravages of entropy," "life disobeyed the Second Law" and so on. Unfortunately, all these statements are *unwarranted*; there exists no law of nature that dictates going "from order to disorder" in the first place. The tendency of entropy to increase applies to an *isolated system*, and certainly not to a living system which is an open system, far from equilibrium.

It is only on page 74 that Schrödinger explicitly relates the Second Law to the behavior of living systems:

> The general principle involved is the famous second law of thermodynamics (entropy principle) and its equally famous statistical foundation. It is avoiding the rapid decay into the inert state of "equilibrium" that an organism appears to be enigmatic; so much so that from the earliest times of human thought some special non-physical or supernatural force (*vis viva*, entelechy) was claimed to be operative in the organism, and in some quarters is still claimed.

Then he asks, "How does the living organism avoid decay? The obvious answer is: By eating, drinking, breathing and (in the case of plants) assimilating. The technical term is *metabolism*." Indeed, this is the obvious answer. The highlight of the book is reached on page 76, where Schrödinger explains:

> What then is that precious something contained in our food which keeps us from death? That is easily answered. Every process, event, happening — call it what you will; in a word, everything that is going on in Nature means an increase of the entropy of the part of the world where it is going on. Thus, a living organism continually increases its entropy — or, as you may say, produces positive entropy — and thus tends to approach the dangerous state of maximum entropy, which is death. It can only keep aloof from it, i.e. alive, by continually drawing from its environment negative entropy — which is something very positive, as we shall immediately see. What an organism feeds upon is ***negative entropy***. Or, to put it less paradoxically, the essential thing in metabolism is that the organism succeeds in freeing itself from all the entropy it cannot help producing while alive.

In my view, Schrödinger dismally failed in his answer to the question he posed.

First, I certainly do not agree that everything that is going on in Nature means an "increase of the entropy." Second, that living things "produce positive entropy." Third, I have no idea what "the dangerous state of maximum entropy" (for a living organism) is. And, finally, that the only way it can stay alive ("keep aloof" from death) is by drawing *negative entropy* from its environment. Of course, I realize that such assertions have been made by many scientists. Unfortunately, all such statements are, in my opinion, meaningless. Entropy, by definition, is a positive quantity. There is no *negative entropy*, just as there is no negative temperature, or negative mass or negative time. Of course, there are processes involving negative *changes* in entropy, but there exists no *negative entropy*!

No doubt Schrödinger's statements are more than just a bad slip of the tongue. Of course, this had no effect on his reputation as a great theoretical physicist. It is unfortunate, however, that many others, scientists as well as nonscientists, fell into the pitfall created by Schrödinger's "negative entropy."

Throughout the book, Schrödinger identifies entropy as a measure of disorder, and the Second Law as the "tendency to go from order to disorder."

Entropy is sometimes correlated with the qualitative notion of order or disorder, but there is no general relationship between entropy and order. Furthermore, the Second Law is not equivalent to the "tendency to go from order to disorder." First, because entropy does not always correlate with disorder; and second, because the tendency of the entropy to increase is true only for an isolated system. Furthermore, it is by no means clear that living organisms are more "ordered" and maintain this "order" — whatever "order" might mean.

On page 78 Schrödinger concludes that "organization is maintained by extracting order from the environment."

> Living organisms ... delay the decay into thermodynamic equilibrium (death), by feeding upon negative entropy, attracting a stream of negative entropy upon itself ... and to maintain itself on a stationary and fairly low entropy level.

Since there is no way of measuring or calculating the "entropy level" of a living system, all these statements are outright meaningless. They certainly do not answer the question posed in the title of his book.

In the last chapter of the book Schrödinger speculates on the possibility that the present laws of physics might not suffice to explain life:

> We must therefore not be discouraged by the difficulty of interpreting life by the ordinary laws of physics. For that is just what is to be expected from the knowledge we have gained of the structure of living matter. We must be prepared to find a new type of physical law prevailing in it. Or are we to term it a nonphysical, not to say a super-physical, law?
>
> No. I do not think that. For the new principle that is involved is a genuinely physical one: it is, in my opinion, nothing else than the principle of quantum theory over again.

The basic idea that we might need to extend the theory of physics in order to understand life was elaborated eloquently by Penrose (1989, 1999). Whether such an extended theory would evolve new concepts which are "physical" or "super-physical" is not clear.

In conclusion, Schrödinger's book was no doubt a very influential one, especially in encouraging many physicists to enter into biology. Most people praised the book, but some expressed their doubts about its content. Perhaps the most famous skeptic of Schrödinger's contribution (to the understanding of life, of course) was Pauling.

Hager's (1995), in his biography of Linus Pauling, writes about Pauling's view about Schrödinger's book:

> Pauling thought the book was hogwash. No one had ever demonstrated the existence of anything like "negative entropy" ... Schrödinger's discussion of thermodynamics is vague and superficial ... Schrödinger made no contribution to our understanding of life.

I fully agree with Pauling's views about Schrödinger's use of thermodynamics in connection with life.

3.7 Fifty Years After *What Is Life?*

Fifty years after the publication of the book *What Is Life?* a conference was held at Trinity College, Dublin, in September 1993, celebrating the anniversary of Schrödinger's lectures. A book titled *What Is Life? The Next Fifty Years*[26] presented the views of scientists from different disciplines on the original book, and on its influence on the ensuing development in the field of molecular biology.

Needless to say, no one offered an answer to this question. Though molecular biology has advanced hugely during those past 50 years, no one at the conference was able to answer this question. I would not venture into claiming that we might, in the near or distant future, have the answer. I believe that no one can predict when we will have the answer. Perhaps we will never have the answer to this question. In the rest of this section, I will present only a few of the highlights of this symposium, where some insightful thoughts were presented.

The first article, by Manfred Eigen, includes some interesting comments on "How is biological information generated?" He starts by listing the three essential characteristics of any living system:[27]

> Self-reproduction — without which the information would be lost after each generation. Mutation — without which the information is "unchangeable" and hence cannot even arise. Metabolism — without which the system would regress to equilibrium, from which no further change is possible (as Erwin Schrödinger already rightly diagnosed in 1944).

Systems which show these properties are predestined to selection. Eigen comments that "it would be meaningless to ask who does the selection" and then claims that "evolution on the basis of natural selection entails the *generation of information*" (italics mine). The information he refers to is the information written by means of the alphabet of the DNA bases (A, C, T, G).

Eigen, following Shannon, emphasizes that SMI *does not* deal with information itself, but rather with a measure of the size of the message — more precisely, the amount of information associated with the probability distribution (or the frequency of occurrence) of the various bases in the DNA. Regarding the answer to the question "What is life?" Eigen speculates:

> The legacy of biological research in this century will be a deep understanding of information-creating processes in the living world. Perhaps this entails an answer to the question "What is life?".

One of the most interesting chapters in that book is the one by J. Maynard Smith and E. Szathmary, titled "Language and Life":

> All living organisms can transmit information between generations. The property of heredity — that like begets like — depends on this transmission of information, and in turn heredity ensures that populations will evolve by natural selection. If we ever encounter, elsewhere in the galaxy, living organisms derived from an origin separate from our own, we can be confident that they too will have heredity, and a language whereby hereditary information is transmitted. The need for such a language was central to Schrödinger's argument in *What Is Life*?

I generally agree with the essence of this statement. I am not sure, however, that if we encounter some creatures from other galaxies, which we recognize as "living organisms," they will necessarily have "heredity, and a language whereby heredity information is transmitted." We are biased by the vast diversity of life on our planet, and all these forms of life have some common characteristics, such as reproduction and heredity. However, it might be possible that in some remote galaxy other creatures that we might encounter will have neither reproduction nor heredity. One can in principle envisage the evolution of a superliving system, which has the capacity to create new copies of itself without the need for reproduction as we know it on this planet. We might also envisage highly developed creatures that have the ability to "heal themselves," or rather to fix themselves, or to replace overused or defective parts. Such creatures could, in principle, live "forever," and there will be no need for evolution to "work" on such creatures to evolve, and therefore no need to reproduce.

Kaufmann, in a chapter titled "What Is Life? Was Schrödinger Right?", discusses and questions the validity of some of Schrödinger's statements. He makes a clear distinction between a *closed* and an *open* system. But then he makes some vague statements about "information" in open and closed systems:

> The critical distinction between a closed system at equilibrium and an open system displaced from equilibrium is this: in a closed system, no information is thrown away. The behavior of the system is, ultimately, reversible. Because of this, phase volumes are conserved. In open systems, information is discarded into the environment and the behavior of the subsystem of interest is not reversible.

It is not clear to me which "information is thrown away" in an open system, but cannot be thrown away in a closed system — "the behavior of the system is, ultimately, reversible." Again, it is not clear which kind of "reversibility" is referred to here. Such vague and perhaps meaningless statements do not add much to answering the question "What Is life?"

Penrose's chapter, titled "Why New Physics Is Needed to Understand the Mind," presents the view that physics, and physical laws as we know them, cannot possibly address the question of mind, including consciousness, and life in general:

> Human consciousness, on such a view, would be such a quality — so it is not simply a manifestation of computation. Indeed, I shall argue so myself; but more than this, I shall argue that those actions which our brains perform in accordance with conscious deliberations must be things that cannot even be simulated computationally — so certainly computation cannot of itself give rise to any kind of conscious experience.

Penrose's ideas are discussed in full detail in his book (1994). In my opinion this chapter, along with his book, contain the most insightful discussion on the problem of consciousness (and by implication also of life phenomena in general), and its possible description by any of the laws of physics known at present.

Schneider and J. Kay discuss the question of "Order from Disorder: The Thermodynamics of Complexity in Biology . . . ".

As I have noted before, I do not believe that invoking *order*, *disorder*, *organization*, etc. in connection with life is useful. Also, the association of "life" with more "order," and more "organized," compared to death — which is associated with the second law — is at best highly speculative.

I should add a personal comment here regarding the association of entropy and the Second Law with the question of life. I believe that if ever a theory of life will be developed, it will rely heavily on information theory, but not on entropy and the Second Law of Thermodynamics. In light of this belief, I think that Schrödinger's book, along with this fifty-years-after book, have unintentionally encouraged people to make a lot of meaningless statements associating entropy and the Second Law of Thermodynamics with life phenomena. A small sample of such statements is scattered throughout the book.

3.8 Entropy and Life

Open any book discussing the question "What is life?" and you are likely to read fancy and grandiose statements ranging from "Life violates the Second Law of Thermodynamics" to "Life emerges from the Second Law of Thermodynamics," to "The Second Law of Thermodynamics explains many aspects of life, and perhaps life itself."

Here is an example from Katchalsky (1965):

> Life is a constant struggle against the tendency to produce entropy by an irreversible process. The synthesis of large and information-rich macromolecules . . . all these are powerful antientropic forces . . . living organisms choose the least evil.

This is a beautiful statement, but devoid of any meaning. No one knows how to *define* the entropy of a living system, how much entropy is produced by a living organism, and what these powerful antientropic forces are.

The description of *life* as an "antientropic" force is very common. A diametrically opposite view was expressed by Volkenstein (2009):

> At least we understand that life is not "antientropic," a word bereft of meaning. On the contrary, life exists because there *is* entropy, the export of which supports biological processes . . . .

This statement is as bereft of meaning as the previous quotation by Katchalsky (1965).

As mentioned earlier, at present we do not have a definition of life. For the purpose of this section we will be more specific and say that we do not know how to characterize the *state* of a living organism. Therefore, the concept of entropy *cannot* be applied to a living system. This is true even if we could have somehow developed a thermodynamic theory which applies to systems far from equilibrium. Such a theory would probably not be applicable to a living organism. Therefore, all statements involving the role of entropy (either that life is "antientropic," or that "life exists because there *is* entropy") in living systems are at best meaningless.

In popular-science books, we find many statements which relate biology to *ordering*, and *ordering* to decrease in entropy. Here is an example[28]:

> The biological trend to order proceeds directly counter to the trend to disorder (entropy growth) that is one of the most fundamental characteristics of macroscopic physical systems.

It is not! First, because the association of "biological trend" with "order" is highly speculative. Second, the "trend to disorder" is a common but erroneous interpretation of the Second Law of Thermodynamics.[29]

Atkins (1984), in the introduction to his book, writes:

> In Chapter 8 we also see how the second law accounts for the emergence of the intricately ordered forms characteristic of life.

This is an unfulfilled promise. The Second Law of Thermodynamics does not account for the emergence of "ordered forms characteristic of life." And, of course, Chapter 8 does not show that.

At the end of Chapter 7, Atkins writes:

> We shall see how chaos can run apparently against Nature, and achieve that most unnatural of ends, life itself.

Again, this is an unfulfilled promise. No one has shown that chaos achieves life itself!

Finally, after discussing some aspects of processes in a living organism, Atkins concludes his book with these words:

> We are the children of chaos, and the deep structure of change is decay. At root, there is only corruption, and the unstemmable tide

> of chaos . . . . This is the bleakness we have to accept as we peer deeply and dispassionately into the heart of the universe.
>
> Yet, when we look around and see beauty, when we look within and experience consciousness, and when we participate in the delights of life, we know in our hearts that the heart of the universe is richer by far. But that is sentiment, and is not what we should know in our minds. Science and the steam engine have greater nobility. Together, they reveal the awesome grandeur of the simplicity of complexity.

I must admit that I very much enjoyed reading these beautiful, poetic words. However, there are several points I failed to understand. I would appreciate it if any of the readers could write to me and explain to me the following:

(1) What is meant by "*we* are the children of chaos"? As far as I know — this is written explicitly on my ID card — I am the son of Samuel and Rachel Ben-Naim. I find the claim that I am the son of chaos both untrue and offensive.
(2) What is the meaning of the phrase "the deep structure of change is decay"?
(3) I have never had the chance to look "deeply into the heart of the universe." Can anyone help me in doing so?
(4) Lamentably, I do not know in *my* heart that "the heart of the universe is richer by far." Perhaps I missed something very beautiful. Can someone help?

To conclude, entropy, which is a *meaningful quantity*, becomes *meaningless* when applied to a living system. It is *a fortiori* true when a *meaningless quantity* such as *negative entropy* is applied to a living system.

3.8.1 *Do we feed on negative entropy?*

In Schrödinger's book *What Is Life?*, which we mentioned earlier, he postulates that an organism feeds on *negative entropy*, which prevents it from decaying into the state of being dead.

I do not know what "negative entropy" means; neither do I accept Brillouin's suggestion to use *"neg-entropy"* instead of "negative entropy," which is supposed to be interpreted as information.[30]

Brillouin goes even further and claims:

> If a living organism needs food, it is only for the neg-entropy it can get from it, and which is needed to make up for the losses due to mechanical work done, or simple degradation processes in living systems. Energy contained in food does not really matter: Since energy is conserved and never gets lost, but neg-entropy is the important factor.

Aha, if that is so, why do all food products have the *caloric* value printed on their labels? They should instead report the more "important factor" of *neg-entropy* in units of calories per degree, or perhaps in bits. Thus, the next time you look at the labels on food products, ignore the "energy value," which "does not really matter." Only the meaningless *neg-entropy* is important!

While I was still baffled by the concept of *negative entropy* — or the shorter version, *neg-entropy* — I was greatly relieved to read the explanation (Hoffmann, 2012) that:

> Life uses a low-entropy source of energy (food or sunlight) and locally decreases entropy (created order by growing) at the cost of creating a lot of high-entropy waste energy (heat and chemical waste).

Thus, in modern books, the meaningless notion of negative entropy (or neg-entropy) is replaced by the more meaningful term "low entropy".

Is it meaningful to claim that we, living organisms, feed on low entropy food? If you are convinced that feeding on low entropy food is the thing that keeps you alive, you should then take your soup (as well as your coffee and tea) as cold as possible. This will ensure that you feed on food with the lowest entropy possible. As for solid food, you should try to eat frozen food (but be careful not to put anything with a very low temperature into your mouth — you might get a cold burn). You can enjoy as much (and as cold) ice cream as you wish. Forget about the *calories* of the ice cream and remember its *entropy* value (calories divided by degrees Kelvin).

While reading about low entropy food, I wonder why food manufacturing companies are not required by law to label the "entropy value" of their products (per 100 gr or per serving), as they usually do with the energy values (in calories).

On "feeding on negative entropy," I would like to add this comment: Entropy, by definition, is a positive quantity. Therefore, "feeding on negative entropy" is like saying "swimming in a negative temperature pool." Such statements can be classified as being either nonsensical or meaningless — you choose. The more meaningful concept is "low entropy food." This certainly should be both delicious and healthy food. It might also boost your longevity — perhaps fighting the deadly effects of entropy.

As we have noted before, the entropy of a living system is not defined — not yet, or perhaps never. The main reason is that we do not know how to define the *state* of a living system — not yet, or perhaps never.

Thus, when we feed on a low entropy food (I mean real food, like eggs or apples) we have no idea what happens to the entropy of the food, or what effect the low entropy food has in our bodies. Even if you believe that our bodies are struggling to lower their entropy (or increase their order), it is not clear that feeding on low entropy food will help the body lower its entropy.

Each of us starts as a tiny baby, and for some time grows into a larger body volume. This volume is measurable and observable. Does eating more voluminous (bigger) apples make us bigger in volume?

We feed on food for its energy content and not its entropy value, or its information value. Without the energy of the chemical bonds in the molecules we eat, we could not move, or digest food, or think and feel, besides many other activities of the body. We also need some specific chemicals, such as minerals, vitamins and, of course, water. These compounds are needed not for their energy — certainly not for their entropy.

We all know that water is indispensable to life. It is an indispensable agent in numerous processes in our bodies, and yet its *caloric* value is zero. If the entropy of our foods is the important factor, then consider this:

If you swallow a cube of ice at 0°C, or drink the equivalent amount of liquid water at 0°C, you will get the same benefit from the water molecules. If you have a choice between the two options, I will recommend that you take the water (with a higher entropy) rather than the ice (with a lower entropy), not because of the entropy difference between the two, but simply because the latter might get stuck in your throat.

The whole idea of feeding on either the meaningless "negative entropy" or the meaningful "low entropy" food is based on the fallacious interpretation of entropy as a measure of disorder or disorganization. We all agree, qualitatively speaking, that a living system has some degree of *order* and *organization*. This is not the same as saying that a living system has a *lower entropy*. The entropy of a living system is not defined! If you believe that eating low entropy food is healthy, you can eat high-ordered and well-organized foods. It is far easier for the layperson to identify a high-ordered food than a low entropy food.

In Seife's book (2007) *Decoding the Universe*, the author identifies entropy with information. He even claims that thermodynamics (which includes the concept of entropy) is a special case of information theory (which includes SMI). If this identification is valid, then we should ask the following question:

3.8.2 *Why shouldn't we feed on information?*

In the previous subsection I wondered why foods are not labeled with their entropy values. The lower the entropy, the higher the information content of the food. I am well aware that most people do not know what entropy means, and even scientists are debating its meaning. However, if entropy is identified with information, why not label all foods with their *informational* values (say in bits per 100 grams or per serving)? Such labels, as well as the foods, will not only be easier to understand — they may also be easier to digest . . . .

Besides, if low entropy food is equivalent to high information food, why can't we feed on *information* itself rather than get it from food? This is not a joke. Experiments were carried out on flatworms which had been taught to perform some simple tasks, such as moving toward a source of light.[31]

In the 1960s, a group of biologists did some interesting experiments on flatworms. The idea of the experiments was to *teach* flatworms some basic things, and after the learning the "smart" flatworms were ground and fed to "uneducated" flatworms. All of a sudden, the once "dumb" worms knew what the "smart" worms had been taught. The scientists concluded that memory, or the learned information of the "educated" worms, had been transferred to the "uneducated" ones.

Realistically, there is not much that humans can teach flatworms. At best, these can be taught how to recognize their food supply, or turning right or left in a simple maze with food as a reward. These experiments were disputed by many scientists. It is inconceivable that a worm or any other living creature which had learned to do some task could pass this information through the digestion of its brain. The moral of these experiments is well summarized by vos Savant (1972):

> Think how wonderful it would be if, instead of being born as a *tabula rasa*, a blank state, we could be fed, along with our infant formula, every scrap of the hard-won information possessed by every generation that has gone before! Instead of having to learn the bittersweet lessons of history all over again, we could start, not from square one, but from somewhere in the middle; square ten, say. (Of course, we might also be drinking in all the misinformation and all the disinformation, prejudices and misconceptions of previous generations, but let's not look at the gloomy side. This is, after all, only a fantasy.) Yet, would having all this predigested information at its tiny, infant fingertips make the baby more intelligent?

Can you imagine human beings being fed by the "intelligent" brains of other human beings?

Remember, the registered information in the brain is in some kind of patterns of neuron connections. Once the brain dies (even before being ground. . . .) these connections are lost, and gone with them is all the information stored in it. You should be aware of these facts before eating an "intelligent" brain!

3.8.3 *Life and the Second Law of Thermodynamics*

Schrödinger asked the question "How do living organisms avoid decay?". More recently, people have claimed that the Second Law of Thermodynamics can *explain* life. In his book *The Second Law*, Atkins (1984) writes:

> In Chapter 8 we also see how the second law accounts for the emergence of the intricately ordered forms characteristic of life.

Furthermore, in a more recent book, Atkins (2007) says:

> The second law is one of the all-time great laws of science, for it illuminates why anything, from the cooling of hot matter to the formulation of a thought, happen at all.
>
> The second law is of central importance ... because it provides a foundation for understanding why any change occurs ... the acts of literary, artistic, and musical creativity that enhance our culture.

Finally, in both of the abovementioned books, Atkins writes on the Second Law:

> ... no other scientific law has contributed more to the liberation of the human spirit than the second law of thermodynamics.

All these quotations are extremely impressive, yet are totally empty statements. The Second Law *does not provide an explanation* for "formulation of a thought," and certainly not the "acts of literacy, artistic and musical creativity." Can anyone tell me how the Second Law "contributed to the liberation of the human spirit"?

In my opinion, such claims not only do not make any sense, but they can actually discourage people from even trying to understand the Second Law of Thermodynamics. Life phenomena involve extremely complicated processes. Everyone knows that life is a complex phenomenon, many aspects of which, such as thoughts, feelings and creativity, are far from being well understood. Therefore, the undeliverable promise of explaining life by the Second Law will inevitably frustrate the reader to the point of concluding that entropy and the Second Law, as life, are hopelessly difficult to understand.

In the introduction to his book, Hoffmann (2012) writes:

> Understanding life is not an easy task . . . . Scientific literature is replete with articles that attempt to explain various aspects of life, yet much is a conjecture; much controversial.

I fully agree with this statement. This is true until we reach the chapter on entropy and life, the Second Law and evolution. At this point, the author slips into the same pitfall he had warned the reader about in the introduction.

As is stated on the cover of Hoffmann's book:

> Rather than relying on some mysterious life force to drive them — as people believed for centuries — life's ratchets harness instead the second law of thermodynamics and the disorder of the molecular storm . . . to tell the story of how the noisy world of atoms gives rise to life itself.

That is not true! The Second Law does not tell the "story of life itself."

On page 69 of the book, Hoffman writes:

> Schrödinger saw a contradiction between the chaos of atoms and the structure of life. But today we know that chaotic motions of atoms and molecules — controlled by life's intricate structure —

> give rise to life's activity. There is no contradiction. Life emerges from the random motions of atoms, and statistical mechanics can capture the essence of this emergence.

This is again a typical statement which is highly impressive but totally vacuous. We do not know how "chaotic motions of atoms and molecules" give rise to "life's activity." Besides, statistical mechanics does not say anything about life's emergence. I have spent over 40 years learning and teaching the extremely powerful theory of statistical mechanics. I also did research on the application of statistical mechanics to some aspects of life, e.g. cooperative binding.[15] I have never seen any example of the application of statistical mechanics to "life's emergence."

Seife (2007) devotes a whole chapter in his book to "Life." Before he discusses life itself he makes the statement that "thermodynamics is, in truth, a special case of information theory." I certainly do not agree with that statement. As we have seen in Chapter 2, entropy may be viewed as a special case of SMI. Clearly, entropy is one concept within thermodynamics. SMI is also one concept, though a central one, in information theory. From this it does not follow that thermodynamics is a special case of information theory.

Then Seife makes the statement that "all matter and energy is subject to the 'laws of thermodynamics,' *including us*. In my view, there is no justification for claiming that we are subject to the laws of thermodynamics — certainly not to the (nonexistent) "laws of information."

Just before he turns to discussing life, another vague, yet very common, statement is made:

> What is life? That answer is quite disturbing.

And what is the disturbing answer? Reading through the whole chapter on life, I could not find the answer to the question "What is life?" Perhaps the following paragraph contains his answer:

> It's a grim picture. Life might be nothing more than information's scheme to duplicate and preserve itself. But even if this is true, it doesn't provide the complete picture.

To the question "How, then, can life exist at all?" Seife's answer is:

> On a purely physical level, is it not too much of a puzzle? Just as a refrigerator can use its engine to reverse entropy locally by keeping its insides colder than the room it is in, the cell has biological engines that are used to reverse entropy — locally — by keeping the information in the cells intact.[32]
>
> The information remains more or less intact, surprisingly well protected from the ravages of time and entropy.

In my opinion, Seife confuses here, as well is in other parts of his book, information with SMI, and SMI with entropy, and then applies entropy or information where these concepts cannot be applied.

The specific sequence of letters in the sentences you are reading now conveys some information in a colloquial sense. The same sequence can be assigned a value of SMI (see Chapter 1). This SMI has nothing to do with the colloquial information conveyed by the sentences. Neither the colloquial *information* nor the SMI of the sentences in this book has anything to do with the *entropy* of the book.

A few months ago I saw a TV series on entropy, life and the universe. The narrator shows us a ruined and desolate city, then the wreckage of an ancient ship, and comments, "This is the result of the ravages of entropy." An awesome, yet meaningless, statement. Entropy is a specific measure of a specific piece of information belonging to some probability distribution of a macroscopic body. It does not ravage the body, just as the *length* of the road from Jerusalem to Tel Aviv does not ravage the road.

Finally, to the question "Where did we come from?" Seife answers, "Here, too, information is yielding surprising answers to this ancient puzzle." Indeed, Seife tells a surprisingly pleasant and most interesting story about the Lemba tribe in Zimbabwe. The tribe claimed to be a lost tribe of Judea and that they were Jews.

However, not many believed their claim as there was very little to establish their link with the ancient Jews. Just like the Jews, they observed the Sabbath, did not eat pork, and circumcised their sons. But similar, traditional practices remained as merely oral traditions, which were often difficult to prove. Recently, however, geneticists from the United States, Israel and England have carried out some tests and analyzed the Y chromosomes of the Lemba males. What makes the Y chromosomes interesting is that they contain a strong marker of the Jewish people's heritage. They contain the "priestly genes."

Based on legend, the priestly class, or the *cohanim,* were descendants of a single male, Aaron, Moses' brother. Handed down from generation to generation to the males was the title of priest, or *cohen*, just like the Y chromosome. Therefore, if the legend is true, all Jewish priests must have the same Y chromosome as Aaron.

It was discovered that the modern-day *cohanim* shared distinctive genetic characteristics. So, regardless of geographical boundaries, the priestly genes were handed down from generation to generation, just like in the case of the Lembas. Although they were separated from their Jewish roots, and were geographically away from their Jewish brothers, they had genetic markers similar to those of other Jewish priests around the world.

This is indeed an extraordinary story, and I salute Seife for bringing it up. I also agree with his final remark:

> Scientists don't know how life began, but the near immortality of information has preserved a story that goes back to the very beginning of life.

This is true. However, it does not answer the question "Where did we come from?" Perhaps it answers the question of the origin of the Lemba tribe, but neither that particular information nor general information theory can tell us where *we* came from, or about the "beginning of life."

There are many other statements connecting life to entropy and the Second Law. Here is another typical example. In his recent book, *From Eternity to Here*, Carroll (2008) writes[33]:

> The arrow of time manifests itself in many different ways — our bodies change as we get older, we remember the past but not the future, effects always follow causes. It turns out that all of these phenomena can be traced back to the Second Law. Entropy, quite literally, makes life possible.

Although I certainly agree with the first part of this paragraph, I do not think that "these phenomena can be traced back to the Second Law." No one has ever shown that remembering the past and not the future has anything to do with the Second Law. And entropy, *quite literally,* does not make life possible. As I have noted before, such statements amplify the mystery enshrouding entropy. If entropy makes life possible, it must be at least as complex as life itself. We all know that life is far from being understood. Therefore, one might rightfully conclude that entropy — which makes life possible — is hopelessly difficult to understand.

On page 180 the author discusses "Information and Life" and says:

> ... our ability to remember the past but not the future must ultimately be explained in terms of entropy, and in particular by recourse to the Past Hypothesis that the early universe was in a very low-entropy state.

I cannot agree with such statements. I do not see how entropy or the Second Law can *explain* the ability to remember the past. I do not accept the Past Hypothesis either. I will comment on this in Chapter 4. Here, I only want to say that the Past Hypothesis is an unfounded, perhaps even meaningless, hypothesis. But whatever the Past Hypothesis means, it certainly cannot explain the ability to remember the past and not the future.

3.9 Is Schrödinger's Cat Alive or Dead?

In the previous sections we suggested refraining from applying either entropy or the Second Law to living systems. The main reason for this is that in order to define the entropy of a living system, we need to specify its *thermodynamic state.* Since this is not possible to do at present, we cannot assign an entropy value to an entire living system.

The same difficulty is encountered when we want to describe the *microscopic state* of a living system, either classically or quantum-mechanically. Perhaps the most dramatic, as well as famous, manifestation of this difficulty is embodied in the so-called Schrödinger's cat paradox.

In connection with the so-called Copenhagen interpretation of quantum mechanics, Schrödinger described a thought experiment in which a cat is held in a steel chamber along with a device which can detonate a bomb or release some poisonous gas. The device is set in such a way that it has probability one half to release the gas (hence causing the death of the cat), and probability one half not to release the gas (hence leaving the cat alive).

In accordance with the Copenhagen interpretation of the wave function of the cat, there is a probability of one half that the cat is alive and a probability of one half that it is dead. As long as we do not open the chamber and look inside it, we assume that the *state* of the cat is a *superposition* of the two states.[34] Let us denote these states as ψ(alive) and ψ(dead). In other words, before we open the chamber, the cat is alive and dead *at the same time*. This principle of superposition of two states works well on an atomic level. For instance, a particle can have a two-state spin: either "up" or "down." As long as we do not make an experiment to determine the state of the particle, it is said to be in a superposition of the states ψ(up) and ψ(down). However, it is not clear what it means when we say that the state of the cat (or any other living organism) is a *superposition* of two states: ψ(alive) and ψ(dead). This is sometimes referred to as the observer's paradox.

Without entering into the debate about the applicability of the principle of superposition of states to a living organism, we can ask whether it is meaningful even to postulate that there exists a wave function which describes the two states of the cat as "alive" and "dead," even when we know that the cat is either alive or dead.[35]

Figure 3.17 represents schematically the two states of the cat. According to the conventional interpretation of quantum mechanics, as long as we do not look inside the chamber the cat is "alive" with probability one half, and "dead" with probability one half. These probabilities are usually inferred from the probabilities of the poison being released or not. However, there is no simple relationship between these probabilities and the probabilities of the cat being alive or dead.[36]

In my opinion, before we contemplate the question of the applicability of the superposition principle to a living system, we should ask a more fundamental question: What is the meaning of the state functions ψ(alive) and ψ(dead)? Everyone can distinguish between a live and a dead cat. However, we do not know how to write *different wave functions* for these two states. In other words, we do not have at present such two different *functions* — one describing the state of being "alive," and the other the state of being "dead." It is rumored that Schrödinger himself said he wished he had never met the "Schrödinger cat."

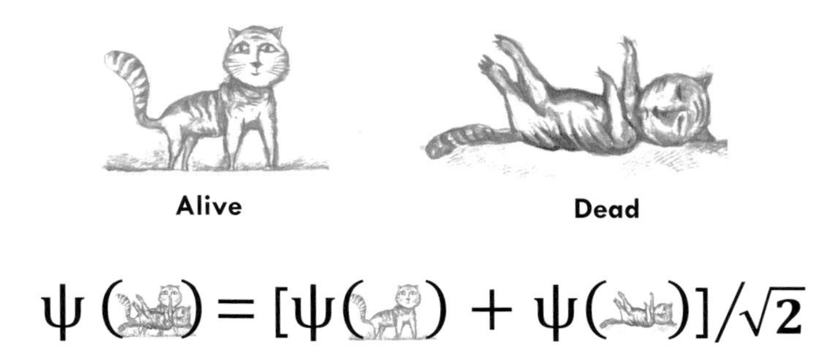

Fig. 3.17 Schrödinger's cat being a superposition of the two states "alive" and "dead."

This problem arises whether we describe a system classically or quantum-mechanically. We do not know whether having all the locations and all the momenta of a living system is enough for the description of the *state* of a living system.

Thus, for the moment we can be sure that the unfortunate cat will not be in a state of superposition. Of course, under the setup of this particular experiment, there is a probability that it will die after some time. But, as long as the bomb does not explode (or the poison is not released), the cat is, to the best of our knowledge, alive.

The Schrödinger cat has been raised in this book not to question the validity of the principle of superposition of states for macroscopic objects. Suppose that it is valid for any object, microscopic or macroscopic. The question still can be asked: Do we have a wave function which distinguishes between the state of "being alive" and of "being dead"? Having such a function would be tantamount to a definition of the states of "being alive" and "being dead." Unfortunately, such a function is not available for any living organism — even the most primitive one.

In fact, the problem is much wider than just distinguishing between the state of being alive and of being dead. What about distinguishing between two states of mind — say, happy and sad, or angry and loving?

Imagine that instead of putting a cat in the chamber, you put a boy. Instead of poison, let us throw into the box a coin which can fall with either face H or face T up, with *equal probability*. In the chamber, there is a robot which reads the result of tossing the coin. If the result is H, then the robot shows a comic movie to the child, and the child is happy. If the robot sees T, it will play a tragic story on the video and the child will be sad.

We, from the outside, know only that the coin could fall on either side with equal probability. This does not imply that the

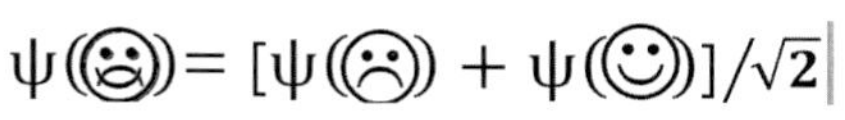

ψ(😧)= [ψ(☹) + ψ(☺)]/√2

Fig. 3.18 A superposition of the two states "happy" and "sad."

boy is either happy or sad with equal probability.[37] Now the question: Is the boy in a superposition of states — happy and sad (Figure 3.18)?

$$\psi(\text{boy}) = \frac{1}{\sqrt{2}}[\psi(\text{happy}) + \psi(\text{sad})].$$

In my view, before we contemplate the question of superposition we should ask whether there is a wave function that describes the boy's mood in the first place. If we find such a function we can still ask whether it will obey any "equation of motion," Schrödinger's equation, or any other, perhaps nonlinear differential equation. In which case the superposition state would not even arise.

It is puzzling that Schrödinger, who had given much thought to the question "What is life?", had failed to see that the superposition principle applies whenever we *have* two well-defined wave functions of the two states. Having no such wave functions for the states "alive" and "dead" dismisses the whole "paradox" associated with Schrödinger's cat.

Hence, for the moment, the state of Schrödinger's cat should not worry Schrödinger, the man. There is no reason for Schrödinger's cat to haunt physicists either. Seife (2007) summarizes this paradox as follows:

> Information can be deadly. This absurd-sounding conclusion seemed to be an unavoidable consequence of the principle of superposition.

It is not the superposition principle that leads to the absurd result, but rather the application of this principle to a living organism.

According to the Copenhagen interpretation of quantum mechanics, as long as we do not look inside the chamber the cat is in a *superposition state* of "being alive" and "being dead." Once we look inside (make a measurement), the wave function *collapses*, and we see that the cat is either alive or dead.

A different interpretation of quantum mechanics is known as the many-worlds (or multi-universes) interpretation. When we do a measurement and find that the cat is alive (or dead), it is indeed alive in "our world" but there is another "copy of the world" in which the observer sees that the cat is dead (or alive). In general, when we do a measurement we see one of the possible outcomes. However, all other possible outcomes exist but in other, parallel worlds. Figure 3.19 shows a caricature of the two "fates" of Schrödinger's cat in different universes.

In connection with the many-worlds theory, Bruce (2004) presents one absurd scenario:

> If you believe in many-worlds, there is an infallible way for you to get very rich. All you need to do is buy a single ticket in a big-money lottery and wire yourself up to a machine that will kill you instantly and painlessly if your ticket does not win. The chance of winning such a lottery is about 1 in 100 million. But the odds do not matter as long as they are finite. If you believe in many-worlds, then you believe that there is literally an infinite number of versions of yourself in universe-variants that are diverging all the time. After the lottery is run, and the machine has killed you (in an infinite number of worlds) or not killed you (in an infinite number

Fig. 3.19 The unfolding of the fate of Schrödinger's cat within the multiuniverse theory.

of others), then all the versions of you still alive will be extremely rich.

I do not recommend anyone to try this particular "trick" in order to get rich. As I have explained, before contemplating the meaning of the superposition state of the cat, or the splitting of the entire world into a world in which the cat is alive and a world in which the cat is dead, we must first determine whether or not quantum mechanics is applicable to a living system.

3.10 The Origin and Evolution of Life

No one knows how life began. We have evidence that life on earth existed between 3.5 and 4.5 billion years ago. This evidence in itself does not pinpoint any "date" for the beginning of life.

In addition, it is not clear whether life began on our planet, or perhaps it was transported to earth from some other planet (this idea is referred to as panspermia). Also, it is not clear whether there was a "date" at which a mixture of chemicals was transformed into a living organism. It is more likely that there was some kind of continuous *evolution* of nonliving chemicals which gradually acquired some attributes which are characteristic of what we recognize today as a living system. This evolution continued smoothly from the "chemical phase" to the "biological phase." A very interesting and illuminating discussion of this process may be found in Pross (2012).

Pross devotes a whole chapter (7) to "Biology Is Chemistry." Most of the chapter deals with the transition from "chemical evolution" to "biological (Darwinian) evolution."

I agree with the author[38] that "Darwinian theory can be integrated into a more general chemical theory of matter, and that biology is just chemistry, or to be more precise, a sub-branch of chemistry — replicative chemistry."

Indeed, most of the chapter describes the smooth passage from evolution in chemistry (replicating nonliving molecules) to biological evolution. Even when one agrees with the author that there is a continuous transition from chemical to biological evolution, the question "What is life?" is still left unanswered. Darwinian evolution is certainly central in biology but not a *defining* characteristic of biology.

The diagram in Figure 3.20 is taken from Pross' book (Figure 6).

There is no problem in envisaging "evolution" in a purely chemical (nonliving) system — including replication, mutation, selection and evolution. This sequence of events characterizes

Fig. 3.20 Schematic transition from chemical evolution to biological evolution.

evolution, and evolution characterizes life. However, it is far from clear that evolution *defines* life.

I agree with the author's contention that one can blur the transition between chemical evolution and biological evolution. I can also accept that biology is essentially chemistry. The question is: What does one include in "biology"? Does it include all *chemical* processes in a biological system which *by definition* is nothing but *chemistry*, or does it also include aspects of biology such as consciousness, thoughts or creativity? Can all of these be explained by chemistry alone?

To the best of my knowledge, no one has offered a convincing answer to this question. It seems to me that Pross believes that such an answer has already been given:

> ... the unification tells us that chemistry and biology are one ... that there is a complexity extension continuum that connects them, that biology is just an elaborate extension of replicative chemistry.

Pross also discusses the intimate connection between the three questions "What is life?", "How did life emerge?" and "How to make life?" Of course, these questions are intimately related. However, it is by no means clear that knowing the answer to one question will provide answers to the others.

These three questions are shown schematically in Figure 3.21. They seem to be initially independent and unrelated. Yet, Pross claims that being able to answer any one of these questions depends on knowing the answer to the other two.

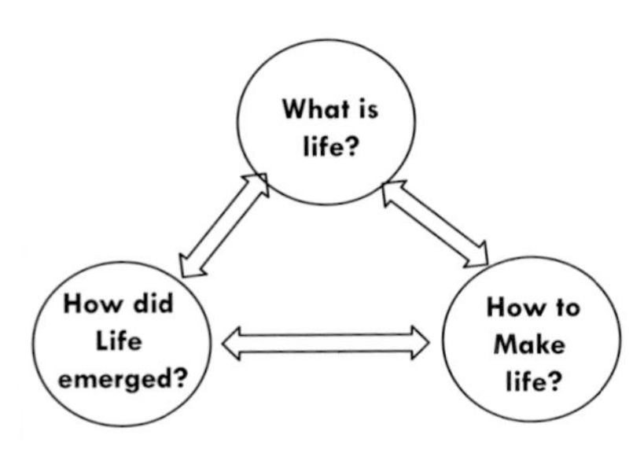

Fig. 3.21 The three key questions in connection with understanding the living systems.

I am not sure this is true. It might be the case that knowing the answer to one question will pave the way for answering the other questions. However, one cannot guarantee that solving one mystery will lead to a solution to the others.

For instance, we might one day understand what life is. This could help in theorizing about the origin of life, and perhaps also in creating life. However, we might never know how life as we know it was actually created on earth. It is also far from clear that understanding what life is, is a sure-fire recipe for making life (by "making life" we mean creating life from inanimate matter; creating new life from a living system is trivial, and depends neither on understanding life nor on knowing its origin).

Thus, although these three questions are certainly intimately related, it is not clear to what extent answering one can shed light on the others.

Leaving aside the unanswered questions about the origin of life, one can still speculate on what came first: the chicken or the egg? Or, in molecular terms, the DNA or the proteins? Equivalently, did the *information* in the DNA come first, and then the proteins, which execute so many functions in the cell (including the replication and translation of the DNA)? Or perhaps the catalytic activity of the proteins evolved first, then the DNA?

An interesting suggestion is the so-called RNA world hypothesis, which essentially claims that the RNA, having both *informational* content and *catalytic* activity, evolved first.

In my opinion, there is something of a continuity between the evolution of chemicals which share some attributes with living organism, and the evolution of living organisms as we know them today. Once a system has acquired the minimal requirements for evolution (in Darwinian sense), we can claim that we understand how life has evolved from the very primitive to the most complex living organism.

In discussing the evolution of life, there seems to be a large degree of confusion between the *theory of evolution*, the *conjecture of evolution* and the *story of creation,* or any other *story of evolution*.

The *theory of evolution* is basically a theory based on common sense. Any system which can reproduce, mutate, metabolize, etc. and exists in a changing environment must either get extinct (if it cannot survive in a new environment), or evolve into one of the mutants which has a "survival advantage" in the new environment. The *conjecture of evolution* states that the evolution of life on our planet occurred by the mechanism described by the *theory of evolution*. This conjecture depends on evidence, and there is abundant evidence in favor of it. However, no matter how much evidence we might have, we can never prove that evolution has actually occurred according to the mechanism of the theory of evolution. There might be other mechanisms that could explain evolution, or one could choose to believe in the *story* of *evolution*, or the story of creation. And there are many stories of creation, one of which is narrated in the Book of Genesis in the Bible.

Shermer (2006), in his delightful book *Why Darwin Matters,* criticizes those creationists who use the Second Law as an argument against the theory of evolution. Clearly, this kind of (very common) argument used by creationists is false. Shermer summarizes the objection to such an argument as follows:

> Species do not evolve from simple to complex, and nature does not simply move from chaos to order. The history of life is checkered with false starts, failed experiments, small and mass extinctions and chaotic restarts. It is anything but the textbook foldout of linear progress from simple cells to humans.

Evolution is usually described as a process which involves transition or evolution from *disorder* to *order*, or to more *organization*, or more *complexity*. In fact, we cannot claim that evolution is associated with a one-way increase in some property — order, organization or complexity. In one environment, the bigger (or the more ordered, the stronger, the more organized, etc.) might have an advantage and therefore will survive. But, in another environment, the bigger (or the more ordered, the stronger, the more organized, etc.) might have a *disadvantage* compared to the smaller, and therefore the *smaller* will be the one which will survive. Thus, in general, we cannot pinpoint any property that changes in evolution in one direction except for evolution itself, which always evolves. This brings us to the (nonexistent) connection between evolution and the Second Law: first, the misconception about evolution that proceeds from *less order* to *more order*; second, the misconception that the Second Law requires that a system (isolated!) go from *more order* to *less order*. The inevitable conclusion is that there is a conflict between evolution and the Second Law. In my opinion, this apparent conflict between evolution and the Second Law is anything but a conflict.

Evolution proceeds according to the *theory of evolution*. It applies to any system which replicates, mutates, selects, etc. The second law applies to a spontaneous process in an isolated system.[39]

The Second Law does not apply to the (Darwinian) process of evolution. These two theories do not share any overlapping regions of applicability.

Another common misconception about entropy and the Second Law is a result of applying the Second Law to the entire universe. We will further discuss this misconception in Chapter 4. In many popular-science books you find the phrases "ravages of time" and "ravages of entropy."

Indeed, there is nothing wrong in using the phrase "ravages of time" in connection with processes of decay, destruction and death. However, this figure of speech is not a law of nature. One can also use the phrase "the beneficial effect of time" in connection with blooming flowers, getting wiser with age, new births of babies or the creation of a novel, a painting or a symphony. In all of these processes *time* is not the *cause* that destroys or creates anything. Processes (of destructions, creations, etc.) occur *in time*, not because of some power of time. Yet, these phrases are acceptable as figures of speech. The phrase "ravages of entropy" cannot be accepted, not even as a figure of speech. In fact, *entropy* is even more *innocent* than time. It does not *create* anything (certainly not life), and it does not *ravage* anything (certainly not life).

Pause and ponder

Here are two statements from Floridi (2010). I will leave it to you to ponder on their meanings:

> The ultimate anti-entropic weapon had been born. Biological life is a constant struggle against thermodynamic entropy.
>
> It suggests that there is something even more fundamental than life, namely being — that is, the existence and flourishing of all entities and their global environment — and something even more fundamental than suffering, namely entropy.

In the next section I will discuss very briefly the oft-repeated association between evolution (the conjecture) and entropy and the Second Law of Thermodynamics. In my opinion, there is no connection between the two. The Second Law cannot be applied either to justify the conjecture of evolution or to negate it.

3.11 Entropy and Evolution

Monod (1997), in his book *Chance and Necessity*, discusses the relationship between evolution and the Second Law. He starts with the following statement:

> No definable or measurable violation of the Second Law has occurred.

The truthfulness of this statement depends on what one means by the *Second Law*. If one means a spontaneous decrease of entropy in an isolated system, then of course no one has observed that. But if the meaning is a reversal of the *state* of the system, then it can be observed in principle — albeit after a very long time.

Another statement by Monod (1997) relating entropy to evolution:

> Evolution in the biosphere is therefore a necessarily irreversible process defining a direction in time — a direction which is the same as that enjoyed by the law of increasing entropy.

I strongly disagree with this statement (see also Section 2.5). First, because evolution is not "irreversible." Mutation that occurs spontaneously can occur in the *reverse* direction. The survival of a mutant or the reverse of this mutant ultimately depends on the changing environment. Second, I do not see any relationship between the process of evolution and the increase in entropy. The author finally concludes:

> Indeed, it is legitimate to view the irreversibility of evolution as an expression of the Second Law in the biosphere.

A similar view was expressed by Volkenstein (2009):

> Biological evolution is irreversible and directed.

In my view such statements are not only not true, but they also add to and deepen the "mystery" and the "incomprehensibility" associated with entropy and the Second Law.

Those who claim that the Second Law prohibits evolution, or that evolution violates the Second Law, use the platitudinous argument that evolution proceeds toward a more ordered or more organized organism. Since entropy is interpreted as a measure of disorder, one concludes that evolution must run against the Second Law, which requires that the entropy (or the disorder . . .) always increase.

This argument is erroneous.

First, it is a misconception that evolution drives toward more order or more organization. Although this is what we would have liked to believe to be true (*we* are more ordered, more organized than primitive organisms), evolution does not *necessarily* proceed toward more ordered, more organized or more complex organisms.

Second, the identification of entropy with disorder is totally unfounded. Indeed, there are examples in which an increase in disorder in a system correlates with the increase in entropy — but there are also examples in which positive change in entropy correlates with increase in order. [Examples were provided in Chapter 2, and more in Ben-Naim (2008).]

Finally, the Second Law applies to isolated systems. Living systems are open systems, and even if we could have defined the "entropy of a living system" (which I doubt is possible), the law of increase in entropy would not apply to such systems.

These arguments are sufficient for rejecting the claim that evolution either violates or defies the Second Law. In an article entitled "Entropy and Evolution," Styer (2008) begins with a question: Does the Second Law of thermodynamics prohibit biological evolution? Then he shows, *quantitatively,* that there is no conflict between evolution and the Second Law. Here is how he calculates the "entropy required for evolution." Suppose that, due to evolution, each individual organism is 1000 times "more improbable" than the corresponding individual was 100 years ago. In other words, if Ω_i is the number of microstates consistent with the specification of an organism 100 years ago, and Ω_f is the number of microstates consistent with the specification of today's "improved and less probable" organism, then $\Omega_f = 10^{-3}\ \Omega_i$. From these two numbers he estimates the change in entropy per evolving organism, and then estimates the change in entropy of the entire biosphere due to evolution. Styer's conclusion:

> The entropy of the earth's biosphere is indeed decreasing by a tiny amount due to evolution, and the entropy of the cosmic microwave background is increasing by an even greater amount to compensate for that decrease.

In my opinion this *quantitative* argument is both misleading and superfluous. It adds more confusion than clarification. In fact, it weakens the *qualitative* arguments I have given above. No one can say that one living organism is less or more probable than the other. No one knows how to calculate the "number of states" (Ω_i and Ω_f) of any living organism. No one knows *what the states* of a living organism are, let alone *count* them. Therefore the author's estimated change in entropy due to evolution has nothing to do with either entropy or evolution — just meaningless numbers.

If you follow me through the entire book, you will see how futile and meaningless the debate over evolution is — whether it is consistent with or contradicts the Second Law.

The Second Law does not state that a system (animate or otherwise) tends to go from order to disorder. It does not state that the entropy of an open system cannot descend or ascend. It does not state anything about a living system. Therefore, all this talk about life, evolution and the Second Law is superfluous.

In Section 2.5, I quoted Boltzmann's reaction to the apparent conflict between the reversibility of the molecular motions and the irreversibility associated with the Second Law. Here,

I will quote Lord Kelvin [Thomson (1874)], who discussed the reversibility paradox, as well as the possibility of *life reversal*:

> If, then, the motion of every particle of matter in the universe were precisely reversed at any instant, the course of nature would be simply reversed for ever after. The bursting bubble of foam at the foot of a waterfall would reunite and descend into the water . . . . Boulders would recover from the mud the materials required to rebuild them into their previous jagged forms, and would become reunited to the mountain peak from which they had formerly broken away. And if the materialistic hypothesis of life were true, living creatures would grow backwards, with conscious knowledge of the future, but no memory of the past, and would become again unborn. But the real phenomena of life infinitely transcend human science . . . . Far otherwise, however, is it in respect to the reversal of the motions of matter uninfluenced by life, a very elementary consideration of which leads to a full explanation of the theory of dissipation of energy.

An interesting discussion about life, entropy and the second law may be found in Denbigh (1989).

3.12 Summary of Chapter 3

Many authors who discuss the question "What is life?" admit that there is no answer to it. Instead, they describe some of the attributes of life, such as reproduction, metabolism, or adaptation to the environment. All these of course are recognized as characteristics of life. However, none of these are necessary or sufficient to define life.

A huge amount of knowledge (or information) has been accumulated on many aspects of life. Yet, there is one aspect of life which is elusive, and that is life itself. We do not know how to define life, how life was created and whether or not life succumbs to the laws of physics and chemistry as we know them today. Specifically, we do not know how to describe the state of being "alive," or being "dead." We can tell when something is alive or not for any living organism, but we cannot specify these states in any of the available physical terms. Thus, there is no point in applying the concept of *entropy*, or of the Second Law, to a living system.

We can still apply the concept of information in both its colloquial sense and its information-theoretical sense. In spite of the many claims in the literature, the *information* we have about life is in general not measurable. On the other hand, we can use SMI for many probability distributions associated with living systems. We can define the probability distribution of compounds in a cell, in an organ, or in the entire organism. We can assign a distribution to the letters in the DNA or the letters of proteins, and so on. For each of these distributions we can define the corresponding SMI. All these SMIs are well-defined quantities, but they are not entropy. Entropy, when viewed as a particular case of a SMI, is defined for specific distributions of a thermodynamic system at equilibrium. We know that a living system is not in an equilibrium state. In addition, we do not know how to define the *state* of a living system. We do not know whether a living system *tends* to an equilibrium state, and whether it will ever reach an equilibrium state. Therefore, as long as a living system is *alive*, it is meaningless to apply to it the concept of entropy, or of the Second Law of Thermodynamics.

Life does not violate the Second Law, nor does it emerge from the second law. The Second Law simply does not apply to a living system.

At this stage of our knowledge of life, we can be satisfied with applying SMI to well-specified distribution functions associated with a living system. Unfortunately, we do not know whether or not SMI or information theory can be applied to life itself. Certainly, it cannot be applied to *explain* aspects of life that are far from being understood, such as consciousness, thoughts, feelings or creativity. Here again statements claiming that information theory can help us with the comprehension of these aspects of life abound in the literature. These statements are no doubt very impressive, but are far from being true.

I would like to add here one final thought about life and the Second Law of Thermodynamics. Entropy is defined for a macroscopic system, and its properties depend on the fact that any macroscopic system consists of an enormous number of particles. A glass of water having chemical composition N, at temperature T and pressure P, can be assigned a value for its entropy. This value of the entropy is the same for *any* glass of water, no matter where it is, how it was prepared and who prepared it, provided that they all have the same T, P and N.

Some aspects of life, such as thoughts, feelings, consciousness, and perhaps life itself, could be *emergent* properties of a highly complex system.[40] If this is true, then all the emergent properties of an individual organism are dependent on both the information contained in its DNA, and all the information that the individual has received and accumulated from the environment throughout its life history. In other words, these emergent properties are highly individual. Unlike the fact that two glasses of water with the same P, T and N will have the same properties, and in particular the same entropy, no two living individuals of the same species can be said to be in the same "state." As I noted in connection with the Schrödinger cat, we still do not know how to characterize the *state* of a living cat, let alone write down its wave function. If we will ever be able to characterize the state of being alive which is distinguished from the state of being dead, then we might be able also to assign an entropy value to each individual. Until that time my suggestion is to refrain from using entropy and the Second Law in connection with a living system.

It is fitting to end this chapter with Woody Allen's witty remark to a Haaretz journalist. When he was asked, "What are your views about death?", he succinctly replied, "I am against it!"

Chapter 4

The Universe

This chapter is the shortest one in the book. The universe is, of course, unimaginably vast. Physicists know a great deal about it — more than can be fitted in one voluminous book. Yet, as in the case of life discussed in the previous chapter, there are many fundamental questions about the universe, the answers to which are unknown, and perhaps will never be known. Is the universe finite or infinite? Can we predict the future of the universe? Can we ever know the exact "beginning" of the universe? Is the total energy and mass in the universe finite? And perhaps the least-talked-about questions: Are the laws of physics as we know them today the same everywhere in the universe, will they be the same in the distant future, and were they the same laws in the distant past?

My knowledge about the physics of the universe is very limited — far less than cosmologists and astrophysicists, and perhaps I should add universologists. However, I feel I know enough about what physicists admit they do not know about the universe to make a few comments about what people say about the *entropy* and *information* of the universe. Here is a typical statement — the opening sentences of Lloyd's (2006) book *Programming the Universe*:

> This book is the story of the universe and the bit. The universe is the biggest thing there is, and the bit is the smallest possible chunk of information.

Indeed, the universe is, almost by definition, the "biggest thing," but the "bit" is *not* the smallest possible chunk of information, just as the centimeter is not the smallest chunk of length, or the second the shortest chunk of time. We have seen in Chapter 1 that the SMI for one binary question can vary between zero and one bit.

The author continues:

> The universe is made of bits. Every atom, and elementary particle, registers bits of information. The universe is a quantum computer. This begs the question: What does the universe compute? It computes itself. The universe computes its own behavior. As soon as the universe began, it began computing . . . .

I cannot agree with any of the above sentences. They all sound impressive and are frequently quoted by many authors. I have no idea what and how the universe "computes," and assuming that it does, I have no idea what is meant by "it computes itself."

The title of Lloyd's book is *Programming the Universe*. It reminds me of Seife's book, *Decoding the Universe*. Both are impressive, but are at best meaningless. One can say that "the universe is a huge metabolizing machine." What does it metabolize? It metabolizes itself. Or perhaps "the universe is a huge creator." What does it create? Of course, it creates itself. You can also say that the universe is a watch, a car or a table. What do all these statements mean? Whatever you like!

In the next section I will start with a simple expansion process. The analysis of this process provides us with a glimpse of the problems involved in any discussion of the entropy of the universe. In Section 4.2, I will present two examples of processes in which the entropy decreases. We will then examine the question of whether the entropy of the universe must increase as a result of these processes. Next, we will discuss the question of the applicability of the concept of entropy to the entire universe. We will touch on the problem of black hole entropy. Though black holes are "smaller" objects compared with the entire universe, the uncertainties about the nature of black holes are not less than the uncertainties about the universe. Therefore, the applicability of entropy and the Second Law to the black holes is as doubtful as its applicability to the entire universe.

4.1 Entropy Change in a Simple Expansion Process

The processes of expansion of an ideal gas from volume V to $2V$ was discussed in Section 2.7 (Figure 4.1). We found that the entropy change in this process is $k_B N \ln 2$, where k_B is the Boltzmann constant, and N the number of atoms — say, of argon (at some temperature where the system may be treated within classical statistical mechanics). The information-theoretical interpretation of this result is very simple; apart from the constant factor $k_B/\log_2 e$, we have $N \log_2 2 = N$, which simply means a change of one bit per particle. Before the expansion, each particle is confined to the left compartment, having volume V. After the expansion, each particle may be in either L or R with equal probability. Therefore, the change in the SMI is one bit per particle. Note that I could have written

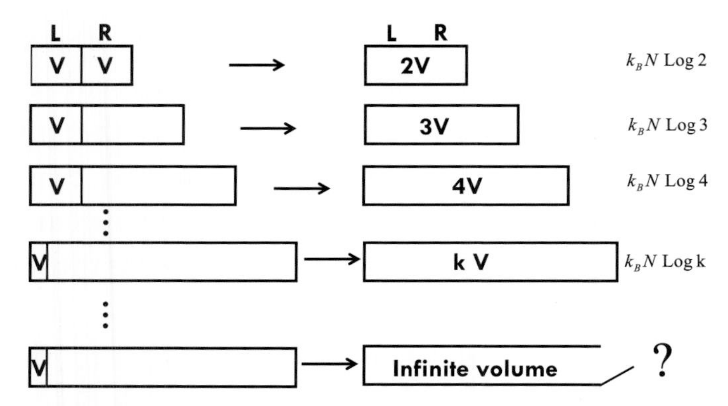

Fig. 4.1 Various processes of expansion of an ideal gas. In each case we start with N particles confined to a volume V. The final volume increases from $2V$ to kV and finally to infinity.

that initially *we* knew that all the particles are in L, and after the expansion *we* do not *know* whether each particle is in L or R. However, such an interpretation might mislead the reader to conclude that ΔS depends on who *knows* the information on the location of the particle. The information-theoretical interpretation of entropy has no traces of subjectivity.

Having firmly secured the objective nature of the change in entropy in this expansion process, consider a few similar processes shown in Figure 4.1. In all of these processes the initial state is the same: all the N particles are in the left compartment L, having volume V. We remove the partition and observe an expansion of the gas. The corresponding entropy change for each process in Figure 4.1 is $k_B N \ln k$, where k is finite (i.e. expansion from V to kV). All the values of ΔS calculated for the various processes can be easily interpreted as changes in the *locational* SMI. For instance, for the expansion from V to $3V$ we simply divide the total volume into three equal volume compartments. Initially, each particle is known to be in L (Figure 4.2). After removal of the partition it can be in any of the regions L, X and Y

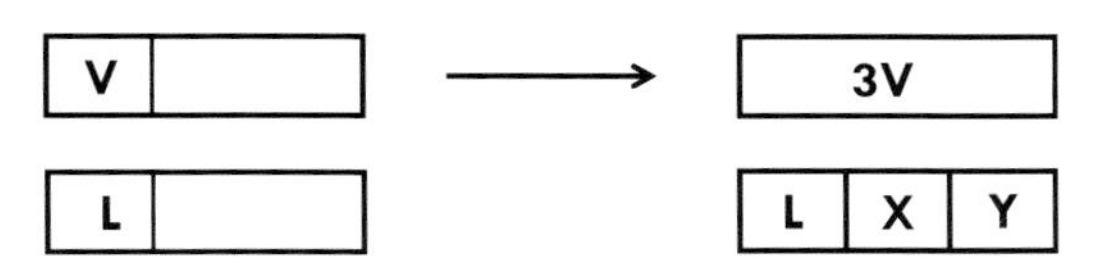

Fig. 4.2 Two expansions from V to $3V$. The informational change in the entropy is the number of questions one must ask in order to determine the location of each particle in L, X or Y.

with equal probability (1/3). Therefore, the change in the SMI is simply $N \log_2 3$, and the corresponding change in entropy is obtained by multiplying by the factor $k_B / \log_2 e$.[1]

What is the change in the entropy for the last expansion process in Figure 4.1, where the particles are free to occupy the entire universe, assuming that the universe is infinite?

An immediate answer would be that ΔS equals infinity. Simply by extrapolating from the series of experiments in Figure 4.1, we can say that as the final volume is larger, the entropy change is also larger. When the final volume is infinity, the change in entropy must also be infinity.

Such reasoning is very plausible. It is also consistent with our estimate of the change in the locational SMI for such a process. However, on second thought we should ask ourselves: How do we know the value of ΔS in this expansion process? Clearly, we cannot measure ΔS in such an experiment. The only way we can calculate ΔS is to use the equation for the entropy (see Section 2.3) and calculate the *difference* between the entropy in the initial and in the final state — assuming that in both the initial and the final state the distribution of locations and of momenta are the *equilibrium* distributions. In our experiment we start with an initial equilibrium state. Once we remove the partition confining the particles to L, the particles will, after a short time, leave the region L. We cannot calculate the locational distribution of the particles, nor can we calculate the velocity distribution of the particle leaving the system. Also, we can say that the locational distribution will never reach an equilibrium distribution. Hence, we can also say that the final state is not a well-defined equilibrium state. Therefore, my answer to the question about ΔS in this process is simple. I do not know the answer.

Pause and think

Consider an experiment which is the same as in Figure 4.1, but carried out in M different systems. In each we start with N particles in a box of volume V. We then open all the boxes and let the particles escape to the vast volume of the universe. Would you conclude that the entropy change for *each* system is $\Delta S = \infty$, and hence the total change in the entropy in the universe is $M \times \infty$?

Another point to think about: If the change in the entropy in a single expansion process from V to infinity is $\Delta S = \infty$, then after one such an experiment the entropy of the system and the whole universe is infinity. In such a universe, what does it mean when we say that the entropy of the universe always increases — increase beyond infinity? What does it mean when we say that the entropy of the universe "tends to a maximum" once we already know that it has reached infinity?

4.2 Two Processes Involving Negative Change in Entropy

In the previous section we discussed a process in which the change in entropy in an expansion process seems to be infinity. In this section we discuss two simple processes for which the

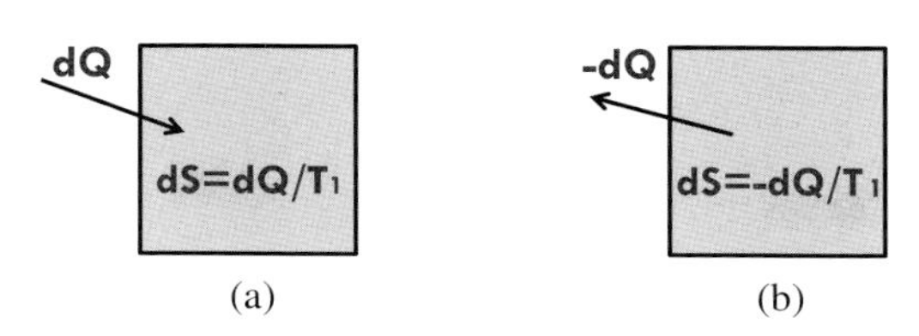

Fig. 4.3 Heat transfer (a) into the system and (b) out of the system, involving positive and negative changes in entropy, respectively.

entropy change of the system is known to be negative. However, it is not clear what the net change in entropy in the entire universe is.

Consider the following experiment. We start with an isolated system at some E, V, N (Figure 4.3). We know that if we *add* a small amount of heat to the system, its entropy will *increase* by the amount $dS = dQ/T_1$ [Figure 4.3(a)]. T_1 is the temperature of the system, defined by the derivative of S with respect to E. If we *extract* a small amount of heat, $-dQ$, from the same system [Figure 4.3(b)], its entropy will *decrease* by the amount of $dS = -dQ/T_1$. Here, we have an example of a system whose entropy has decreased. Most textbooks would go on to explain that when we extract heat from the system, the system is no longer isolated. Its entropy has indeed decreased, but the entropy of the surroundings must have increased by a larger amount, and so the total change in the entropy of the entire universe has increased.

How do we know this? Most people would answer that this is exactly what the Second Law states. In Section 2.7, we saw that when two bodies at different temperatures are brought into thermal contact, heat will flow from the hot to the cold body. We also saw that this process involves a positive change in entropy. In Figure 4.4 we do a slightly modified experiment. Suppose that we take the system that we had in Figure 4.3,

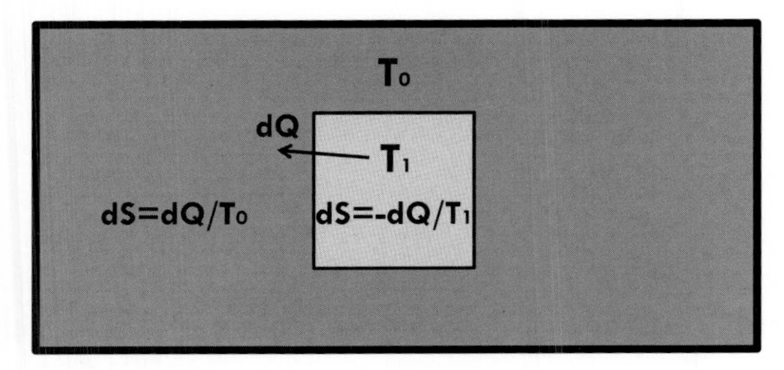

Fig. 4.4 Heat transfer from a heat reservoir at constant temperature T_c. The system and the heat reservoir are isolated from the rest of the world.

bring it into contact with a very large heat reservoir — say, a thermostat at a lower temperature T_0 ($T_0 < T_1$). We let the system exchange heat with the reservoir for a very short time, such that a small amount of heat is transferred from the system into the reservoir. The system and the reservoir are isolated from the entire universe. We can repeat the calculation as we did in Section 2.7 and conclude that the change in entropy in the system is negative: $dS_1 = -dQ/T_1$. The change in entropy in the reservoir is positive: $dS_0 = dQ/T_0$. Since $T_1 > T_0$ we conclude that the sum $dS_1 + dS_0$ is positive. This is indeed a result of the Second Law.

Next, we start with the same system as in Figure 4.4, having temperature T_1. We bring it into thermal contact with the same reservoir, initially at temperature T_0 ($T_0 < T_1$), for a short time. But now the reservoir is not isolated from the entire universe (Figure 4.5). We calculate the entropy change of the system and, as before, it will be negative: $dS_1 = -dQ/T_1$. What is the change in entropy of the entire universe? The answer is that we do not know. In the previous experiment (Figure 4.4), we had a well-defined reservoir at a well-defined temperature, T_0. Therefore, we could calculate its entropy change. We found that the sum of the entropy changes in the *system* and in the *reservoir*

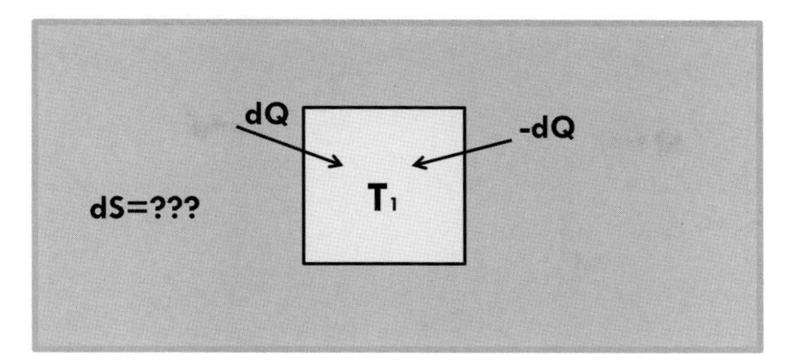

Fig. 4.5 The same process as in Figure 4.4, but now the "heat reservoir is in contact with the universe."

is positive. We cannot say the same thing for the experiment in Figure 4.5. Here, the temperature of the entire universe is not defined, and hence we cannot claim that the entropy of the entire environment has increased. In fact, even the entropy of the system might change with time.

Suppose that we start the experiment as in Figure 4.4 and open the environment (the gray rectangle in Figure 4.4) to the universe. If the walls of the system are made of poor heat-conducting materials, then we can assume that the system is initially at equilibrium at temperature T_1. However, when the temperature of the environment changes, heat flows in and out of the system. If this flow of heat is very slow, the system will go through a sequence of equilibrium states, much as in a quasistatic process (sometimes referred to as a reversible process). In this case the temperature of the system will also change with time, and as a result the entropy of the system will also change with time. For instance, if we follow the change in the entropy of the system during a whole year, we might find that the change in the entropy of the system is negative in the winter (when the ambient temperature is lower than that of the system, heat will flow out of the system into the environment), but the change in the entropy is positive in the summer (when the ambient

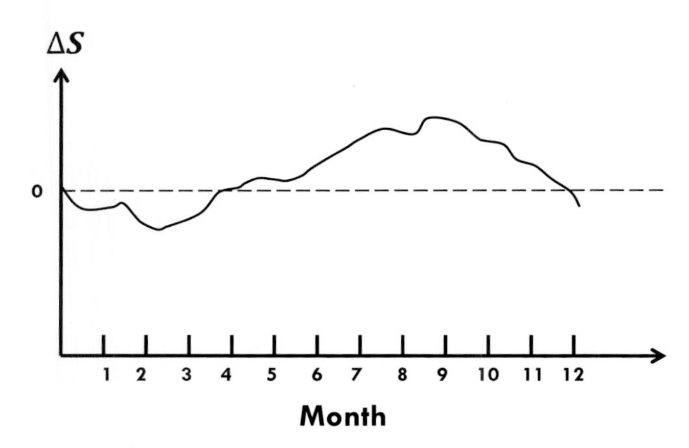

Fig. 4.6 Possible changes in the entropy of the system during a whole year.

temperature is higher than that of the system, heat flows into the system).

Thus, even in this very careful and slow process occurring in the system, we cannot say in which direction its entropy will change. Sometimes it might increase, sometimes decrease; a schematic curve of ΔS as a function of the months of the year is shown in Figure 4.6.

Clearly, if the walls of the system are made of good thermal conducting materials, then we expect flow of heat in and out of the system from different sides of the system, and the entropy of the system might not even be defined. In any case, we can say nothing about the change in the entropy of the entire universe.

To conclude, we can say that in the experiment of Figure 4.4, the entropy of the combined system plus the reservoir will increase. This is the manifestation of the Second Law applied to the combined system plus a reservoir which is presumed to be isolated. We cannot say the same thing for the experiment of Figure 4.5. In this case the system plus the universe is not an isolated system, and the Second Law (in terms of entropy) does not apply.

Another example is the effect of the solubility of argon in water. Starting with pure water at temperature T and pressure P, we add a small amount of argon to the water. The experimental finding is the following: as the argon dissolves into the water, a small amount of heat is released from the water into its surroundings, and hence the change in the entropy of the system ($dS = -dQ/T$) is negative. This is well-explained in terms of a structuring effect of argon on the water. [For more details, see Ben-Naim (2009).]

Without going into the details of what happens in the water (which is a fascinating story in itself), we have here a case of a decrease in the system's entropy. If the system is in a well-defined heat reservoir at temperature T (Figure 4.7), we can calculate the total entropy change of the system (water plus argon) and the reservoir, and find that it is positive. However, if the heat flows into the "universe" we cannot claim that the entropy in the universe either increases or decreases.

In both of these examples it is very common to invoke the Second Law and conclude that the entropy change in the universe must be positive and larger (in absolute magnitude) than the change in the entropy of the system. Such a conclusion is very common. However, we in fact do not know how to measure

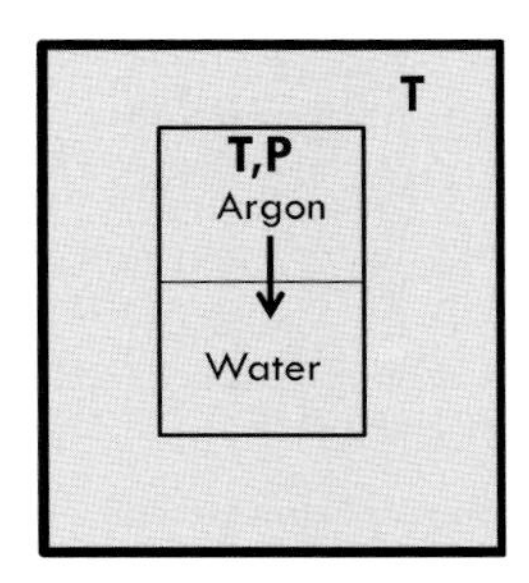

Fig. 4.7 Argon dissolves in water. The whole system is in a heat reservoir at a given temperature T.

or calculate the change in the entropy of the universe in such experiments.

It should be emphasized that in all of the examples given above, my objection is to the application of entropy and the Second Law to the *entire* universe.

4.3 Entropy of the Universe?

Perhaps the most well-known and well-quoted statement regarding the entropy of the universe is the one made by Clausius himself (1865):

> The energy of the universe is constant. The entropy of the universe tends to a maximum.

I can easily accept the first part of this quotation. Although we cannot be absolutely certain that the energy of the entire universe is conserved, I am comfortable with the general principle of conservation of energy, not only because we have never observed violations of this principle; in fact, we did observe when Einstein discovered that mass can be converted into energy. This finding has violated both the principle of conservation of energy and the principle of conservation of mass. But we resolved that violation by reformulating a new principle of conservation of the total energy and mass.

The reason I am comfortable with the first statement is that, as far as we know, the energy does not depend on whether the system is at equilibrium or not. If the universe is infinite, then perhaps the energy is also infinite, in which case the *conservation* of an infinite energy becomes fuzzy, as much as the statement that the number of points in the real interval between zero and one is the same as between zero and ten, or between zero and infinity.

What I have said about the total energy might be drastically changed if a new form of energy is discovered in the future.

The situation is very different for the entropy of the universe. We have no way of either measuring or calculating the entropy of the universe. I doubt that the entropy of the universe can be defined. Therefore, the second statement by Clausius is unwarranted. Here is a relatively rare statement by Corning (2002), with which I fully agree:

> Equally dubious is the claim that the general trend in the universe is toward increased entropy. Indeed, entropy has often been portrayed as a dark force which somehow governs the fate of our species and dooms our progeny to oblivion — in the eventual "heat death" of our universe. The practice of making such cosmic claims for entropy dates back to Clausius. In his classical text, *Abhandlungen über die mechanische Wärmetheorie*, Clausius (1864) wrote: "The energy of the universe is constant; the entropy of the universe tends to a maximum" [(quoted in Harold 1986, p. 8)]. Clausius also coined the term "heat death" (*Wärmetod*).
>
> This dour vision has long since become the conventional wisdom of the western scientific establishment.

Atkins (2007) devotes a whole book to the "Four Laws that drive the Universe." The author means, of course, the Four Laws of Thermodynamics. In the book's preface, he states:

> The Second Law is one of the all-time great laws of science, for it illuminates why anything — anything from the cooling of hot matter to the formulation of a thought — happens at all.

I flatly disagree with this.

On page 50, the author writes:

> All our actions, from digestion to artistic creation, are at heart captured by the essence of the operation of a steam engine.

This is a wild and meaningless sentence. The essence of a steam engine has nothing to do with artistic creation!

On page 62, the author writes:

> The entropy of the universe increases in the course of any spontaneous change. The key word here is *universe*; it means, as always in thermodynamics, the system together with its surroundings. There is no prohibition of the system or the surroundings *individually* undergoing a decrease in entropy provided that there is a compensating change elsewhere.

"The key word here is *universe*." In my opinion, this "key word" invalidates the quoted statement of the Second Law. We know that entropy increases in a spontaneous process in an *isolated* system. In this formulation of the Second Law, the "key word" — "universe" — does not appear, and therefore is superfluous.

There are actually two issues between which most writers on the entropy of the universe do not make a clear distinction. One is the *value* of the entropy of the universe, and in which direction it is expected to change in millions or billions of years from now. The second is the *fate* of the universe. What will happen? Is the universe doomed to reach its "thermal death"? These are two distinctly different issues.

Regarding the issue of the entropy of the universe, all the conclusions reached in the previous chapter regarding the *entropy of life* are *a fortiori* true for the entropy of the universe. Why?

First, because life is *included* in the universe. Since it is meaningless to talk about the entropy of all living organisms in the universe, it is also meaningless to talk about the entropy of the entire universe. Here, by "all living organisms of the universe," I mean those we know about, as well as those that have not been discovered yet, those that once lived and are now extinct, and those that might evolve in the near and the distant future. Of course, the total mass of all living systems in the universe is small compared with the vast size of the universe. However, this fact does not change my argument in principle. Besides, we really do not know about *all life* in the universe.

Second, in addition to having life within the universe, we still do not know whether the universe is finite or infinite. We certainly know that the universe is not in an equilibrium state. In addition, we have no idea whether the universe will ever reach an equilibrium state.

As I noted in Chapter 3, we do not know how to describe the *state* of a living system, either classically or quantum-mechanically, and we do not know whether the laws of physics govern living organisms. This statement was made regarding a single living organism. When we talk about the entire universe, or even the entire planet on which we live, we must also ask ourselves whether we know how to define the *interactions* between two or more living organisms.

We know much about the interactions between atoms and molecules. We also know how these interactions affect the entropy of the interacting particles. [See Ben-Naim (2008).] Qualitatively, we also know what we mean by "interaction" between people: they communicate, they love or hate, they admire or disdain, and so on. But we do not even know how to describe the interactions between people in physical terms. Certainly, we have no idea how these interactions affect the entropy of all living organisms — if such an entropy is meaningful.

In Chapter 3, I argued that because the state of a living system is not yet definable, we cannot assign an entropy value to any single organism. The situation is of course much more complicated when we have many *interacting* organisms (colloquially as well as physically). Even if we find a way to define the entropy of a single organism someday, it will be far from clear what the entropy of two or more interacting organisms is.

Does the entropy increase or decrease when a baby is born? Does the entropy increase or decrease when two people exchange information? Does the entropy increase or decrease when one animal eats another animal? You could go on and on, and ask these questions which at the moment are unanswerable.

From the conclusion we reached in the previous section on the process of expansion into an infinite volume, we can conclude that all statements about the entropy of the universe are premature at best, and more likely are superfluous.

The second issue concerns the ultimate *fate of the universe.* Since we do not know whether the universe will ever reach an equilibrium state, authors should refrain from frightening people about the universe reaching a *thermal death*, or that the *ravaging* power of entropy is a harbinger of the end of life and everything that life has created on our planet or in the entire universe.

Some authors who write about the ravages of entropy give the "bad news" first, saying that the universe is doomed. Then they add the "good news," telling the reader not to worry about it since it will only happen billions of years from now, thus posing no threat to our generation, or to any of the succeeding millions of generations. This is indeed "soothing" news. However, fortunately or unfortunately, both the "bad" and the "good" news are unwarranted.

Note that I do not claim that the universe might not "die" on a rainy day some billions of years from now. This might happen (whatever this might mean). My objection is to the prediction of the death of the universe *based on* the Second Law of thermodynamics, and that this fate is inevitable. If thermal death is to happen, it will not happen because of the "ravages of entropy," or because of the Second Law.

The phrase "ravages of entropy" has become commonplace in recent popular-science books. People who talk about the "ravages of entropy" also talk about the "ravages of time" (this becomes almost the same if one identifies entropy or the Second Law with the arrow of time).

Colloquially, I myself oftentimes use the expression "ravages of time." This is a very common figure of speech used by scientists, as well as nonscientists. There is nothing wrong with this, as long as you mean that some processes seem to lead to deterioration, decay or death. However, there is no such general rule, or a law of nature that states that things are ravaged by time. One can equally say that many phenomena, like flourishing, blooming and births, are the "creation of time." In fact, the history of life on earth shows that more lives are created in the direction of the arrow of time than deaths. Thus, for some processes one can say figuratively that they are the result of the ravages of time, but for others one can also say that they are the result of the "creative power of time" — again only figuratively. Time does not ravage anything, nor does it create anything.

Entropy, on the other hand, is more "innocent" than time. It does not ravage anything, nor does it create anything. It does not create "order out of disorder," or "disorder out of order." In short, entropy does not *do* anything, not even figuratively, just as the *length* of this book, the number of pages, or the number of letters in this book, does not ravage anything or create anything.[2]

Yet, you read in many popular-science books about the ravages of entropy, giving the impression that the almighty entropy has the power to destroy anything while it is growing . . . .

Such statements result from the fundamental misconception of the Second Law as a tendency toward disorder, and the association of entropy with the arrow of time.

All that I have said above about the distant *future* of the universe applies to the distant *past* of the universe. People deduce from the fact that the entropy of the universe always increases, that the value of the entropy must have been very low in the early times of the universe. This is known as the Past Hypothesis.[3] Briefly, the Past Hypothesis states:

> When it comes to the past, however, we have at our disposal both our knowledge of the current macroscopic state of the universe, plus the fact that the early universe began in a low-entropy state. That one extra bit of information, known simply as the "Past Hypothesis," gives us enormous leverage when it comes to reconstructing the past from the present.

Do we have the "current macroscopic state of the universe"? I doubt the veracity of this statement. The "fact" that the early universe began in a low entropy state is not a fact! At best, this is a wild hypothesis, probably even a meaningless one. Even if we had known the exact macroscopic state of the universe, I doubt that the Past Hypothesis could help in reconstructing the past from the present. On the same page we find another "punch line"[4]:

> The punch line is that our notion of free will, the ability to change the future by making choices in a way that is not available to us as far as the past is concerned, is only possible because the past has a

> low entropy and the future has a high entropy. The future seems open to us, while the past seems closed, even though the laws of physics treat them on an equal footing.

The truth of the matter is that we have no idea about the entropy of the past, or about the entropy of the future. Even if we would have known these, I doubt if that these entropies would have anything to do with "free will."

Here are more quotations from Carroll's book (2010):

> If everything in the universe evolves toward increasing disorder, it must have started out in an exquisitely ordered arrangement. This whole chain of logic, purporting to explain why you can't turn an omelet into an egg, apparently rests on a deep entropy, very high order.
>
> The arrow of time connects the early universe to something we experience literally every moment of our lives. It's not just breaking eggs, or other irreversible processes like mixing milk into coffee or how an untended room tends to get messier over time. The arrow of time is the reason why time seems to flow around us, or why (if you prefer) we seem to move through time. It's why we remember the past, but not the future. It's why we evolve and metabolize and eventually die. It's why we believe in cause and effect, and is crucial to our notions of free will.

The fact is that in a controlled experiment carried out in a laboratory in an isolated system, the entropy increases. From these experiments we cannot conclude that the universe started out in an "exquisitely ordered arrangement," and besides an untended room does not get messier over time! These statements involve typical misinterpretations of entropy in terms of disorder, and applying the concept of entropy (or disorder) where it does not apply.

Penrose (1989) asks the question about the origin of the "low entropy" of the universe. In my opinion, this is a meaningless question, as long as the entropy of the universe is not defined.

Penrose also asks about the "origin of entropy." In my opinion this question is as meaningless as it is to ask about the origin of "probability," or the origin of length or mass. It is meaningful to ask about the origin of matter, the origin of energy, and even the origin of life and of the universe. We will always get back one level, and once we find that A is the origin of B, we can ask what the origin of A is, and so on. In my opinion, since entropy is not a substantial thing, it is not meaningful to ask about its *origin*. Perhaps one day we will know the origin of the universe, and we will know its state at that time. In such a case, we could assign meaningful entropy to that state. But since we do not know the precise state of the universe at any point of time, we cannot talk about the entropy of the universe — and therefore it would be meaningless to ask about the "origin of the entropy," or the origin of the "low entropy" of the universe.

Here are some more statements about the Past Hypothesis. I urge the reader to read these quotations carefully and critically.

On page 317, Penrose (1989) has a section titled "The Origin of Low Entropy in the Universe":

> We shall try to understand where this "amazing" low entropy comes from in the actual world that we inhabit. Let us start with ourselves. If we can understand where our low entropy came from, then we should be able to see where the low entropy of the gas held by the partition came from, or in the water glass on the table, or in the egg held above the frying pan, or the lump of sugar held over the coffee cup .... It was, to a large extent, some small part of the low

> entropy in ourselves which was actually made use of in setting up these other low-entropy states.

In my opinion, the title of that section might already be misleading, as it might lead the reader to think that the entropy "has an origin" much as any substance does. With all due respect to Penrose, whose writings I admire, I cannot agree with the whole paragraph quoted above. In my view it is meaningless to talk about "our own low entropy," and certainly one cannot claim that the "entropy in ourselves" sets the low entropy states in the systems mentioned.

In contrast to the idea of the low entropy past, we find in Greene (2004), page 100, the following:

> Thus, not only is there an overwhelming probability that the entropy of a physical system will be higher in what we call the future, but there is the same overwhelming probability that it was higher in what we call the past.

In my opinion such statements create the impression that somehow the entropy of a system is a well-known function of time (as is actually illustrated in Greene's book, Figure 6.2). This is not so. When the same argument is applied to the *whole universe*, it turns from being wrong to meaningless. Since the entropy of the universe is not defined, we cannot say anything about its value in the past, or in the future.

Furthermore, on page 173, Greene (2004) writes:

> Thus, unlike ordinary stars, black holes stubbornly hold on to all the entropy they produce: none of it can escape the black hole's powerful gravitational grip.

Again, this is a highly impressive statement, but it is meaningless. The argument given by Greene is even more puzzling:

> This makes good intuitive sense: high entropy means that many rearrangements of the constituents of an object go unnoticed. Since we can't see inside a black hole, it is impossible for us to detect any rearrangement of its constituents — whatever those constituents may be — and hence black holes have maximum entropy.

First, entropy does not mean that "many *rearrangements* of the constituents of an object go unnoticed." Second, we do not know what the "constituents" of black holes are. Third, the entropy does not depend only on the "constituents," but also on the interaction energies between them. Fourth, "maximum entropy" — maximum with respect to what?

Finally, Greene repeats the same erroneous interpretation of entropy as disorder. On page 333 he writes:

> Entropy is a measure of disorder or randomness. For instance, if your desk is cluttered high with layer upon layer of open books, half-read articles, old newspapers, and junk mail, it is in a state of high disorder, or high entropy. On the other hand, if it is highly organized with articles in alphabetized folders, newspapers neatly stacked in chronological order, books arranged in alphabetical order by author and pens placed in their designated holders, your desk is in a state of high order or, equivalently, low entropy.

Instead of commenting on this paragraph, I will refer the reader to the two figures at the beginning of Chapter 2.

If one interprets entropy as a measure of disorder (as most people do), then one must conclude that "in the beginning" the universe was in a highly "ordered state" — whatever this might mean. I have once (ironically) commented that this conclusion "contradicts" what is written in Genesis[5]:

> In the beginning God created the heaven and the earth. And the earth was unformed, and void.

א בְּרֵאשִׁית, בָּרָא אֱלֹהִים, אֵת הַשָּׁמַיִם, וְאֵת הָאָרֶץ.
ב וְהָאָרֶץ, הָיְתָה תֹהוּ וָבֹהוּ, וְחֹשֶׁךְ, עַל-פְּנֵי תְהוֹם; וְרוּחַ אֱלֹהִים, מְרַחֶפֶת עַל-פְּנֵי הַמָּיִם.

Fig. 4.8

The original Hebrew version includes the expression "*Tohu Vavohu*," instead of "unformed" and "void." The traditional interpretation of "*Tohu Vavohu*" is total chaos, or total *disorder* — if you prefer, the highest entropy!

Personally, I do not subscribe to either the Past Hypothesis or the "past chaos" views. I simply do not know, and I believe no one knows, and perhaps no one will ever know. The most one might say regarding the *state* of the universe is that it might have been in *some state*. Regarding the entropy, I can say more categorically that not only do we not know the *value* of the entropy, but the entropy of the universe is simply undefined.

Carroll (2010), in discussing the Past Hypothesis, makes the analogy between the *expansion of the universe* and the *increase in entropy*. This analogy is invalid for the following reason:

We observe the expansion of the universe. We can extrapolate back in time and conclude that some 14–15 billion years ago the whole universe was highly condensed. This extrapolation is valid provided that the universe had always expanded, and that the laws of physics as we know them today were the same billions of years ago. With these assumptions we may speculate about the big bang theory. Figure 4.9 shows schematically what happens

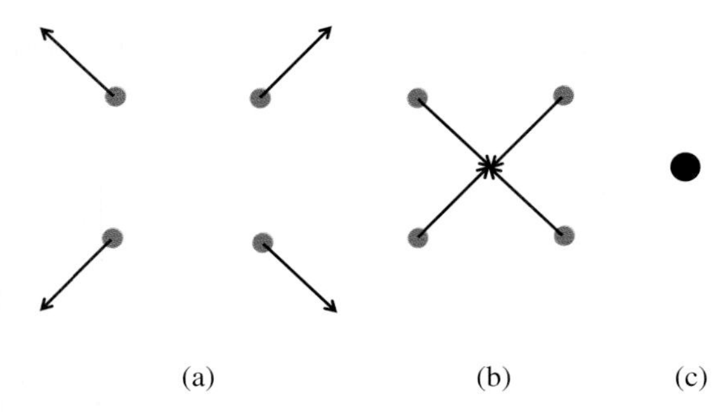

Fig. 4.9 Schematic: (a) expanding of the universe; (b) reversing the velocities of all particles to condensing in a very small region (c).

when we reverse all the speeds of all the objects in the universe, and extrapolate to what might have happened in the very early beginning.

The situation is very different when we discuss the Past Hypothesis. Here, we do not have any *evidence* that the entropy of the universe is increasing. It is doubtful if the entropy of the universe is even definable. Thus, there is nothing to be extrapolated back in time to reach a conclusion on the *value* of the entropy at the big bang.

The big bang theory is based on some speculative assumptions. The Past Hypothesis is based on the misconception of the "entropy of the universe." Therefore, it is a baseless and meaningless hypothesis.

Recently, some physicists have promoted the *idea* about multiverses, or parallel universes, which is essentially an alternative to the Copenhagen interpretation of quantum mechanics. Of course, no one knows anything about these parallel universes, or whether or not they are similar to "our universe," or perhaps completely different universes with different sets of laws of physics, chemistry or biology. Needless to say, we cannot say anything about the entropy of these universes.

A witty and succinct statement by Woody Allen sums it all up:

> There is no doubt that there is an unseen world. The problem is how far is it from midtown and how late is it open?

4.4 Information and the Universe

Having dismissed the entropy of the universe, there still remains the question of the applicability of SMI to the universe. Of course, there is no problem in applying SMI to any well-known probability distribution. It could be the SMI of the distribution of locations, and velocities of particles, the distribution of matter or energy in any part of the universe. But reference to the SMI of the universe is meaningless *unless we specify the relevant distribution*. Unfortunately, the literature is replete with statements about the entropy of the universe, the information of the universe, and the SMI of the universe. I have no idea what these statements mean. I will quote some here and leave the reader to ponder about them, and if anyone finds or knows the meaning of any of these, please write to me.

Carroll (2010):

> The entropy of the universe has increased by an enormous amount since the days when everything was smooth and featureless.
>
> Our universe may be entirely shaped by the information.

Seife (2007), page 262:

> As the universe expands and evolves, the entropy of the cosmos increases.

All these are impressive, perhaps even dramatic, statements. Unfortunately, they lack any meaningful content.

Campbell (1982):

> Dreams may seem to be highly random, entropic and out of control.

Does anyone know what an "entropic dream" means[6]?

> Entropy is a protean concept. It has to do with probability, since the most probable state of a system in constant, random motion is for all its contrast to be smoothed out. (*Page 49*)
>
> In remembering, the brain does not like randomness. It looks for ways to lower the entropy of a collection of items by reducing the number of ways in which they can be arranged. It cuts down the value of W in Shannon's equation, $S = K \log W$, and therefore reduces the entropy S. (*page 215*)

This is very interesting! In *remembering*, the brain does not like *randomness*. It follows that in *forgetting* the brain *likes* randomness. I wonder what the brain likes when it is idle. Besides, can anyone explain to me how the brain, while it is busy remembering, reduces the number of ways a collection of items can be arranged?

Seife (2007):

> The laws of information are giving physicists a way to understand the darkest mysteries that humanity has ever pondered. Yet, those laws are painting a picture that is as grim as it is surreal.
>
> The laws of information have sealed our fate, just as they have sealed the fate of the universe itself.
>
> Information theory ... governs the behavior of objects on many different scales. It tells how atoms interact with each other and how black holes swallow stars. Its rules describe how the universe will die; and they illuminate the structure of the entire cosmos.

> Even if there were no such thing as a computer, information theory would still be the third great revolution of the twentieth-century physics.
>
> This new theory of information was an idea as revolutionary as quantum theory and relativity ... the ideas of information theory ... they describe the behavior of the subatomic world, all life on Earth, and even the universe as a whole.

This is not a mere exaggeration of an existing theory. It is an over-overexaggeration of a nonexisting theory. Whatever one means by "information theory," it does not explain *anything*, it does not govern *anything*, certainly not "life on Earth" and absolutely not the universe itself.

It is clear that Seife (2007) refers to Shannon's "theory of communication" as the foundation of "information theory."

Here are some more fancy statements from Seife (2007):

> Thermodynamics is just a special case of information theory. (*Page 72*)

It is not!

> Understand the relationship among entropy, energy and information and you might begin to understand how computers and humans think. (*Page 81*)

I believe I understand the relationship between entropy, energy and information, but I fail to understand how humans think!

> Information theory consumed thermodynamics. The problem in thermodynamics can be solved by recognizing that thermodynamics is, in truth, a special case of information theory.

It is not!

> All matter and energy is subject to the laws of information. Including us.

This is not even wrong!

> Why must we die? We don't. We are immortal. The catch is that "we" in question is not our bodies or our minds; it is the bits of information that reside in our genes. (*Page 100*)

So we are immortal! What a great revelation.

> Yet, the information in our genes is able to resist the ravages of time and entropy, the arrow of time.

The information in our genes does not resist anything — certainly not the meaningless "ravages of time and entropy".

All these statements and many others in the book can be classified as being either meaningless, absurd, or simply nonsense.

4.5 Entropy and Information of Black Holes

Black holes (BHs) are objects in space which cannot be seen directly but their existence is implied indirectly through their gravity. A BH is characterized by a very strong gravitational field, so strong that nothing can escape from it, including light. BHs cannot be seen, and hence they are referred to as "black."[7]

There is a vast literature on BHs. Yet, not much can be said about a BH. There are essentially three physical quantities that characterize BHs: mass, charge and angular momentum. This means that two BHs having the same mass, charge and angular momentum would be indistinguishable.

Bekenstein (1973) noticed that there are a number of *similarities* between BH physics and thermodynamics. More specifically, the horizon surface area can never decrease with time, and when two BHs collide and merge into one BH, the surface area of the new BH is greater than the sum of the areas of the two original BHs. This property of BHs smacks of entropy. As Bekenstein noted:

> This is reminiscent of the second law of thermodynamics, which states that changes of a closed thermodynamic system take place in the direction of increasing entropy.
>
> The above observation suggests that something like entropy may also play a role in it.

"Something like entropy" is not necessarily entropy.

Another kind of circular argument is the following[8]:

> What is the real meaning of the black hole entropy? ... the entropy reflects the number of microscopic quantum states .... The number of such states is of the order of $\exp[S_{BH}]$.

Here, the entropy of a BH is defined in terms of the number of states of the BH, and the number of states is defined in terms of the entropy of the BH!

Since we do not know how to calculate W, it follows also that the Boltzmann entropy cannot be calculated for a BH.

In my opinion, the very assumption that a BH has entropy is questionable. I should remind the reader that although we know how to calculate the entropy of ideal gases, of perfect solids, and some simple models of systems of interacting particles, we still do not know what BHs consist of, or what the interaction energies between the particles constituting BHs are. Therefore, at this stage I doubt that our knowledge of BHs allows us even to talk about BH entropy, let alone calculate its value.

It seems to me that the mere analogy between BH physics and the Second Law of Thermodynamics is not sufficient to assign entropy to BHs. There have been several skeptical comments regarding the validity of the arguments leading to defining BH entropy.[9] We will not enter into this highly technical issue.

In fact, Bekenstein proposed a generalization of the second law[10]:

> It is thus natural to conjecture that the second law is not really transcended provided that it is expressed in a generalized form: *The common entropy in the black hole exterior plus the black hole entropy never decreases.* This statement means that we must regard black-hole entropy as a genuine contribution to the entropy of the universe.

Since, in my opinion, the "entropy of the universe" is not defined, it is very clear that BH entropy is not *thermodynamic* entropy.[11,12]

In a paper title: "Black Holes and Entropy," Bekenstein (1973) starts from the following observations:

> There are a number of similarities between black-hole physics and thermodynamics. Most striking is the similarity in the behaviors of black-hole area and of entropy: Both quantities tend to increase irreversibly.

The author continues:

> We show that it is natural to introduce the concept of black-hole entropy as a measure of information about a black-hole interior which is inaccessible to an exterior observer.

The existence of a "similarity" between BH physics and thermodynamics does not necessarily mean that there should be a connection between the two. From these similarities, Bekenstein suggests:

> We already have a concept of energy in black-hole physics, and the above observation suggests that something like entropy may play a role in it.

What could that something *like* entropy be?

First, one can rule out thermodynamic entropy as defined by Clausius. One cannot make any measurements on a BH; specifically, one cannot determine its heat capacity (it is not clear that such a quantity is definable for BHs). Indeed, Bekenstein admits that BH entropy is not thermodynamic entropy.[11]

The second possibility is to define Boltzmann's entropy for BHs. This requires some knowledge of the total number of accessible states, W. Unfortunately, it is not clear how to calculate the energy states of a BH; nor is it clear *which states* to include in W.[11] Therefore, Boltzmann's entropy is also ruled out.

What remains is that BH entropy is not *entropy*, but SMI, which is referred to as entropy. Of course, one can always define an SMI for various distributions, if such distributions for BHs are available.

In fact, Bekenstein uses arguments from information theory, which he refers to as Shannon's entropy. As we have noted in Chapter 2, SMI is in general not entropy. Indeed, Bekenstein alludes to this by stating:

> Although there can be little doubt that black-hole entropy corresponds closely to a phenomenological entropy, its deeper meaning has remained mysterious.

> Does it stand for information lost in the transcendence of the hallowed principle of unitary evolution? I would claim that at this stage the usefulness of any proposed interpretation of black-hole entropy turns on how well it relates to the original "statistical" aspect of entropy as a measure of disorder, missing information, multiplicity of microstates compatible with a given macrostate, etc.

In my view, all these interpretations of entropy should *follow* the definition of BH entropy, not *precede* it. Since thermodynamic entropy is not definable for BHs, there is no justification to speculate about the *information* of BHs, information *lost* in BHs, information *swallowed* by BHs, etc., as is common in popular-science books.

Denbigh (1981) has also expressed his doubts about BH entropy:

> No doubt this simple theory, put forward in the springtime of an entirely new field of research, may not prove to be correct.

Denbigh's statement is mild compared to Lavenda's criticism[9] of the concept of BH entropy, concluding that

> "the expression for the 'entropy' of a black hole is anything but entropy."

The involvement of "information" in BH entropy has led to many statements referring to colloquial information and BHs. I will conclude this section with some quotations from recent popular books, and leave the reader to ponder on their meanings:

Seife (2007):

> Although there are very strong reasons for believing that information is conserved, the information in that lump of matter is gone.

> The information is inaccessible. But has it been destroyed? Is this information erased without any trace?
>
> Nobody knows. But there's reason to believe that it isn't — that information survives even the ultimate torture of falling into a black hole.

Seife (2007), p. 240:

> Perhaps a black hole really is the ultimate computer, the ultimate processor of information . . . . Information . . . . is supreme. It might even reveal the existence of hidden universes.

Perhaps! In my opinion, since we know so little about BHs, it is meaningless to say that a BH is the "ultimate computer." It is as meaningless as to say that a BH is the ultimate living system, or the ultimate smartphone, or whatever you wish.

Seife, p. 242:

> The frontiers of information theory are providing a very, very disquieting picture of our universe — and of the ultimate fate of life in the cosmos.

And on page 265, Seife writes:

> What is life? Using the tools of information theory, scientists are beginning to get answers to all of these questions. But at the same time, those tools of information theory have revealed our ultimate fate. We will die, as will all the answers we have to these questions — all the information our civilization has gathered. Life must end, and with it will end all consciousness, all ability to understand the universe. Using information, we may find the ultimate answers, yet those answers will be rendered worthless by the laws of information.
>
> This precious information that may well illuminate the darkest mysteries about the universe carries in it the seeds of its own destruction.

All these statements are the result of confusing *information* with SMI, then confusing SMI with *entropy*, and finally applying entropy to predict the fate of life and of the universe.

Besides, the "tools of information theory" do not, and cannot, provide any answer to the question "What is life?"

If you follow me throughout the book you will see that it is meaningless to talk about the entropy of life and the entropy of the universe. It is *a fortiori* meaningless to predict the fate of life or of the universe by using information instead of entropy (see also the Epilogue).

Carroll (2010):

> If instead of throwing the book into a fire, we had thrown it into a black hole, the story would be different. According to classical general relativity, there is no way to reconstruct the information; the book fell into a black hole, and we can measure the resulting mass, charge, and spin, but nothing more. We might console ourselves that the information is still in there somewhere, but we can't get to it.

How does Carroll know that the information is "somewhere there" if we cannot get to it?

Finally, a quote from von Baeyer (2003):

> Talk of the information contained in black holes sounds highly metaphysical, but it turns out that it can actually throw light on some mundane, everyday problems.
>
> If you stretch your imagination to shrink this cube into a black hole, and use the known formula for its entropy to estimate the theoretical maximum amount of information it can store, you get an ultimate limit of about 10^{65} bits per cc, a bound which no memory can exceed. The gap of forty-five orders of magnitude between what might be technically achievable, and

what fundamental theory allows, seems so ridiculously large that it renders the black hole irrelevant.

As I have emphasized several times in this book, the road from SMI to colloquial information is very slippery. One can easily muddle up the two. In my opinion all those references to "information" *of* a BH, *in* a BH or *outside* a BH are at best highly speculative. They certainly have nothing to do with the entropy!

A BH is still far from being a well-defined and well-characterized object. Trying to estimate its entropy based on its mass, energy or the area of its horizon is, at best, making an estimate of the entropy of the BH if that would be definable. It does not provide any *additional information* beyond that.

Finally, having read a few articles on BH entropy, I was wondering: If so little is known about both the microscopic and the macroscopic state of a BH, why bother about its entropy? It seems to me that with the present knowledge of BHs, very little can be gained by estimating the BH entropy. It only allows writers of popular-science books to write all kinds of fanciful things about entropy and information of BHs which have little (if anything) to do with reality.

4.6 Summary of Chapter 4

The entropy of the universe always increases.

This statement has been quoted innumerable times. Unfortunately, there is no basis, either experimental or theoretical, to either justify it, or to lend any meaning to it. The universe is not a well-defined thermodynamic system. It is not clear whether the universe is finite or infinite, and whether or not it will ever reach an equilibrium state. Therefore, it is meaningless to speculate about the changes in the entropy of the universe.

Clearly, the thermodynamic entropy of the universe cannot be measured. We cannot make any experiment on the entire universe. The same is true of Boltzmann's entropy. Can we calculate the number of states of the universe?

What remains is to use SMI, which could be applied to well-defined distributions. However, these SMIs have nothing to do with thermodynamic entropy.

All the statements regarding the entropy of the universe, including some quoted in this chapter, arise from confusing entropy with SMI, and SMI with information.

Of course, one can have a lot of information *about* the universe, or the BH. However, this information is not subject to any law of physics. The information about the weather in Jerusalem or New York is not subject to Newton's laws. It is not subject to the Second Law of Thermodynamics either.

My general conclusion in this chapter is the same as in Chapter 3. We do not know how to define the thermodynamic *state* of being alive. Therefore, we cannot talk about the entropy of a living organism. This conclusion is *a fortiori* true for the entire universe, which not only includes life but also because we have no idea of the thermodynamic *state* of the universe.

Epilogue

The Future Hypothesis

I have discussed two hypotheses in this book: the Future one and the Past one. Both are based on the Second Law of Thermodynamics. The first is not considered to be a hypothesis. It is based on the "well-tested" and "never-violated" Second Law. Therefore, extrapolating back in time, it is considered to be as solid as a rock, a well-established and well-accepted conclusion based on the Second Law. Namely, that the universe is *doomed to death*. The second is referred to as the Past Hypothesis. It is based on the premise that the entropy of the universe always increases both in the present and in the future. Therefore, extrapolating back in time, one concludes that in the distant past the entropy of the universe must have been very low (and if you subscribe to the order–disorder interpretation of entropy, you would conclude that the early universe must have started in a *highly ordered* state).

I have challenged both of these conclusions in Chapters 2 and 4 on the basis of the simple fact that no one has ever measured or calculated the entropy of the universe. In fact, I claimed that the entropy of the universe is undefinable, not in the past, not in the present, and not even in the future. Perhaps the entropy of the universe will never be defined. Therefore, it is meaningless to talk about the thermal death of the universe, the low entropy state of the past, the "origin" of the low entropy of the universe, and so on.

The immediate consequence of my conclusion is that you (and I) can relax. We have no idea whether the universe is doomed to die or not.

In order to balance this dismal fate of the universe predicted by almost anyone who writes about the Second Law, I would like to offer my own *Future Hypothesis*, which is far brighter and better founded than the future (hypothetical) prediction based on the Second Law.

My Future Hypothesis is not based on the Second Law, but on logical deduction. Of course, this is only a hypothesis or a conjecture. I cannot prove it, but I believe it is far more plausible than the conventional Future Hypothesis based on the Second Law.

We know that in the past few billions of years life emerged on our planet. Ever since it began (whenever that was), it has flourished, proliferated, and of course evolved. It is reasonable to expect that life will continue to evolve, and I believe that at some stage — it could be thousands, millions or billions of years from now — evolution of humans, in the sense of *Darwinian evolution*, will slow down and perhaps come to a halt. Human life could still evolve, but not necessarily according to the natural selection mechanism.

The reason I am saying this is that Darwinian evolution is based on some basic premises: reproduction, metabolism, mutation, survival, etc. It is not far-fetched to foresee that in millions of years from now, humans will gradually evolve into machines — first by replacing some mechanical parts (muscles, bones, the heart, etc.), then also the chemical factories, such as the liver and other glands, and eventually the brain.

I believe that all these material parts of the body will eventually be replaced by long-enduring and self-correcting parts. This scenario is different from the science fiction predictions that the world will be "conquered" by computers, robots or machines. It is *we*, or some future variation of us, that will evolve into something that resembles robots and machines. [For some interesting thoughts about the future of life, see Mautner (2009).]

Millions of years from now, all our tissues and organs, even the brain, will be replaced with equivalent parts. At that time, some parts of our bodies might be redundant. As an example, if those human–robots, or superhumans, find a better way to use energy, they might do away with all the metabolism machinery, and perhaps also the circulation of blood. For those creatures, oxygen, water and other essential chemicals for life would no longer be necessary.

These superhumans could produce as many copies of themselves as they wish. Eventually, there might not even be a need for all the reproduction machinery. In this sense, the superhuman will be "simpler" than the "organic" human.

What I am not sure about is what the last step will be. Will replacements for the "mind," the "soul" or the "*neshama*," that feature of life which distinguishes between the living and the nonliving, or whatever you might want to call it, be found? What will these new creatures "feel"? Perhaps the mind is an emergent attribute and it will be carried over from the present organic human to the superhuman. Perhaps we will discover what the mind is, once we have replaced everything.

We might or might not find out what this thing that we call the "mind" or "soul" or "*neshama*" is. If we do, then one can speculate "from here to infinity" about the potential of the mind. It will certainly not obey any presently known physical laws. This "mind" might create or discover new energies, new matter, or new "somethings" that we still have no words for. We might discover new things which we would be able to understand, or perhaps would never be able to understand. What we call the "mind" might itself be a new form of energy. It might travel at an infinite speed — much as our imagination *can* travel at any speed we want, unrestricted by the finite speed of light. This mind would be able to travel to distances far beyond those from which we presently receive signals sent about 15 billion years ago, beyond what we presently call "our universe," to other parallel universes or even to parallel multiuniverses, or multi-multiuniverses and beyond. On reaching other universes, we might find completely different — presently inconceivable and unimaginable — realities, different laws of physics, different elements, different chemistry and biology, and different forms of life.

At that time, we — or should I say these supercreatures — will be able to produce copies of themselves or other creatures "in their image," as in the story of Genesis. Perhaps there will no longer be any need for reproduction in the conventional sense, and evolution in the Darwinian sense will come to a halt. This does not mean, however, that the supercreatures will not evolve further, but if they do, it will not necessarily be in accordance with the Darwinian doctrine.

In this new world the law of conservation of energy, or an extension of it, will still apply. However, the almighty entropy — so many people fear its ravages — might one day become harmless. Supercreatures could control the random behavior of atomic particles, and the statistical nature of the Second Law of Thermodynamics would be an obsolete theory.

If new energies and perhaps new entities will be discovered, there is no reason why the supercreatures will not live forever, and colonize more and more parts of the universe. They could create new flora and fauna, or perhaps completely new forms of life which will live in peaceful coexistence with the supercreatures. At that time, we will see the dawn of a new world and new beginnings, and therefore there will no longer be a reason to worry about a doomed universe.

My friend Diego Casadei, who read this epilogue, commented that perhaps these eternal superrobot creatures will be lonely and bored. At some point in time they might entertain the idea of creating old-style humans (like us with feelings, a need for reproduction, etc.). In such a scenario the story of life in the universe will reiterate itself internally.

Notes

Chapter 1

Note 1 Yes, I know that some people will object to this statement, maintaining that information cannot be destroyed. I do not agree with this. I believe that such a statement results from confusing the general concept of *information* with the measure of information as used in information theory. I will further discuss this aspect of information in Chapters 1 and 2.

Note 2 Aumann (2005).

Note 3 A more dramatic version of this story can be found in Ben-Naim (2010). Briefly, the ten people in the party are replaced by ten women in Baghdad. Instead of colored stickers, each woman could see what all the other husbands were doing behind the backs of their wives, except the deeds of their own husbands. A judge declared that if a wife discovered that her husband was not faithful she must kill him the next morning. He also announced that there was at least one unfaithful husband. After two days, the two wives concluded that their husbands were unfaithful and they killed their husbands.

Clearly, every woman *knew* that there was at least one unfaithful husband, i.e. each woman had that *information*. Why had no one acted before the announcement?

The information that "there was at least one unfaithful husband" was known to each of the women before and after the announcement. The new information that was added was that each woman knew that each of the other women knew the information that "there was at least one unfaithful husband." This *additional* information could be used by logical reasoning to deduce and obtain new *information*; specifically, the two unlucky women could find out the *information* on their own husbands' infidelity.

Note 4 Hofstadter (1981).

Note 5 Hofstadter (1981,1985).

Note 6 Note that the negation of "All Cretans are liars" is "Not all Cretans are liars." This leaves the possibility that *some* Cretans are not liars, and some are.

Note 7 Sainsbury (2009).

Note 8 The paradox is that the barber shaves himself if and only if he does not shave himself, i.e. the barber shaves only those who do not shave themselves.

Note 9 Another clever statement the poor man could make: "I will be eaten cooked." If they cooked him, then his statement was true, and therefore they should eat him uncooked. But if they eat him uncooked, then his statement was false, and therefore they should cook him first. Unfortunately, the poor man could be eaten (cooked or uncooked) before they contemplated the truthfulness of his statement.

Note 10 Besides the technical difficulty of retrieving the information from the ashes and smoke. There is a more fundamental reason for the impossibility of retrieving this information. Once the molecules are in the gaseous phase they are mixed with other molecules of the same kind. Because of the indistinguishability of the molecules, there is no way to tell which molecules originated from the letters in the book. This is a result of the indistinguishability of the molecules. See also Section 1.13. [For more details, see Ben-Naim (2008)].

Note 11 Prove that $\sqrt{2}$ is not a rational number.

An irrational number is defined as a number that cannot be expressed as a ratio, m/n, of two integers ($n \neq 0$).

Assume that $\sqrt{2}$ is a rational number. Then we can write

$$\sqrt{2} = \frac{m}{n}, \tag{1}$$

where m and n are two integers.

Without loss of generality we can assume that m and n have no common factor. If they have one, then we can divide both numerator and denominator by that factor. For instance, if $m = 3$ and $n = 6$, we can divide by 3 and rewrite 3/6 as 1/2.

Take the square of the equation (1) to get

$$2 = \frac{m^2}{n^2}. \tag{2}$$

Or $2n^2 = m^2$. This means that 2 is a factor of m^2. It follows that 2 is also a factor of m itself (if 2 did not divide m, then the factorization of m *into* prime numbers would not contain 2, and therefore also the square of m would not be divided by 2).

Thus, we can write m as $m = 2x$, where x is an integer. Squaring the last equality gives $m^2 = 4x^2$, which means that $2n^2 = 4x^2$, or $n^2 = 2x^2$. Therefore, 2 divides n^2, and also must divide n.

We see that m and n have a common factor (2), in contradiction to our assumption. Therefore, $\sqrt{2}$ cannot be a rational number. We know that all rational numbers are infinite.

The totality of all the numbers in the segment (0, 2) is infinite, but of a higher order.

Note 12 This experiment is discussed in great detail in Ben-Naim (2014).

Note 13 For details, see Ben-Naim (2014).

Note 14 We are talking here about the correlation between two events — say, outcome "6" on one dice, and outcome "4" on the second dice. In the theory of probability we also define the correlation between the two random variables [see Ben-Naim (2008) for more details].

Note 15

בראשית פרק-כט פסוק-יח

וַיֶּאֱהַב יַעֲקֹב אֶת רָחֵל וַיֹּאמֶר אֶעֱבָדְךָ שֶׁבַע שָׁנִים בְּרָחֵל בִּתְּךָ הַקְּטַנָּה:

Note 16 Hofstadter (1985).

Note 17 Most people would answer, "I don't know." But let me give you a hint. I know that you know the name of at least one US president who is not buried in North America.

If you still maintain that you do not know, you either do not know who the US president *is*, or your default assumption is so strong that you cannot brush it aside.

Note 18 Shannon (1948).

Note 19 Consider the following information: "This book is about the theory of information." This information is about the contents of the book you are reading now. Does any of the plausible conditions (1), (2) and (3) apply to this information?

Not only do these conditions not *apply*, but they are not even *defined* for such information!

Note 20 In the interpretation of H as uncertainty and unlikelihood, we started with the interpretation of each term, $-\log p_i$, then reached the interpretation of the sum of this term. The informational interpretation of H does not arise from the informational interpretation of each of the terms, $-\log p_i$. It seems to me that some authors start from the informational interpretation of H, then "split" this interpretation and assign informational value to each of the events.

Another way of saying this is that knowing p_i is equivalent to having information on the probability of the event i. Here, we use "information" in its colloquial meaning. In this sense we have full information on the value of p_i. However, p_i is not the SMI associated with the event i. (See also Subsection 1.29.)

Note 21 For details, see Ben-Naim (2008, 2011).

Note 22 For some research on how young children play the 20Q game, see Ben-Naim (2008).

Note 23 For some simulation of the 20Q game, see *ariehbennaim.com, book, Entropy Demystified, simulated games.*

Note 24 For more details, see Ben-Naim (2008).

Note 25 This assertion needs a proof. Here, we offer only a qualitative argument. [More details can be found in Ben-Naim (2008)].

Note 26 Note that this is a maximum since

$$\frac{d^2F}{dp_i^2} = \frac{-1}{p_i} \leq 0,$$

i.e. the function H is concave downward.

The value of H for the distribution $p_i^*, \ldots, p_n^*$ is

$$H\left(p_i^*, \ldots, p_n^*\right) = -\sum p_i^* \log p_i^* = -\sum \frac{1}{n} \log \frac{1}{n} = \log n.$$

Note 27 Here, we calculate the average number of steps in case A of Figure 1.16. [For more details see Ben-Naim (2008)].

As can be seen in the diagram, the number of steps in case A of Figure 1.16 is also 2. The reductions in the SMI at each step are:

$$\text{First step: } g_1 = H\left(\frac{1}{2}, \frac{1}{2}\right) = 1;$$

$$\text{Second step: } g_2 = \frac{1}{2} H\left(\frac{1}{2}, \frac{1}{2}\right) + \frac{1}{2} H\left(\frac{1}{2}, \frac{1}{2}\right) = 1.$$

Hence, the total reduction in the SMI is

$$g_1 + g_2 = 2.$$

Clearly, we have two steps. At each step we gain 1 bit of information.

Let us calculate the probabilities of getting the required information at each step. Denote by G_i and N_i gaining and not gaining the information at the ith step. We have for the first step the following probabilities:

$$P(G_1) = 0,$$
$$P(N_1) = 1.$$

This simply means that at the first step we cannot gain the required information.

At the second step, we have

$$P(G_2, N_1) = P(G_2|N_1)P(N_1) = 1 \times 1 = 1.$$

This means that the probability of gaining the information *and* not gaining it at the first step is unity.

Thus, the average number of steps is

$$0 \times 1 + 1 \times 2 = 2,$$

which is the same as the total SMI. Note that at the second step, whatever answer we get, we will know where the coin is.

Note 28 At the first step, in case B of Figure 1.16, in contrast to the previous strategy, we *can* find the coin in one step. The probability of finding the coin in one step is 1/4. The SMI at the first step is

$$g_1 = H\left(\frac{1}{4}, \frac{3}{4}\right) = 0.8113.$$

We can refer to this quantity as the *average* gain of information at the first step. At the second step we have

$$g_2 = \frac{1}{4}H(1) + \frac{3}{4}H\left(\frac{1}{3}, \frac{2}{3}\right) = 0.6887,$$

and at the third step we have

$$g_3 = \frac{3}{4}\left[\frac{1}{3}H(1) + \frac{2}{3}H\left(\frac{1}{2}, \frac{1}{2}\right)\right] = \frac{1}{2}.$$

Clearly, the total amount of information gained at the three steps is

$$g_1 + g_2 + g_3 = 2.$$

Thus, the *total* SMI acquired is the same. However, the average number of questions is different in the two methods.

The average number of questions in the second method is obtained from the following probabilities. If we ask "Is the coin in box a?" and obtain the answer "yes", the game is ended. This happens with probability 1/4, and the "no" answer is obtained with probability 3/4:

$$P(G_1) = \frac{1}{4},$$
$$P(N_1) = \frac{3}{4}.$$

To gain the information at the second step, we need to get a "no" answer at the first step, and "yes" at the second step. Hence,

$$P(G_2, N_1) = P(G_2|N_1)P(N_1) = \frac{1}{3} \times \frac{3}{4} = \frac{1}{4},$$
$$P(N_2, N_1) = P(N_2|N_1)P(N_1) = \frac{2}{3} \times \frac{3}{4} = \frac{2}{4},$$
$$P(G_3, N_1, N_2) = P(G_3|N_1, N_2)P(N_1, N_2) = 1 \times \frac{2}{4} = \frac{2}{4}.$$

The average number of steps in this case is

$$\frac{1}{4} \times 1 + \frac{1}{4} \times 2 + \frac{2}{4} \times 3 = \frac{1+2+6}{4} = \frac{9}{4} = 2\frac{1}{4},$$

which is slightly larger than the number of steps in the first method.

Note 29 Note that Jaynes uses the term "maximum entropy," which, more appropriately, should be maximum SMI."

Note 30 Katz (1967).

Note 31 Brillouin (1967).

Note 32 Denoting by BR the experiments B *and* R, the assumption of independence of the two events is

$$\Pr^{(BR)}(i, j) = \Pr^{(B)}(i) \times \Pr^{(B)}(j). \quad (1)$$

The corresponding joint SMI of the two experiments is defined by

$$\mathrm{SMI(BR)} = -\sum_{i=1}^{N}\sum_{j=1}^{M} \Pr^{(\mathrm{BR})}(i,j) \log \Pr^{(\mathrm{BR})}(i,j). \qquad (2)$$

When the two experiments are independent, i.e. the equation (1) holds for each pair of outcomes i and j,

$$\begin{aligned}
\mathrm{SMI(BR)} &= -\sum_{i=1}^{N}\sum_{j=1}^{M} \Pr^{(\mathrm{BR})}(i,j) \log[\Pr^{(\mathrm{B})}(i)\Pr^{(\mathrm{R})}(j)] \\
&= -\sum_{i=1}^{N} \Pr^{(\mathrm{B})}(i) \log[\Pr^{(\mathrm{B})}(i)] \\
&\quad -\sum_{j=1}^{M} \Pr^{(\mathrm{R})}(j) \log[\Pr^{(\mathrm{R})}(j)] \\
&= \mathrm{SMI(B)} + \mathrm{SMI(R)}. \qquad (3)
\end{aligned}$$

Note 33 For more details, see Ben-Naim (2008, 2011).

Note 34 This follows from the identity

$$2^{2N} = (1+1)^{2M} = \sum_{i=0}^{2N} \binom{2N}{i} > \frac{(2N)!}{N!N!}. \qquad (4)$$

The term on the right hand side is only one term in the sum of positive numbers. Hence,

$$2N \log 2 > \log\left[\frac{(2N)!}{(N!)^2}\right].$$

Note 35 Brillouin (1967) defines "absolute information" as "any piece of information available to any human being on earth, this information being counted only once, whatever the number of people knowing it might be."

In addition, he defines "distributed information" as the product of the amount of absolute information and the number of people who share that information.

Both of these concepts refer to the *information* itself, i.e. the C-information, and not to the SMI associated with the information. It is interesting to note that in both definitions Brillouin refers to the information "available to" or "shared by" people. Does this include information stored in books, records, diskettes, etc. which is not available to people? Personally, I doubt the usefulness of these two concepts of information.

Note 36 Seife (2007), p. 216.

Note 37 Colloquially speaking, we may say that the bath at 25°C can only transmit information on that temperature. However, the *information* reached at the brain by the two hands is different information. One hand feels cold, while the other feels hot. The reason is that the information transmitted by the hand is not about the absolute temperature but the *difference* between the temperature of the hand and the temperature of the bath. Therefore, one hand transmits the information "colder," the other "hotter" — these two different pieces of information arrive at the brain, which interprets (erroneously) that as one hand is in a hotter bath, the other hand is in a colder bath.

Of course, this erroneous feeling does not last too long. After some time the hands get used to the new temperature of 25°C, and they "forget" their original temperature. From that point the two hands will transmit the same information to the brain.

Note 38 Remember that we used the Stirling approximation in the form

$$\ln N_i! \simeq N_i \ln N_i - N_i.$$

A better approximation would be

$$\ln N_i! \simeq N_i \ln N_i - N_i + \ln \sqrt{2\pi N_i}.$$

In this approximation the expression for the probability Pr is

$$\Pr(p_i, \dots, p_M) \cong \left(\frac{1}{M}\right)^N \frac{2^{NSMI(p_i,\dots,p_M)}}{\sqrt{(2\pi N)^{M-1} \prod_{i=1}^{M} p_i}}.$$

For the uniform distribution we have

$$p_i = \frac{1}{M}.$$

Hence,

$$\Pr\left(\frac{1}{M}, \dots, \frac{1}{M}\right) \approx \left(\frac{1}{M}\right)^N \frac{2^{N \log_2 M}}{\sqrt{(2\pi N)^{M-1} M^{-M}}}$$
$$= \sqrt{\frac{M^M}{(2\pi N)^{M-1}}}.$$

Note that this maximal value of Pr decreases as N increases. This result seems to be paradoxical, since we expect that the probability Pr of the uniform distribution should tend to unity. Indeed, if we take a small interval about the uniform distribution — say, $0.5-\delta \leq p_i \leq 0.5 + \delta$, with δ small compared with N — we will obtain probability 1 for the uniform distribution with a small vicinity about it. [For more details, see Ben-Naim (2008, 2011).]

Note 39 Seife (2007).

Chapter 2

Note 1 This is a highly schematic drawing of a heat engine. Note that the insulator and the heat source should both envelop the entire vessel. Here, these are shown at the bottom of the vessel.

Note 2 Denote by $\Delta Q(\text{Hot})$ and $\Delta Q(\text{Cold})$ the heat flow inside and outside the engine, respectively. The efficiency is defined by

$$\eta = \frac{\Delta W}{\Delta Q(\text{Hot})} = \frac{\Delta Q(\text{Hot}) - \Delta Q(\text{Cold})}{\Delta Q(\text{Hot})} = 1 - \frac{\Delta Q(\text{Hot})}{\Delta Q(\text{Cold})},$$

where ΔW denotes the useful work done by the system (lifting the weight).

For a Carnot engine working between the two temperatures T_2 and T_1 ($T_2 > T_1$), the efficiency is given by

$$\eta = \frac{T_2 - T_1}{T_2} = 1 - \frac{T_1}{T_2} \leq 1.$$

It is 0 when $T_1 = T_2$, and it approaches 1 when $T_1/T_2 \to 0$.

Note 3 It should be noted that spontaneous *demixing* can occur *spontaneously* in a different experimental setup. (See Section 2.7.)

Note 4 Quoted by Cooper (1968). For some comments on Cooper's comments, see Ben-Naim (2008).

Note 5 Actually, Einstein's equation $E = mc^2$ combines the conservation of mass (m) and energy (E) into one law of conservation of mass–energy. c is the speed of light.

Note 6 For further discussion of the mystery associated with the term "entropy," see Ben-Naim (2007, 2008).

Note 7 See Ben-Naim (2011b), Appendix Q.

Note 8 We use the convention that the heat flow into the system is positive, and the work done *on* the system is positive. In some textbooks the first law is written as $\Delta E = Q - W$. In this case work done on the system is negative.

Note 9 The expansion from V_1 to V_2 at constant pressure (P) is calculated from

$$W = -\int_{V_1}^{V_2} PdV = -P(V_2 - V_1).$$

The negative sign of W means that work is done *by* the system. (Here, W is negative.) In the process depicted in Figure 2.4 the pressure is not *constant*, and in fact it is not even defined along the way from the initial to the final state. The trick is to perform the process along a sequence of small steps. (See Figure 2.6.)

In the limit of very small steps we can calculate the work of expansion of an ideal gas as follows:

$$W = -\int_{V_1}^{V_2} PdV = -Nk_BT \int_{V_1}^{V_2} \frac{dV}{V} = -Nk_BT \ln 2.$$

And the corresponding entropy change:

$$\Delta S = \int_{V_1}^{V_2} \frac{P}{T} dV = \int_{V_1}^{V_2} \frac{Nk_B}{V} dV = Nk_BT \ln 2.$$

Note 9 We note here that once we have the *entropy function* (see Section 2.3), we do not need to devise a quasistatic process to move from the initial to the final state. For the particular process of expansion (Figure 2.4), we can simply calculate ΔS by taking the difference in the entropy function.

Note 10 Callen (1985).

Note 11 For more details, see Ben-Naim (2012).

Note 12 Details are provided in Ben-Naim (2008, 2012).

Note 13 The formal problem we pose is to find the maximum of the SMI,

$$H(1D \text{ location}) = -\int_0^L f(x)\log f(x)dx, \tag{1}$$

subject to the constraints

$$\int_0^L f(x)dx = 1, \tag{2}$$

where $f(x)dx$ is the probability of finding the particle in an interval dx at x, (Figure 2.4). The solution to this problem is given in the text. Note that $f(x)$ itself is a density distribution. However, we shall often refer to $f(x)$ as a distribution.

Note 14 The formal mathematical problem is to find the maximum value of the SMI,

$$H(\text{momentum in } 1D) = -\int_{-\infty}^{\infty} f(v_x)\log f(v_x)dv_x,$$

subject to the two constraints

$$\int_{-\infty}^{\infty} f(v_x)dv_x = 1,$$

$$\int_{-\infty}^{\infty} \frac{mv_x^2}{2} f(v_x)dv_x = \frac{m\langle v_x^2\rangle}{2} = \frac{m\sigma^2}{2},$$

where σ^2 is the variance of the distribution $f(v_x)$. The solution is given in the text.

Note 15 The average kinetic energy of a particle moving in a one-dimensional system at equilibrium is

$$\frac{m\left\langle v_x^2\right\rangle}{2}=\int_{-\infty}^{\infty}\frac{mv_x^2}{2}\frac{\exp\left(-v_x^2/2\sigma^2\right)}{\sqrt{2\pi\sigma^2}}dv_x=\frac{m\sigma^2}{2}.$$

The average kinetic energy of a particle moving in three dimensions is given by

$$\frac{m\left\langle v^2\right\rangle}{2}=\frac{m\left\langle v_x^2\right\rangle}{2}+\frac{m\left\langle v_y^2\right\rangle}{2}+\frac{m\left\langle v_z^2\right\rangle}{2}=3\frac{m\left\langle v_x^2\right\rangle}{2},$$

where v is the *speed* of the particles (this is defined by $v^2 = v_x^2+v_y^2+v_z^2$). From the kinetic theory of gases we have the relation between the absolute temperature and the average kinetic energy:

$$k_BT=\frac{2}{3}\frac{m\left\langle v^2\right\rangle}{2}.$$

From the above equations we identify σ^2 as k_BT/m.

Note 16 The one-dimensional density distribution of momenta is

$$f^*\left(p_x\right)=\frac{\exp\left(-p_x^2/2mk_BT\right)}{\sqrt{2\pi mk_BT}}.$$

The corresponding SMI in one dimension is

$$H_{\max}\left(p_x\right)=\frac{1}{2}\log\left(2\pi e\,mk_BT\right).$$

And in three dimensions

$$H_{\max}\left(p_x,p_y,p_z\right)=\frac{3}{2}\log\left(2\pi e\,mk_BT\right).$$

For a system of N independent particles the SMI is simply the sum of the SMI of each particle, and hence we have

$$H_{\max}(\boldsymbol{p}^N) = \frac{3N}{2}\log(2\pi e m k_B T),$$

where $\boldsymbol{p}^N = (\boldsymbol{p}_1, \ldots, \boldsymbol{p}^N)$.

Note 17 The *speed* of the particles is defined by the positive square root

$$v = \sqrt{v_x^2 + v_y^2 + v_z^2}.$$

The *speed* is the absolute magnitude of the velocity of a particle having the components of velocities v_x, v_y and v_z along the three axes.

Since the motions along the three axes are independent, we have

$$\begin{aligned} f^*\left(v_x, v_y, v_z\right) &= f^*(v_x)f^*(v_y)f^*(v_z) \\ &= \sqrt{\frac{m}{2\pi k_B T}}\exp\left[\frac{-m\left(v_x^2 + v_y^2 + v_z^2\right)}{2k_B T}\right] \\ &= \sqrt{\frac{m}{2\pi k_B T}}\exp\left(\frac{-mv^2}{2k_B T}\right). \end{aligned}$$

$f^*(v_x, v_y, v_z)\, dv_x, dv_y, dv_z$ is the probability of finding a molecule with velocities between v_x and $v_x + dv_x$, between v_y and $v_y + dv_y$, and between v_z and $v_z + dv_z$. The *speed* v as defined above can be obtained with infinitely many combinations of v_x, v_y, and v_z.

The distribution of the speeds is obtained by transforming to spherical polar coordinates and integrating over all the angles.

The result is

$$f^*(v) = \left(\frac{m}{2\pi k_B T}\right)^{3/2} 4\pi v^2 \exp\left(\frac{-mv^2}{2k_B T}\right).$$

Here, $f^*(v)dv$ is the probability of finding a particle with a *speed* between v and $v+dv$ (Figure 2.14). Note carefully the difference between the distribution of *velocities* and the distribution of *speeds*. The velocity v_x can be either positive or negative; its distribution is normal (Figure 2.12), and centered at $v_x = 0$. It is a symmetric distribution, i.e. the particle has the same probability of moving with a velocity v_x as with $-v_x$. On the other hand, the *speed* distribution (Figure 2.14) is not symmetric.

Note 18

$$\begin{aligned} H^D(1,2,\ldots,N) &= H_{\max}(\text{locations}) + H_{\max}(\text{momenta}) \\ &\quad - I(\text{uncertainty principle}) \\ &\quad - I(\text{indistinguishability}) \\ &= \left(N \log V\right) + \left[\frac{3N}{2}\log\left(2\pi emk_B T\right)\right] \\ &\quad - \left(3N \log h\right) - \left(\log N!\right) \\ &= N \log\left[\frac{V}{N}\left(\frac{2\pi mk_B T}{h^2}\right)^{3/2}\right] + \frac{5N}{2}. \end{aligned}$$

Note 19 The thermodynamic limit is obtained by letting $N \to \infty$ and $V \to \infty$, but keeping the ratio N/V constant.

Note 20 See Ben-Naim (2009).

Note 21 A function can have more than one maximum, in which case one needs to define each maximum *locally*. The mathematical *conditions* for a maximum of the function

$y = f(x)$ at the point x^* are

$$\left.\frac{\partial f(x)}{\partial x}\right|_{x=x^*} = 0,$$
$$\left.\frac{\partial^2 f(x)}{\partial x^2}\right|_{x=x^*} < 0.$$

This means that the *slope* of the function at $x = x^*$ is zero, and the *curvature* is negative.

By generalization, if we are given a function $y = f\left(x, y, z\right)$ and we say that there exists a maximum at some point $\left(x^*, y^*, z^*\right)$, we mean that y has the largest value when x, y and z are varied in the neighborhood of the point x^*, y^*, z^*.

Note 22 There are other fundamental functions, such as $G(T, P, N)$ and $A(T, V, N)$, where G is the Gibbs energy and A the Helmholtz energy.

Note 23 Jaynes, E. T. (1963), Information theory and statistical mechanics, in *Proc. Brandeis Summer Institute 1962: Statistical Physics*. W. A. Benjamin, New York.

Note 24 Lewis, G. N. (1930), *Science* **71**: 569.

Note 25 Denbigh, K. (1981), *Chem. Brit.* **17**: 168.

Note 26 Jaynes E. T. (1965), *Am. J. Phys.* **33**: 391.

Note 27 In this section, we have dealt with ideal gases only. However, one can show that intermolecular interactions always introduce mutual information which reduces the entropy of the ideal gases. [See Ben-Naim (2008).] The reader should note that our procedure for obtaining the equation for the entropy of an ideal gas differs in a fundamental way from the Sackur–Tetrode method.

Sackur and Tetrode estimated the *number of states* W of an ideal gas contained in a volume V. Once you have W you can calculate the entropy by using the Boltzmann relationship $S = k_B \ln W$. This procedure provides the correct entropy of an ideal gas, but does not offer any *interpretation* of the entropy. In our approach we did not calculate W, then the entropy. Instead, we calculated directly the SMI of an ideal gas (with respect to locational and momentum distribution) at equilibrium. We then showed that this SMI is identical (up to a constant) with the entropy of an ideal gas. With this identification the entropy we have obtained has the same meaning and the same interpretation as the SMI.

Note 28 For details, see Ben-Naim (2012).

Note 29 This is the formulation of the Second Law for an *isolated* system. There are equivalent formulations of the Second Law for systems under constant temperature and volume or under constant temperature and pressure.

Note 30 See any book on probability, e.g. Papoulis and Pillai, 4th edition (2002). A shorter discussion is also available in Ben-Naim (2008) and Ben-Naim (2014). It should be said that the high probability of an event does not mean that the "probability" is the cause of an event. Cause and conditional probability are often confused [see Ben-Naim, (2014)]. When we are considering a very large probability ratio — say, between two events — we can safely use the frequency definition of probability to infer that if you start from any initial event, and repeat the experiment many times, the event having the higher probability will occur most of the time. There is an awkward statement in Brillouin's book (1962): "The probability has a tendency to increase and so thus entropy." I believe this is a slip of the tongue. The state of the system changes from low probability states to high probability states. The probabilities do not change with time.

Note 31 The condition for the maximum of $W(n)$ is

$$\frac{\partial \ln W(n)}{\partial n} = \frac{1}{W(n)}\frac{\partial W(n)}{\partial n} = -\ln n + \ln (N-n) = 0.$$

The solution to this equation is

$$n^* = \frac{N}{2}.$$

This is a maximum, since

$$\frac{\partial^2 \ln W(n)}{\partial n^2} = \frac{-1}{n} - \frac{1}{N-n} = \frac{-N}{(N-n)n} < 0.$$

Thus, the function (n), as well as $P_N(n)$, has a maximum with respect to n (keeping $E, 2V, N$ constant).

Note 32 It is easy to examine how **Pr** (n^*) changes with N, by using the Stirling approximation of the form

$$j! \cong \left(\frac{j}{e}\right)^j \sqrt{2\pi j}.$$

Thus, the maximum probability is

$$P_N\left(n^* = \frac{N}{2}\right) \approx \sqrt{\frac{2}{\pi N}}.$$

To calculate the probability of finding the system in the neighborhood of n^*, we use the De Moivre–Laplace theorem, and then integrate the Gaussian function [for details see Ben-Naim (2008)]. The result is

$$P_N(n^* - \delta N \le n \le n* + \delta N) = \sum_{n=n^*-\delta N}^{n^*+\delta N} P_N(n)$$
$$\approx \int_{n+n^*-\delta N}^{n=n^*+\delta N} \frac{1}{\sqrt{\pi N/2}} \exp\left(\frac{-(n-\frac{N}{2})^2}{N/2}\right) dn.$$

This is the error function $\text{erf}(\delta\sqrt{2N})$. The function is shown in Figure 2.22.

Note 33 For any distribution (p, q) we can define the SMI as

$$\text{SMI}(p, q) = -p \log p - q \log q = (-p \ln p - q \ln q)/\ln 2.$$

(Note that we have used both the natural logarithm, ln and the logarithm to the base 2, $\log_2$.)

This is the SMI per particle in the system. It measures the uncertainty in the location of each particle, with respect to being in L or R.

When $N \to \infty$, we can use the Stirling approximation to rewrite the number of states as

$$\ln[W(p, q)] = \ln \binom{N}{pN} \approx N(p \ln p + q \ln q) - \frac{1}{2} \ln(2\pi Npq)$$
$$= N \times \text{SMI}(p, q) \ln 2 - \frac{1}{2} \ln(2\pi Npq),$$
$$\text{for large } N.$$

Hence, in this approximation

$$\Pr(p, q) \approx \left(\frac{1}{2}\right)^N \frac{\exp[N \times \text{SMI}(p, q) \ln 2]}{\sqrt{2\pi Npq}}$$

Or, equivalently,

$$\Pr(p,q) = \left(\frac{1}{2}\right)^N \frac{2^{N\times \mathrm{SMI}(p,q)}}{\sqrt{2\pi Npq}}.$$

Note 34 Greene (2004).

Note 35 Today, there are at least five different "arrows of time." We will discuss here only the one arrow associated with the Second Law of Thermodynamics.

To the best of my knowledge, there exists neither a proof, nor a convincing argument relating the positive change in entropy to the "arrow of time." There are many things or concepts which always increase with time (for example, the accumulated number of events occurring in the universe always increases with time). This does not make these concepts *identical* with the arrow of time.

We perceive time as having a direction. We believe that this direction is absolute. We cannot perceive time changing in an opposite direction (whatever this might mean). On the other hand, as was understood by Boltzmann, the direction of unfolding of events is not absolute, only highly probable. Therefore, we cannot claim that the increase in entropy throughout the universe *defines* the arrow of time.

Even if one could prove that the Second Law is absolute, and that entropy *always* increases, one could not conclude that this defines the arrow of time. Such a relationship must be proven.

Note 36 Hawking, S. W. (1988).

Note 37 Carroll (2010), *From Eternity to Here: The Quest for the Ultimate Theory of Time.*

In my opinion there exists no theory of time, just as there is no theory of "length." I am not sure what the "quest for the ultimate theory of time" means.

Note 38 Boltzmann (1896).

Note 39 Brillouin (1967).

Note 40 Seife (2010), page 55.

Note 41 I have no idea what the change in the entropy of the Robot is. I have also no idea what the change in the entropy of the universe is. Regarding the use of information, I doubt that the Robot used any information. If you do believe that the Robot did use information, you will have a hard time explaining how the Robot could achieve three different results in (a), (b) and (c) using the same amount of information. After all, the Robot did exactly the same job of transferring N molecules from the right to the left in the three cases.

Note 42 We always assume that there are no external fields. Although we cannot avoid the presence of gravitational fields, we assume that their effect on the processes is negligible. In some processes it is important to take into account the effect of gravity on the distribution of density in a vertical column of air. [See for example Chapter 4 of Ben-Naim (2012).]

Note 43 The total differential of the entropy viewed as a function of the variables E, V, N is

$$dS = \frac{\partial S}{\partial E}dE + \frac{\partial S}{\partial V}dV + \frac{\partial S}{\partial N}dN = \frac{1}{T}dE + \frac{P}{T}dV + \frac{\mu}{T}dN,$$

where T is the temperature, P the pressure of the gas, and μ the chemical potential.

Since in this expansion process there is no change in the energy of the system, or in the number of particles, we write

$$dS = \frac{P}{T}dV.$$

Assuming that the system is an ideal gas, we write the equation of state as

$$PV = k_B NT,$$

where k_B is the Boltzmann constant. Integrating from V to $2V$ leads to

$$\Delta S = \int_V^{2V} \frac{P}{T}dV' = \int_V^{2V} \frac{k_B N}{V'}dV' = k_B N \ln \frac{2V}{V}$$
$$= k_B N \ln 2.$$

Thus, the entropy change for the expansion process in Figure 2.32 is

$$\Delta S(\text{expansion}\, V \rightarrow 2V) = k_B N \ln 2.$$

In thermodynamics one uses the gas constant R, which is related to the Boltzmann constant k_B by $R = k_B N_{\text{av}}$, where $N_{\text{av}} = 6.023 \times 10^{23}$ is the number of atoms or molecules in one *mole* of the particles.

In the thermodynamic calculation of ΔS we calculate the integral from V to $2V$ based on a process for which at each stage of the integration the system is well defined, and hence the entropy function is also a well-defined function of the variable V. Such a process is sometimes referred to as a *reversible* process. We will use the term "*quasistatic*" for such a process. The term "reversible" has many definitions and we will refrain from using it here. This is discussed in Ben-Naim (2013).

Note 44 For details, see Ben-Naim (2008).

Note 45 Many textbooks reach this conclusion. Unfortunately, this conclusion is not true. We have already seen an example of a mixing process with no change in entropy. We will soon show an example of a *demixing* process with an *increase* in entropy (which means a process of mixing with a *decrease* in entropy).

Note 46 See Ben-Naim (2008) for a proof that this is always positive.

Note 47 For details, see Ben-Naim (2008, 2012).

Note 48 Note that we can use the same equation to "predict" the direction of heat flow. Knowing that the entropy must increase in this spontaneous process, i.e. $dS > 0$, we can conclude that dQ must be positive.

Note 49 In the initial state we have two systems having velocity distributions $g_1(v)$ and $g_2(v)$ at temperatures T_1 and T_2, respectively. The variance of the initial system as a whole is

$$\sigma_i^2 = \int_{-\infty}^{\infty} (v - \overline{v})^2 \left(\frac{g_1(v) + g_2(v)}{2} \right) dv = \frac{\sigma_1^2 + \sigma_2^2}{2}.$$

For simplicity, we can take v to be the x component of the velocity. The three components of the velocities are independent, and hence their variances are equal.

We now apply Shannon's theorem for the continuous distribution. The theorem says that of all the distributions for which the variance is constant, the normal (or Gaussian) distribution has a maximum SMI. Since the total kinetic energy of the system before and after the process must be conserved, and since the average kinetic energy is proportional to the temperature, the variance in the final state must be equal to the variance in the initial state, i.e.

$$\sigma_f^2 = \int_{-\infty}^{\infty} (v - \overline{v})^2 g(v) dv = \sigma_i^2.$$

From this it follows that the final temperature T must be, in this particular process,

$$T = \frac{T_1 + T_2}{2}.$$

The change in the SMI in this process can be easily calculated from the SMI of an ideal gas:

$$\begin{aligned}\mathrm{SMI}(f) - \mathrm{SMI}(i) &= \frac{3}{2}(2N)\log T - \frac{3}{2}N\log T_1 - \frac{3}{2}N\log T_2\\ &= \frac{3}{2}N\log\frac{T}{T_1} + \frac{3}{2}N\log\frac{T}{T_2},\end{aligned}$$

where the two terms correspond to the changes in the SMI when the temperature changes from T_1 to T and from T_2 to T, respectively. Since T is the arithmetic average of T_1 and T_2, we have

$$\begin{aligned}\Delta S &= \frac{3}{2}N\log\frac{T_1 + T_2}{2T_1} + \frac{3}{2}N\log\frac{T_1 + T_2}{2T_2}\\ &= \frac{3}{2}N\log\frac{(T_1 + T_2)^2}{4T_1T_2} = \frac{3}{2}N\log\frac{\left(\frac{T_1+T_2}{2}\right)^2}{T_1T_2}\\ &= 3N\log\frac{\frac{T_1+T_2}{2}}{\sqrt{T_1T_2}} \geq 0.\end{aligned}$$

The last inequality follows from the inequality about the arithmetic and the geometric average. [For details, see Ben-Naim

(2008).]

$$\frac{T_1 + T_2}{2} > \sqrt{T_1 T_2}.$$

Note 50 This is true when all the energy entering into the system is used to increase the kinetic energy of the particles. If there are two phases at equilibrium — say, water and vapor — and we add thermal energy to the system, the average kinetic energy does not change. Instead, water molecules are transferred from the liquid to the gaseous phase.

Note 51 For proof, see Ben-Naim (2008).

Note 52 Ben-Naim (2008, 2012).

Note 53 See Ben-Naim (2007).

Note 54 Thims (2014) has published an article entitled "Thermodynamics ≠ Information Theory: Science's Greatest Sokal Affair." The author repeatedly criticizes those equating thermodynamics with information theory. This issue has already been raised in the first part of the title. These two fields are obviously different. Thus, there is no point in devoting a whole article to this issue. Of course, there are many who equate one concept from thermodynamics with one concept from information theory. The second part of the title, alluding to the Sokal affair, is also inappropriate. I do not know of any author in these fields who intentionally wrote a meaningless article, mocking those who write a seemingly meaningful article, which in fact is a meaningless article.

Here is part of the abstract for Thims' article:

> This short article is a long-overdue, seven decades — 1940 to present — delayed, inter-science departmental memorandum —

> though not the first — that INFORMATION THEORY IS NOT THERMODYNAMICS and thermodynamics is not information theory. We repeat again: information theory — the mathematical study of the transmission of information in binary format and/or the study of the probabilistic decoding of keys and cyphers in cryptograms — is not thermodynamics! This point cannot be overemphasized enough.

This highly repetitious warning on a trivial matter goes on and on throughout the rest of the very long article. In my view this is a superfluous criticism. In another section of the article, Thims writes:

> Ben-Naim is so brainwashed by the idea that thermodynamics needs to be reformulated in terms of information theory, that . . . he states "I would simply say that I shall go back to Clausius' choice of the term, and suggest that he should not have used the term entropy in the first place." This is an example of someone afflicted by the Shannon syndrome.

This quotation is highly misleading, and shows that my message in the book *A Farewell to Entropy* (2008) was totally misunderstood. First, I do not know what the "Shannon syndrome" is. If the author means "Shannon's measure", then yes, I admire Shannon for developing this measure. I have also shown how the entropy can be derived from the SMI. Second, if the author means "brainwashed" in a derogatory sense, then of course I reject that description. However, if he means it in the sense that I have "washed my brain" of any missconceptions, and thoroughly convinced myself (not by someone else), that what I wrote in my book (2008) is true, then I wholeheartedly embrace his compliment.

Indeed, I am convinced that the entropy — the thermodynamic entropy — may be interpreted, up to a multiplicative constant, as a particular case of SMI. Not with the general concept of information, not with information theory, and not with *any* other measure of information, but with the specific measure of information defined by Shannon.

Note 55 From Kurzweil (2005): In his highly praised book *The Singularity Is Near*, he discusses the exponential growth of, among other things, information.

> The ongoing acceleration of technology is the implication and inevitable result of what I call the law of accelerating returns, which describes the acceleration of the pace of the exponential growth of the products of an evolutionary process. These products, and in particular, information ... their acceleration extends substantially beyond the predictions made by what became known as Moore's Law. The singularity is the inexorable result of the law of accelerating returns.

This introductory paragraph summarizes the theme of the entire book. I am not convinced that there exists such a law as the "law of accelerating returns," and whether the growth of information will continue at the same rate is highly speculative. The rest of the book is extremely repetitious of this highly speculative idea.

Here are a few meaningless quotations from this book.

On page 130 the author writes:

> When a bit of information is erased, that information has to go somewhere. According to the Laws of Thermodynamics, the erased bit is essentially released into the surrounding environment, thereby increasing its entropy.
>
> ... entropy, which can be viewed as a measure of information (including apparently disordered information.)
>
> This results in a higher temperature for the environment (because temperature is a measure of entropy).

In my view, it is meaningless to talk about "information has to go somewhere" — certainly not information which is erased. If Bob stops loving Linda, does his love go somewhere, perhaps released into the surrounding environment? Second, the Second Law of Thermodynamics does not apply to information. Third, while it is true that entropy can be viewed as a *measure* of information (see Section 2.2), I do not know what "disordered information" is. Finally, temperature is *not* a measure of entropy!

Chapter 3

Note 1 Genesis 1:26:

וַיֹּאמֶר אֱלֹהִים, נַעֲשֶׂה אָדָם בְּצַלְמֵנוּ כִּדְמוּתֵנוּ

וַיִּבְרָא אֱלֹהִים אֶת-הָאָדָם בְּצַלְמוֹ, בְּצֶלֶם אֱלֹהִים בָּרָא אֹתוֹ

Note 2 Genesis 2:7:

וַיִּיצֶר יְהוָה אֱלֹהִים אֶת-הָאָדָם, עָפָר מִן-הָאֲדָמָה, וַיִּפַּח בְּאַפָּיו, נִשְׁמַת חַיִּים; וַיְהִי הָאָדָם, לְנֶפֶשׁ חַיָּה

Note 3 Ezekiel 37: 5–37:

כֹּה אָמַר אֲדֹנָי יְהוִה, לָעֲצָמוֹת הָאֵלֶּה: הִנֵּה אֲנִי מֵבִיא בָכֶם, רוּחַ—וִחְיִיתֶם

וְנָתַתִּי עֲלֵיכֶם גִּידִים וְהַעֲלֵתִי עֲלֵיכֶם בָּשָׂר, וְקָרַמְתִּי עֲלֵיכֶם עוֹר, וְנָתַתִּי בָכֶם רוּחַ, וִחְיִיתֶם

וְרָאִיתִי וְהִנֵּה-עֲלֵיהֶם גִּדִים, וּבָשָׂר עָלָה, וַיִּקְרַם עֲלֵיהֶם עוֹר, מִלְמָעְלָה; וְרוּחַ, אֵין בָּהֶם

וַיֹּאמֶר אֵלַי, הִנָּבֵא אֶל-הָרוּחַ; הִנָּבֵא בֶן-אָדָם וְאָמַרְתָּ אֶל-הָרוּחַ

כֹּה-אָמַר אֲדֹנָי יְהוִה, מֵאַרְבַּע רוּחוֹת בֹּאִי הָרוּחַ, וּפְחִי בַּהֲרוּגִים הָאֵלֶּה, וְיִחְיוּ

וְהִנַּבֵּאתִי, כַּאֲשֶׁר צִוָּנִי; וַתָּבוֹא בָהֶם הָרוּחַ וַיִּחְיוּ, וַיַּעַמְדוּ עַל-רַגְלֵיהֶם--חַיִל, גָּדוֹל מְאֹד-מְאֹד

Note 4 There were many attempts to weigh the soul by weighing the body before and after death. No definite conclusions were reached.

Note 5 Alexander (2012). There are many stories of this kind. No one can argue about the "reality" of the experience of people who visited heaven (or perhaps hell) and came back. What is especially funny about this book is not the story of the journey to heaven, but rather the conclusion that the author reaches: "Today, Alexander is a doctor who believes that true health can be achieved only when we realize that God and the soul are real and that death is not the end of personal experience but only a transition."

Do not argue with the author. He is a neurosurgeon and has a proof of heaven!

Dr. Alexander tells the story of his near-death experience (NDE). While in a coma, he journeyed beyond this world into the realm of superphysical existence. There he met and spoke with the divine source of the universe itself.

On the back cover of the book, it states, "This story would be remarkable no matter who it happened to. That it happened to Dr. Alexander makes it revolutionary. No scientist or person of faith will be able to ignore it. Reading it will change your life."

I read this book out of curiosity. No one can prove whether his story is true or false. You either believe it or not. Personally, I trust that the author tells his story faithfully. I do not agree, though, with his interpretation of that experience. Contrary to what is written on the cover of the book, reading the book did not cause any change in my life . . . .

Note 6 Shermer (2011).

Note 7 Classical mechanics originated from Newton's laws of mechanics. Take any living system — a bacterium or cat or human being. Suppose that you know all the locations and all the velocities of all the particles in that living system.

If you also know the interactions between the particles, you can write the Hamiltonian function, and the corresponding (classical) equations of motion. The question is whether such a *Hamiltonian* can distinguish between the state of being alive and of being dead. Perhaps we will need to add one or more parameters to *describe* the system. Perhaps the equations of motion, if they exist, will not obey the classical Newton or Langrange, or Hamiltonian equations of motion. Perhaps a whole new function should be created to describe the equations of motion of such a system.

We will face the same problem if we want to describe the system in quantum-mechanical language. We can write down the *Hamiltonian operator* of the specific living system. The question is again: Is all the information on all the particles involved in a living system enough to describe and distinguish between the *state* of being alive and the *state* of being dead? Perhaps one would need one or more parameters to write down the Hamiltonian operator of the system, and if such a Hamiltonian exists it is by no means clear that the wave function (if existent) of the living system will obey a linear differential equation, or any other differential or integral equation. Therefore, until we settle the question of the *existence* of a wave function that has two values in the states "alive" and "dead," there is no point in speculating about the meaning of the superposition of these two states.

Note 8 Looking at any living system, we see that it uses the available resources from its environment to its advantage. This is true of any living system. A small bacterium can sense the gradient of the concentration of some nutrients, then propels itself toward the source of the nutrients. In higher animals, we observe many complex behaviors, from exploiting resources to changing their environments to their advantage. All these can be interpreted as manifestations of the purposefulness of life. Whether purposefulness can serve as a definition of life is not clear.

The whole question of "purpose is nature" is still a vigorously debated one. Are changes driven by a physical cause or a purpose? Teleology (from the Greek *telos* — "purpose") asserts that there is a purpose in nature. We certainly feel that there is purpose that drives everything we do. Some believe that *vitalism* is purposefulness. Still others believe that all changes can ultimately be explained by the laws of physics and chemistry.

Recently, I enjoyed many discussions with Naftali Tishby. His view of life is that any living system uses *information* from the past to gain some survival advantages for the future. This is certainly a characteristic of living systems; however, I doubt that it can serve as a definition of life. In this view, if one could teach a robot to exploit any information from its environment to its advantage, it would be considered alive. I wonder how many scientists would consider such a robot as a living system.

Note 9 The folding of the protein into a precise 3D structure has been the prevalent paradigm for many years. Recently, many functioning proteins were discovered which do not have a well-defined structure. They are referred to as "intrinsically disordered proteins." [See Lehninger (2014).]

Note 10 For more details, see Ben-Naim (2013).

Note 11 Kennedy and Norman (2005).

Note 12 Anfinsen (1973).

Note 13 For details, see Ben-Naim (2010, 2013). It is also known that the same sequence can have different *structures*. Also, note that the *structure* of the protein is not a *unique* sequence of coordinates of all the atoms of the protein, even for those proteins that have a well-defined 3D structure. This structure is not unique. Furthermore, under different solvent compositions one can have different structures (e.g. hemoglobin and allosteric, regulatory enzymes). It is also known that many proteins which have an important *function* in the cell do not have a structure at all. They are referred to as intrinsically disordered proteins.

Note 14 We discuss here a spontaneous folding of a protein *in vitro*. It is known that *in vivo*, i.e. in the cells of a living organism, there might be some proteins, referred to as chaperones, which help the protein in the folding process.

Note 15 Ben-Naim (2001).

Note 16 Hager (1995).

Note 17 Gatlin (1972).

Note also that, according to Gatlin, viruses are not considered to be alive; they *store* information but they do not *process* information.

Note 18 Gatlin defined two SMIs for pairs of bases. One is for the *actual* distribution of a pair of bases,

$$\mathrm{SMI}_D^{(2)} = -\sum \Pr(i,j)\log[\Pr(i,j)],$$

where $\Pr(i,j)$ is the probability of finding the specific pair i and j as nearest neighbors. The other SMI is for the pair of letters, assuming that the second letters in the sequence are independent:

$$\mathrm{SMI}_{\mathrm{Ind}}^{(2)} = -\sum \Pr(i)\Pr(j)\log[\Pr(i)\Pr(j)].$$

The summation is over all the four letters of the alphabet.

Note 19 While I was writing my book *Entropy Demystified*, I searched the literature on "entropy." Among others, I found Sanford's book *Genetic Entropy and the Mystery of the Genome*. Since I am interested in both "entropy" and the "evolution theory," I immediately ordered the book from Amazon. As soon as I got the book, I leafed through it, and I was deeply disappointed upon realizing that "entropy" is mentioned only in a few places in the book. Moreover, I found out that what the author says about entropy has nothing to do with entropy. Here is what the author says on page 144 (note that page 144 is 10 pages shy of the end of the book):

> For decades biologists have argued on a philosophical level that the very special qualities of natural selection can easily reverse the biological effects on the second law of thermodynamics. In this way, it has been argued, the degenerative effects of entropy in living systems can be negated — making life itself potentially immortal. However, all of the analyses of this book contradict that philosophical assumption. Mutational **entropy** appears to be so strong within large genomes that selection cannot reverse it. This makes eventual extinction of such genomes inevitable. I have termed this fundamental problem **Genetic Entropy**. Genetic Entropy is not a starting axiomatic position — rather, it is a logical conclusion derived from careful analysis of how selection really operates.

Clearly, the author has no idea what entropy means, and what the Second Law of Thermodynamics states, and yet he had the guts to use "entropy" in the book's title. The whole paragraph quoted above is sheer nonsense. Although disappointed, I was still curious to read what the author has to say on the "mystery of the genome." While I was reading the book, my disappointment turned into disgust, specifically when I read page 116, on which the author insinuates a linkage between the concepts of "Darwinism," "evolution theory," "eugenics" and "Hitler's racism." Here is what the author writes:

> Darwin repeatedly pointed to human efforts in animal and plant breeding as a model for such men-directed selection ... Darwin went further and contended that there is a need for superior races (i.e. the white race) to replace the "inferior races." This ushered in the modern era of racism, which came to a head in Hitler's Germany. Before World War II, many nations including America had government-directed eugenics programs .... Ever since the time of Darwin, essentially all his followers have been eugenicists at heart, and have advocated the genetic improvement of the human race. When I was an evolutionist, I was at heart a eugenicist .... However, after the horrors of WWII, essentially all open discussions of eugenics were quietly put aside.
>
> Eugenics has from its inception been a racist concept, and has always been driven by the Primary Axiom.

These words smack of Hitler's propaganda rather than belong to a scientific text. Clearly, they have nothing to do with evolution theory. Thus, the author has nothing to say on either "entropy" or "evolution theory." It is a shower of nonsensical assaults on what he refers to as the "Primary Axiom."

The book is full of obnoxious ideas. The horrific crimes committed in WWII have nothing to do with Darwin's theory of evolution.

There is no connection between the theory of evolution and "racism" or "eugenics." Only a perverted mind can claim that such a connection exists, or *use* such a purported connection to commit such crimes.

The foreword by John Baumgardner says, "*The Mystery of the Genome* is a brilliant exposé on the un-reality of the Primary

Axiom."

And what is the "Primary Axiom"? On page v of the prologue, the author explains:

> Modern Darwinism is built most fundamentally upon what I will be calling "the Primary Axiom." The Primary Axiom is that man is merely the product of random mutations plus natural selection.

The truth of the matter is that there exists no "Primary Axiom," or any other axiom in modern Darwinism. Modern Darwinism is built on *experimental evidence* (and supported by logical deductions), not on *axioms*. To the best of my knowledge, the theory of evolution is the best theory that science can offer.

Here are a few "pearls" of the author's work, on page v:

> Within our society's academia, the Primary Axiom is universally taught and almost universally accepted . . . . It is very difficult to find any professor on any college campus who would even consider (or should I say — dare) to question the Primary Axiom.

The author continues on page vi:

> Late in my career, I did something which for a Cornell professor would seem unthinkable. I began to question the Primary Axiom. I did this with great fear and trepidation. By doing this, I knew I would be at odds with the most "sacred cow" within modern academia. Among other things, it might even result in my expulsion from the academic world.

The above quotation is nothing but slander. I have been teaching in the university for many years and have visited many universities all over the world. In all of these places, academia's teaching encourages teachers and students alike to question any axiom, conjecture or theory. What the author insinuates in the quotations above reminds me of only one place where questioning of the "fundamental axiom" is forbidden. And that is the religious school which I attended as a boy. Nothing like that is practiced in any university that I am aware of.

The author concludes on page vii:

> Furthermore, every form of objective analysis I have performed has convinced me that the axiom is clearly false. So now, regardless of the consequences, I have to say it out loud: "**The Emperor has no clothes.**"

Indeed, it should be said loud and clear: **The author has nothing to say, yet he says it screamingly loud.**

Note 20 The relationship between Pr and ΔG is $\Pr = C \exp(-\Delta G/k_B T)$, where ΔG is the Gibbs energy of the binding, T the temperature, k_B the Boltzmann constant, and C a normalization constant. [For details, see Ben-Naim (2001).]

Note 21 Jacob, F. and Monod, J. (1961), *J. Mol.* **3**: 318.

Note 22 Jacob, F. (1966), *Science* **152:** 1470.

Note 23 Ptashne (2004), *The Genetic Switch.*

Note 24 Seife (2010).

Note 25 Schrödinger, E. (1969), *What Is Life? The Physical Aspects of the Living Cell.* Cambridge University Press.

Note 26 Murphy and O'Neill (1995).

Note 27 Many others have provided lists of *characteristics* of life. It is clear, however, that these characteristics of life are neither necessary nor sufficient for life. One could envisage creatures that are alive that do not "reproduce–mutate–metabolize." On the other hand, one can also envisage some kind of highly sophisticated robots that reproduce–mutate–metabolize but will not be recognized (by us) as being alive.

Note 28 Ellis (1994).

Note 29 For more details, see Ben-Naim (2008, 2012).

Note 30 Brillouin (1967).

Note 31 vos Savant (1972), page 17.

Note 32 Seife (2006), pages. 100–111.

Note 33 Carroll (2010), page 38 .

Note 34 The superposition of the two state functions $\psi(\text{alive})$ and $\psi(\text{dead})$ would be written as

$$\psi(\text{cat}) = \frac{1}{\sqrt{2}}[\psi(\text{alive}) + \psi(\text{dead})].$$

Note 35 The superposition principle is simply a result of the fact that Schrödinger's equation is linear and homogenous. This means that if $\psi_1, \ldots, \psi_n$ are solutions to the Schrödinger equation, then any linear combination of these functions is also a solution.

When this principle is applied to a living organism, one gets the absurd result that as long as an observer does not look into the chamber, the cat is in a superposition of the two states: alive and dead. Most physicists believe that this state of the cat is nonsense. The reason given is that the superposition principle does not apply to a macroscopic system. My view is different. Even if the superposition principle applies to a macroscopic system, it is far from clear that it will apply to a *macroscopic living system*, not because of the validity of the superposition principle, but because it is not clear that there exist wave functions which describe the states of being alive and being dead.

Note 36 It should be noted that even when the device which releases the poisonous gas is designed in such a way that the probabilities of being released or not released (or the bomb detonating or not) are *exactly* half and half, one cannot infer that the probabilities of the cat being alive or dead are also half and half.

The gas could be released (with probability 1/2) and yet the cat could still survive. On the other hand, the gas could not have been released and yet the cat could have died from other causes. Thus, the probabilities of being alive or dead could not be inferred from the known probabilities that the poisonous gas was or was not released.

Note 37 As we have noted in connection with the fate of the Schrödinger cat, here also we cannot infer from the equation

$$\psi(\text{boy}) = \frac{1}{\sqrt{2}}[\psi(\text{happy}) + \psi(\text{sad})],$$

the probabilities of being happy and being sad. Even if the probabilities of the outcomes of the coin are exactly half and half, we cannot say anything about the probabilities of the state of mind of the boy. The boy can watch a funny movie but can be sad (or in any other mood). He can watch a sad movie but can be happy (or in any other mood).

Note 38 Pross (2012), page 127.

Note 39 This formulation is in terms of increase in entropy in an isolated system. Other formulations apply in a nonisolated system. For instance, in a spontaneous process in a system characterized by the temperature, pressure and composition, the Gibbs energy decreases.

Note 40 Sometimes the term "holism" is used in connection with emergence. The idea is that the "whole" is more than the sum of its parts. It is believed that once a system becomes very complex, new properties emerge from the totality of the parts.

Chapter 4

Note 1 Note that in the expansion of a gas from volume V to $3V$ the entropy change is $k_B N \ln 3$. The corresponding change in SMI is $N \log_2 3$. This is *larger* than the change in SMI for the process of expansion from V to $2V$, which is $N \log_2 2 = N$. The informational interpretation of these results is as follows:

For the expansion from V to $2V$, we have to ask *one* question per particle to determine in which *half* of the system the particle is.

For the expansion from V to $3V$, we have to ask *more* than one question to determine in which third of the system the particle is.

On the other hand, if we are interested only in the location of the particle, either in L (of volume $1/3V$), or on the right of L (i.e. either X or Y of volume $2/3V$), then the number of questions is *less* than one per particle (i.e. $-1/3 \log_2 1/3 - 2/3 \log_2 2/3 < 1$).

Note 2 Yes, I know I have said similar things in Chapter 3 about the ravages of entropy for life. Whatever I said there about life is *a fortiori* true for the entire universe.

Note 3 This term was coined by Albert (2000) and it is discussed at great length by Carroll (2010).

Note 4 Carroll (2010).

Note 5 When I first quoted the first two sentences from the Bible, a few reviewers of my book Ben-Naim (2007) misunderstood my message as being a "scientific evidence" against the Past Hypothesis. No, my argument against the Past Hypothesis is not "based" on the Bible, but rather on the lack of any supporting evidence.

Note 6 The adjective "entropic" here refers to a disordered, chaotic sequence of events or images. "Entropic" is used in thermodynamics in connection with the driving force of a process. One writes the Gibbs energy change as $\Delta G = \Delta H - T\Delta S$, then interprets the two contributions to the "driving force" ΔG as the enthalpic (or energetic) contribution, ΔH, and the entropic contribution, $T\Delta S$. This usage of the term "entropic" is appropriate. It does not imply any involvement of "disorder" or "chaos" in the process.

Colloquially, we may use the adjective "energetic" to describe a very active boy or a high speed bullet. But an "energetic dream" sounds to me meaningless.

In the case of the adjective "entropic," it would be meaningless to refer to an "entropic boy" or an "entropic bullet" — and certainly to an "entropic dream." Therefore, my suggestion is to refrain from using "entropic" as an adjective for any object.

Note 7 Black holes are sometimes depicted as the ultimate vacuum cleaner — nothing can escape from them.

Note 8 Susskind and Lindesay (2005).

Note 9 Lavenda (1991), and private communication.

Note 10 Bekenstein (1973).

Note 11 Bekenstein (private communication).

Note 12 The black hole (BH), or the Bekenstein–Hawking, entropy is given by the formula

$$S_{\mathrm{BH}} = \left(\frac{k_B c^3}{G\hbar}\right)\frac{A}{4}$$

where k_B is the Boltzmann constant, c the speed of light, G the Newton gravitational constant and $\hbar$ the Planck constant divided by 2π. A is the horizon's surface area of the black hole.

For a spherical black hole, the surface area is related to the total mass of the black hole, and hence the entropy may be expressed as

$$S_{\mathrm{BH}} = \left(\frac{2\pi k_B G}{\hbar c}\right) m^2.$$

Thus, the BH entropy is proportional to the square of its mass (m) — or, equivalently, using Einstein's formula $E = mc^2$, the BH entropy would be proportional to the square of its energy. The latter result makes the entropy a homogenous function of second order, whereas the thermodynamic entropy is a homogenous function of first order (referred to as an extensive property of the entropy). In addition, the temperature as defined in thermodynamics is an *intensive* quantity (or a homogenous function of order 0). The thermodynamic temperature, defined as $\partial S/\partial E = 1/T$, would be proportional to the inverse of the energy, and hence would not be an intensive property. We would face the same problem had we defined the chemical potential of whatever the particles of the black hole are. We would get a chemical potential which is not a homogenous function of order 0, as in thermodynamics.

References

Abbot, D., Davies, P. C. W., and Pati, A. K., eds. (2008), *Quantum Aspects of Life*. Imperial College Press, London.

Adami, C. (2004), Information theory in molecular biology, *Physics of Life Reviews* **1**: 3–22.

Albert, D. Z. (2000), *Time and Chance*, Harvard University Press.

Alexander, E. (2012), *Proof of Heaven: A Neurosurgeon's Journey into the Afterlife*. Simon and Schuster, New York.

Amit, D. J., Gutfreund, H., and Sompolinsky, H. (1987), Statistical mechanics of neural networks near saturation, *Ann. Phys.* **173**: 30–67.

Anfinsen, C. B. (1973), Principles that govern the folding of protein chains, *Science* **181**: 223.

Arnheim, R. (1971), *Entropy and Art: An Essay on Disorder and Order*. University of California Press.

Atkins, P. (2003), *Galileo's Finger: The Ten Great Ideas of Science*. Oxford University Press, London.

Atkins, P. (2007), *Four Laws That Drive the Universe*. Oxford University Press.

Aumann, R. J. (2005), Musings on information and knowledge, in *Symp. Information and Knowledge in Economics, Econ. J. Watch* **2**(1): 88–96.

Avery, J. (2003), *Information Theory and Evolution*. World Scientific.

Baierlein, R. (1971), *Atoms and Information Theory: An Introduction to Statistical Mechanics*. W. H. Freeman and Company, San Francisco.

Baierlein, R. (1994), Entropy and the second law: a pedagogical alternative, *Am. J. Phys.* **62**: 15.

Bar-Hillel, Y., and Carnap, R. (1954), Semantic information, *J. Philos. Sci.* **9**: 12.

Battino, R. (2007), "Mysteries" of the first and second laws of thermodynamics, *J. Chem. Educ.* **84**: 753.

Ben-Naim, A. (1987), Is mixing a thermodynamic process? *Am. J. Phys.* **55**: 725.

Ben-Naim, A. (2001), *Cooperativity and Regulation in Biochemical Processes*. Kluwer Academic/Plenum.

Ben-Naim, A. (2006), *A Molecular Theory of Solutions*. Oxford University Press, London.

Ben-Naim, A. (2007), *Entropy Demystified: The Second Law of Thermodynamics Reduced to Plain Common Sense*. World Scientific.

Ben-Naim, A. (2008), *A Farewell to Entropy: Statistical Thermodynamics Based on Information*. World Scientific.

Ben-Naim, A. (2009), An informational-theoretical formulation of the second law of thermodynamics, *J. Chem. Educ.* **86**: 99.

Ben-Naim, A. (2010), *Discover Entropy and the Second Law of Thermodynamics: A Playful Way of Discovering a Law of Nature*. World Scientific.

Ben-Naim, A. (2011a), *Molecular Theory of Water and Aqueous Solutions, Part II: The Role of Water in Protein Folding, Self-Assembly and Molecular Recognition*. World Scientific.

Ben-Naim, A. (2011b), Entropy: order or information, *J. Chem. Educ.* **88**: 594.

Ben-Naim, A. (2012), *Entropy and the Second Law: Interpretations and Misinterpretations*. World Scientific.

Ben-Naim, A. (2013), *The Protein Folding Problem and Its Solutions*. World Scientific.

Bent, H. A. (1965), *The Second Law*. Oxford University Press, New York.

Bekenstein, J. D. (1973), Black holes and entropy, *Phys. Rev. D* 7: 2333–2346.

Bekenstein, J. D. (1980), Black hole thermodynamics, *Phys. Today*, pp. 24–31.

Bekenstein, J. D. (2003), Information in the holographic universe, *Sci. Am.*, pp. 58–69.

Blum, H. F. (1968), *Time's Arrow and Evolution*, Princeton University Press.

Boltzmann, L. (1877), *Vienna Academy, "Gesammelte Werke"* **42**: 193.

Boltzmann, L. (1896), *Lectures on Gas Theory*, transl. by S. G. Brush, Dover, New York (1995).

Bricmont, J. (1996), Science of chaos or chaos in science? In *The Flight from Science and Reason, Ann. N.Y. Acad. Sci.* **775**: 131.

Brillouin, L. (1962), *Science and Information Theory*. Academic, New York.
Brissaud, J. B. (2005), The meaning of entropy, *Entropy* 7: 68.
Bruce, C. (2004), *Schrödinger's Rabbits: The Many Worlds of Quantum*. Joseph Henry Press, Washington, D.C.
Callen, H. B. (1960), *Thermodynamics*. John Wiley and Sons, New York.
Callen, H. B. (1985), *Thermodynamics and an Introduction to Thermostatistics*, 2nd edition. John Wiley and Sons, New York.
Campbell, J. (1982), *Grammatical Man: Information, Entropy, Language and Life*. Simon and Schuster, New York.
Carroll, S. (2010), *From Eternity to Here: The Quest for the ultimate Theory of Time*. First Plume Printing, New York.
Chaisson, E. J. (2001), *Cosmic Evolution: The Rise of Complexity in Nature*. Harvard University Press.
Cooper, L. N. (1968), *An Introduction to the Meaning and the Structure of Physics*. Harper and Row, New York.
Cover, T. M., and Thomas, J. A. (1991), *Elements of Information Theory*. John Wiley and Sons, New York.
Corning, P. A. (2002), Thermodynamics: beyond the second law, *J. Bioecon.* **4**: 57.
Crick, F. H. C. (1967), *Of Molecules and Men*. University of Washington Press.
Crick, F. H. C. (1981), *Life Itself: Its Origin and Its Nature*. Simon and Schuster, New York.
Crick, F. H. C. (1994), *The Astonishing Hypothesis: The Scientific Search for the Soul*. Simon and Schuster, New York.
Clausius, R. (1865), Presentation to the Philosophical Society of Zurich.
Dawkins, R., (1996), *The Blind Watchmaker: Why the Evidence of Evolution Reveals a Universe Without Design*. W. W. Norton & Company, New York.
Denbigh, K. (1981), How subjective is entropy? *Chem. Brit.* **17**, 168–185.
Denbigh, K. G., and Denbigh, J. S. (1985), *Entropy in Relation to Incomplete Knowledge*. Cambridge University Press.
Denbigh, K. (1989), Note on entropy, disorder and disorganization, *Brit. J. Philos. Sci.* **40**: 323–332.
Deutsch, D. (1997), *The Fabric of Reality*. Penguin, New York.
Dugdale, J. S. (1996), *Entropy and Its Physical Meaning*. Taylor and Francis, London; entropysite.oxy.edu
Dyson, F. (1985), *Origins of Life*. Cambridge University Press.

Eisenberg, D., and Crothers, D. (1979), *Physical Chemistry with Applications to the Life Sciences*. Benjamin/Cummings, California.

Ellis, G. F. R. (1993), *Before the Beginning, Cosmology Explained*. Boyars/Bowerdean, London.

Elsasser, W. M. (1958), *The Physical Foundations of Biology*. Pergamon, New York.

Erill, I. (2012), Information theory and biological sequences: insights from an evolutionary perspective, *Information Theory: New Research*, eds. P. Deloumeaux and J. D. Gorzalka, Nova Science, Chap. 1.

Fast, J. D. (1962), *Entropy: The Significance of the Concept of Entropy and Its Applications in Science and Technology*. Philips Technical Library, The Netherlands.

Floridi, L. (2010), *Information: A Very Short Introduction*. Oxford University Press, London.

Frank, S. A. (2012), Natural selection, *J. Evol. Biol.*, **25**: 2377.

Freeman, D. (1985), *Origin of Life*, Cambridge University Press.

Gatlin, L. L. (1972), *Information Theory and the Living System*, Columbia University Press.

Georgescu-Roegen, N. (1971, 1999), *The Entropy Law and the Economic Press*. Harvard University Press.

Gibbs, J. W. (1906), *Collected Scientific Papers of J. Willard Gibbs*. Longmans Green, New York.

Gibbs, J. W. (1906), *The Scientific Papers of J. W. Gibbs*. Longmans Green, London.

Glansdorff, P., and Prigogine, I. (1971), *Thermodynamic Theory of Structure, Stability and Fluctuations*. Wiley-Interscience, London.

Glansdorff, P., and Prigogine, I. (1979), *Thermodynamic Theory of Structure, Stability and Fluctuations*, John Wiley and Sons.

Gleick, J. (1987), *Chaos: Making a New Science*. Viking Penguin.

Gleick, J. (2011), *The Information: A History, a Theory, a Flood*. Pantheon, New York.

Goldstein, S. (2001), *Boltzmann's Approach to Statistical Mechanics*. Published in arXiv:condmat/0105242,v1, 11 May 2001.

Goodsell, D. S. (2009), *The Machinery of Life*. Copernicus, New York.

Greene, B. (1999), *The Elegant Universe: Superstrings, Hidden Dimensions, and the Quest for the Ultimate Theory*. Vintage, New York.

Greene, B. (2004), *The Fabric of the Cosmos: Space, Time, and the Texture of Reality*. Alfred A. Knopf, New York.

Greene, B. (2011), *The Hidden Reality: Parallel Universes and the Deep Laws of the Cosmos*. Alfred A. Knopff, New York.

Greven, A., Keller, G., and Warnecke, G. (2003), editors, *Entropy*. Princeton University Press.

Guggenheim, E. A. (1949), Statistical basis of thermodynamics, *Research* **2** : 450.

Hager, T. (1995), *Force of Nature: The Life of Linus Pauling*. Simon and Schuster, New York.

Harper, M. (2010), *The Replicator Equation of an Inference Dynamics*, arXiv: 0911.1763v3.

Hawking, S. W. (1988), *A Brief History of Time: From the Big Bang Theory to Black Holes*. Bantam, New York.

Hazen, R. M., and Singer, M. (1997), *Why Aren't Black Holes Black?* Anchor, Doubleday, New York.

Hill, T. L. (1960), *Introduction to Statistical Mechanics*, Addison-Wesley, Reading, Massachusetts.

Hoffman, P. M. (2012), *Life's Ratchet: How Molecular Machines Extract Order from Chaos*. Basic, New York.

Hofstadter, D. R. (1979), *Gödel, Escher, Bach: An Eternal Golden Braid*. Basic, New York.

Hofstadter, D. R. (1980), Metamagical themas, *Sci. Am.*, pp. 22–32.

Hofstadter, D. R., and Dennett, D. C. (1981), *The Mind's I: Fantasies and Reflections of Self and Soul*. Basic, New York.

Hofstadter, D. R. (1985), *Metamagical Themas: Questing for the Essence of Mind and Pattern*. Bantam, New York.

Holland, J. H. (1998), *Emergence: From Chaos to Order*. Oxford University Press, London.

Hoyer, B. H., and Roberts, R. B. (1967), In *Molecular Genetics*. Academic, New York.

Jacob, F., and Monod, J. (1961), *J. Mol. Biol.* **3**: 318.

Jacob, F. (1966), *Science* **152**: 1470.

Jaynes, E. T. (1957a), *Phys. Rev.* **106**: 620.

Jaynes, E. T. (1957b), *Phys. Rev.* **108**: 171.

Jaynes, E. T. (1965), Gibbs vs Boltzmann entropies, *Am. J. Phys.* **33**: 391.

Katz, A. (1967), *Principles of Statistical Mechanics: The Informational Theory Approach*. W. H. Freeman, London.

Kåhre, J. (2002), *The Mathematical Theory of Information*, Kluwer Academic, The Netherlands.

Kaufmann, S. (1995), *At Home in the Universe: The Search for Laws of Self-Organization and Complexity*. Oxford University Press, London.

Kennedy, D., and Norman, C. (2005), What don't we know? *Science* **309**: 75.

Khinchin, A. I. (1957), *Mathematical Foundation of Information Theory*. Dover, New York.

Kline, M. (1980), *Mathematics: The Loss of Certainty*. Oxford University Press, USA.

Kondepudi, D., and Prigogine, I. (1998), *Modern Thermodynamics from Heat Engines to Dissipative Structures*. John Wiley and Sons, UK.

Kozliak, E. I., and Lambert, F. L. (2005), "Order — to disorder" for entropy change? Consider the numbers! *Chem. Educ.* **10**: 24.

Kurzweil, R. (2005), *The Singularity Is Near: When Humans Transcend Biology*. Penguin, New York.

Lambert, F. L. (1999), Shuffled cards, messy desks and disorderly dorm rooms, *J. Chem. Educ.* **76**: 1385.

Lambert, F. L. (2002), Entropy is simple, qualitatively, *J. Chem. Educ.* 79: 1241.

Lambert, F. L. (2006), A modern view of entropy, *Chemistry* **15**: 13.

Landsberg, P. T. (1961), *Thermodynamics with Quantum Statistical Illustrations*. Interscience, New York.

Lavenda, B. H. (1991), *Statistical Physics: A Probability Approach*. John Wiley, New York.

Lebowitz, J. L. (1993), Boltzmann's entropy and time's arrow, *Physica Today* **46**: 32.

Lebowitz, J. L. (1999), Microscopic origins of irreversible macroscopic behavior, *Physica* A **263**: 516.

Leff, H. S. (1966), Thermodynamics entropy: the spreading and sharing of energy, *Am. J. Phys.* **64**: 1261.

Leff, H. S. and Rex, A. F. (1990), eds., *Maxwell's Demon: Entropy, Information, Computing*. Princeton University Press.

Leff, H. S. and Rex, A. F. (2003), eds., *Maxwell's Demon 2: Entropy, Classical and Quantum Information, Computing*. Institute of Physics Publishing, Philadelphia.

Leff, H. S. (2007), Entropy, its language and interpretation, *Found. Phys.* **37**: 1744.

Leff, H. S., and Lambert, F. L. (2010), *J. Chem. Educ.* **87**: 143.

Lehninger, A. L. (1971), *Bioenergetics: The Molecular Basis of Biological Energy Transformation*, 2nd edition. W. A. Benjamin, Menlo Park, California.

Lewis, G. N. (1930), The Symmetry of time in physics, *Science* **71**: 569.

Lindley, D. V. (1965), *Introduction to Probability and Statistics*. Cambridge University Press.

Lloyd, S. (2006), *Programming the Universe: A Quantum Computer Scientist Takes on the Cosmos*. Alfred A. Knopf, New York.

Lloyd, S. (2008), Quantum mechanics and emergence, Chap. 2, in Abbott, Davies and Pati (2008).

Mahon, B. (2003), *The Man Who Changed Everything, The Life of James Clerk Maxwell*. John Wiley & Sons, UK.

Margulis, L., and Sagan, D. (1995), *What Is Life?* Simon and Schuster, New York.

Mautner, M. N. (2009), Life-centered ethics, and the human future in space, *Bioethics* **23**: 433.

Mensky, M. B. (2010), *Consciousness and Quantum Mechanics: Life in Parallel Worlds*. World Scientific.

Monod, J. (1971), *Chance and Necessity: An Essay on the Natural Philosophy of Modern Biology*. Alfred A. Knopf, New York.

Morange, M. (2008), *Life Explained*, transl. by M. Cobb and M. DeBevoise, Yale University Press.

Morowitz, H. J. (1968), *Energy Flow in Biology: Biological Organization as a Problem in Thermal Physics*. Academic, New York.

Morowitz, H. J. (2002), *The Emergence of Everything*. Oxford University Press, London.

Morris, H. (1972), *The Troubled Water of Evolution*. Creation Life, San Diego, California, p. 110.

Murphy, M. P., and O'Neill, L. A. J. (1995), *What Is Life? The Next Fifty Years: Speculations on the Future of Biology*. Cambridge University Press.

Nelson, D. L. and Cox, M. M. (2014). *Principles of Biochemistry*, 6th ed. WH Treeman & Co., New York.

Nordholm, S. (1997), In defense of thermodynamics — an intimate analogy, *J. Chem. Educ.* **74**: 273.

Patel, A. D. (2008), Towards understanding of the origin of genetic languages, in *Quantum Aspects of Life*, ed. by D., Abbott, P. C. W. Davies, and A. K., Pati, Imperial College Press, World Scientific.

Pauling, L. (1935), *Proc. Natl. Acad. Sci.* U.S.A. **21**: 186.

Peacocke, A. R. (1983), *An Introduction to the Physical Chemistry of Biological Organization*. Clarendon, Oxford.

Penrose, R. (1989), *The Emperor's New Mind: Concerning Computers, Minds, and the Laws of Physics*. Penguin, USA.

Penrose, R. (1994), *Shadows of the Mind: An Approach to the Missing Science of Consciousness*. Oxford University Press, London.

Pierce, J. R. (1980), *An Introduction to Information Theory: Symbols, Signals and Noise*, 2nd edition. Dover, New York.

Polanyi, M. (1967), *Chem. Eng. News* **45**: 54.

Pross, A. (2012), *What Is Life? How Chemistry Becomes Biology*. Oxford University Press, London.

Ptashne, M. (1992), *A Genetic Switch: Phage λ and Higher Organisms*, 2nd edition. Cell Press and Blackwell, Massachusetts.

Ptashne, M. (2004), *A Genetic Switch, 3rd ed.: Phage λ Revisited.* Cold Spring Harbor Laboratory Press, New York.

Rawlinson, G. E. (1976), *The Significance of Letter Position in Word Recognition*. P.h.D thesis, Psychology Department, University of Nottingham.

Roy, B. N. (2002), *Fundamentals of Classical and Statistical Thermodynamics*. John Wiley & Sons, UK.

Sackur, O. (1911), *Annalen der Physik* **36**: 958.

Sainsbury, R. M. (2009), *Paradoxes*. Cambridge University Press.

Sanford, J. C. (2005), *Genetic Entropy and the Mystery of the Genome*. Ivan Press, a division of Elim Publishing, New York.

vos Savant, M. (1972), *Brain Building: Exercising Yourself Smarter*, Bantam, Toronto.

vos Savant, M. (1996), *The Power of Logical Thinking*. St. Martin's Press, New York.

Schopf, J. W. (ed.) (2002), *Life's Origin: The Beginnings of Biological Evolution*. University of California Press, Los Angeles.

Schrödinger, E. (1969), *The Physical Aspects of the Living Cell.* Cambridge University Press.

Scully, R. J. (2007), *The Demon and the Quantum: From the Pythagorean Mystics to Maxwell's Demon and Quantum Mystery*. Wiley-VCH, Weinheim.

Seife, C. (2007), *Decoding the Universe: How the New Science of Information Is Explaining Everything in the Cosmos, from Our Brains to Black Holes*. Penguin, USA.

Sethna, J. P. (2006), *Statistical Mechanics: Entropy, Order Parameters and Complexity*. Oxford University Press, London.

Shannon, C. E. (1948), A mathematical theory of communication. *Bell Sys. Tech. J.* **27**: 379.

Sheehan, D. P., and Gross, D. H. E. (2006), Extensitivity and the thermodynamic limit: why size really does matter, *Physica A* **370**: 461.

Shermer, M. (2006), *Why Darwin Matters: The Case Against Intelligent Design*. Times, Henry Holt and Company, New York.

Shermer, M. (2011), *The Believing Brain from Ghosts and Gods to Politics and Conspiracies How We Construct Beliefs and Reinforce Them as Truths*. Henry Holt and Company, New York.

Siegfried, T. (2000), *The Bit and the Pendulum: From Quantum Computing to M Theory — The New Physics of Information.* John Wiley and Sons, New York.

Sigmund, K. (1993), *Games of Life: Explorations in Ecology, Evolution and Behavior*. Oxford University Press, London.

Smith, J. M., and Szathmary, E. (1995), *Language and Life.* In: Murphy and O'Neil.

Sokal, A., and Bricmont, J. (1998), *Fashionable Nonsense: Postmodern Intellectuals' Abuse of Science*. Picador, New York.

Sommerfeld, A. (1956), *Thermodynamics and Statistical Mechanics*. Academic, New York.

Stent, G. S. (1963), *Molecular Biology of Bacterial Viruses*. W. H. Freeman and Co., San Francisco.

Stewart, I. (1998), *Life's Other Secret: The New Mathematics of the Living World.* John Wiley and Sons, New York.

Styer, D. F. (2000), Insight into entropy, *Am. J. Phys.* **68**: 1090.

Styer, D. F. (2008), Entropy and evolution, *Am. J. Phys.* **76**: 1031.

Susskind, L., and Lindesay, J. (2005), *An Introduction to Black Holes, Information and the String Theory Revolution. The Holographic Universe*. World Scientific.

Susskind, L. (2006), *The Cosmic Landscape, String Theory and the Illusion of Intelligent Design*. Back Bay, New York.

Szilard (1929), *Z. Phys.* **53**: 840.

Tanford, C., and Reynolds, J. (2001), *Nature's Robots: A History of Proteins*. Oxford University Press, New York.

Tetrode, H. (1912), *Annalen der Physik* **38**: 434.

Thomson, W. (1874), *Proc. R. Soc. Edinburgh* **8**: 325.

Tribus, M., and McIrvine, E. C. (1971), Entropy and information, *Sci. Am.* **225**: 179.

Velan, H., and Frost, R. (2007), Cambridge University Versus Hebrew University; the impact of letter transposition on reading English and Hebrew, *Psychon. Bull. Rev.* **14**: 913.

Volkenstein, M. V. (2009), *Entropy and Information*, transl. by A. Shenitzer and A. G. Burns, Birkhauser, Berlin.

Wan-Ho, M. (2008), *The Rainbow and the Worm: The Physics of Organisms*, 3rd edition. World Scientific.

Woolfson, M. (2009), *Time, Space, Stars and Man: The Story of the Big Bang*. Imperial College Press, London.

Wheeler, J. A. (1990), Information, physics, quantum: the search for links. In: W. Zurek, *Entropy and the Physics of Information*. Addison-Wesley, California.

Wheeler, J. A. (1994), *At Home in the Universe, Masters of Modern Physics* **9**: 296. American Institute of Physics.

Wheeler, J. A. (2000), *Geons, Black Holes and Quantum Foam: A Life in Physics*. Norton and Company.

Wigner, E. P. (1967), *Symmetry and Reflections*. Indiana University Press.

Yaglom, A. M., and Yaglom, I. M. (1983), *Probability and Information*, transl. from Russian by V. K. Jain and D. Reider, Boston.

Index